Opioids

Silvia L. Cruz
Editor

Opioids

Pharmacology, Abuse, and Addiction

Springer

Editor
Silvia L. Cruz
Department of Pharmacobiology
Cinvestav
Mexico City, Mexico

Opioids Working Group
Global Studies Seminar
Faculty of Medicine
National Autonomous University of Mexico (UNAM)
Mexico City, Mexico

ISBN 978-3-031-09938-0 ISBN 978-3-031-09936-6 (eBook)
https://doi.org/10.1007/978-3-031-09936-6

This Springer imprint is published by the registered company Springer Nature Switzerland AG
The registered company address is: Gewerbestrasse 11, 6330 Cham, Switzerland

Preface

The history of opioids closely intertwines with the history of humankind. Indeed, opioids can affect how we live and die due to their therapeutic properties, psychoactive effects, and addiction potential.

The writing of this book took approximately 1 year, during which 80,800 people died only in the United States due to overdoses caused by prescription and non-prescription opioids. At the same time, countless patients benefit from opioid medications for pain management or palliative care. How can these drugs cause such profound effects on persons and societies? How can we take advantage of the desirable opioid effects while minimizing addiction?

Addressing these questions requires a multidisciplinary approach to understanding the complex phenomenon of addiction and how opioids produce their actions. Such knowledge provides the foundations for implementing much-needed evidence-based interventions to reduce opioid-induced harm. As a neuroscientist who has studied the neurobiology of addiction for decades and a lecturer in a graduate school, I have experienced firsthand the difficulty of finding comprehensive, accessible, and updated information on the social and pharmacological aspects of drug use in a single book. The same is true for colleagues working on philosophy, sociology, or economics that want to get familiar with the pharmacology of opioids or the complex chemical names of new psychoactive drugs.

This book is divided into two parts. The first one addresses key social aspects related to opioid use and users. It analyzes the current challenges to providing adequate pain management while avoiding the diversion of opioids to the black market. It also reviews the current international regulations to control drug use, the need to adapt them to different regional realities, the impact of opioid use disorders (OUDs) in people's lives, and the structural interventions effective for prevention, harm reduction, and treatment. The second part covers the physiological, cellular, and molecular actions of main opioids, the relationship between opioid use and alterations in the immune system, tolerance development, abstinence, addiction, and medication-assisted treatment for OUDs.

It was my privilege to work with an amazing group of professionals on addiction from different fields; my gratitude and recognition to them for their generous and high-quality work. I hope that the readers will enjoy the process of learning about opioids as much as we enjoyed putting this book together.

Mexico City, Mexico Silvia L. Cruz

Introduction

Part I: Social Aspects of Opioid Use

Men and women of all ages have coexisted with opioids since the dawn of civilization. From pain relief to significant advances in understanding how these drugs exert their actions, opioids have deeply influenced social, medical, and economic aspects of people's lives. Countries have fought to control opioid markets. Societies have developed laws and treaties in an attempt to counteract the adverse consequences of nonmedical opioid use. At the current time, countless physicians and patients benefit from opioid medications, but also numerous persons are afflicted with opioid use disorders and lack treatment. Throughout history, addiction has been considered in changing ways: a moral failure, a personality disorder, a pattern of maladaptive substance use that leads to significant clinical deterioration, a social problem, and a brain disease. Chapter 1 reviews some milestones in the history of opioids and the evolution of concepts associated with opioid use disorders (OUDs).

Opioid use poses many challenges. Two billion people worldwide lack access to opioid medications for pain management, mainly in low- and middle-income countries. At the same time, OUDs continue to increase. In 2019, opioids caused more than two-thirds of deaths associated with substance use. That same year, 20% of the population consumed more than 80% of the morphine legally available for pain management, and low- and middle-income countries had no access to it. Chapter 2 addresses the importance of understanding these disparities and drug market supply and demand forces. It also emphasizes the need to adopt policies that protect human rights, decriminalize drug use, fight stigma, and provide evidence-based treatment to people with OUDs.

Chapter 3 reviews opioid markets, vulnerable populations, and how countries are affected by production, transit, and opioid consumption. Both the appearance of opioids among new psychoactive substances and drug adulteration mark new trends in opioid use. Vulnerable populations face specific challenges in reaching treatment for OUDs, including women, persons who experienced violence in childhood, adolescents, and older adults. In addition, both legal and illegal markets are being affected by an increase in life expectancy, aggressive marketing practices, and increasing prevalence of chronic diseases. More recently, the COVID-19 pandemic brought significant changes in production, marketing, and drug use.

The international drug control regime comprises principles, norms, rules, and processes with which signing states of international treaties must comply or be subject to sanctions. Chapter 4 comments on the implications and consequences of applying the three United Nations treaties, signed in 1961, 1971, and 1988, that are the basis of the current drug control regime. It also mentions the drug control agencies and the agreements derived from United Nations General Assembly (UNGASS) meetings, especially the most recent one, which was held in 2016. The authors proposed a new drug policy agreement that should align with the UN sustainable development goals and centered on persons rather than substances. This new agreement should treat drug use as a public health problem rather than a criminal activity, respect human rights, provide alternatives to incarceration, and facilitate access to medication-assisted treatment to individuals with OUDs.

Half a million people died due to opioid overdose from 1999 to 2019 in the USA. The current opioid crisis in the USA is the most conspicuous but not the only one. In the USA and Canada, the crisis began with heavy promotion of opioid use by pharmaceutical companies, which marketed them as safe medications, practically devoid of addiction liability, and effective for long-term treatment of chronic pain. The unprecedented increase in opioid prescription led to the development of OUDs among patients that began using them as analgesics but later migrated to heroin. More recently, fentanyl and its analogs have become responsible for the third peak in opioid-related overdoses. Chapter 5 details how this epidemic evolved; presents data on opioid misuse in Asia, Europe, and Africa; and briefly reviews some effective measures to prevent or diminish the impact of the global opioid crisis.

With a humane person-centered perspective, Chapter 6 analyzes the living conditions of people who inject drugs (PWIDs) in Tijuana, a border city in northern Mexico, and some differences and commonalities with similar populations in other parts of the world. First-person narratives illustrate the social and emotional problems faced by marginalized people with OUDs, including poverty, poor health care, low education, stigma, police brutality, violence (sexual and otherwise), and few job opportunities. These voices illustrate the precarious day-to-day conditions of vulnerable populations and highlight the need for structural interventions to address the opioid crisis.

Factors such as socioeconomic status, inequality, public policies, and political trends are social determinants of health. Chapter 7 explains the crucial role played by structural interventions in preventing and managing nonmedical opioid use. These interventions involve law reforms, implementing or adapting administrative procedures, and community organization. Recommendations to implement effective structural interventions include:

- Involving persons with lived experience in treating OUDs
- Providing dignified housing programs
- Facilitating access to health care and medication-assisted programs
- Implementing syringe exchange programs and safe consumption services

To reduce fatal opioid overdose, naloxone, the opioid antagonist that reverses opioid-induced respiratory depression, should be widely available, and drug-checking should be a common practice to detect drug adulteration.

Part II: Pharmacological and Medical Aspects of Opioid Use

Opioids have been responsible for many advances in pharmacology and medicine. While searching for morphine-like molecules devoid of adverse effects, hundreds of new compounds were synthesized, which led to the search for endogenous ligands with morphine-like effects, opioid receptor characterization, and the development of effective medications. Chapter 8 reviews the effects of morphine—the opioid prototype—on the nervous, gastrointestinal, cardiovascular, and immune systems. It also describes the endogenous opioid peptides; the distinction between agonists, partial agonists, inverse agonists, or antagonists at different receptor subtypes; and the pharmacological characteristics of clinically relevant opioids. The tables and illustrations contained in this chapter help identify chemical groups, effects, doses, and administration routes of main opioids.

Opioid receptors couple to G-proteins, which convey inhibitory signals, thus decreasing neuron responsiveness. Chapter 9 provides an in-depth look at opioid receptors' structure and intracellular signaling pathways. When agonists bind to the orthosteric site of an opioid receptor, they stabilize it in one or another conformation, which can preferentially activate G-protein or beta-arrestin pathways (biased agonism). Allosteric modulators that bind to receptor sites different from those occupied by opioids can increase or decrease the agonists response (positive or negative modulators, respectively). Phosphorylation, single-nucleotide polymorphisms, the splicing of opioid receptor genes, receptor dimerization, and epigenetic factors modulate opioid effects.

Pain is an unpleasant experience that acts as an alert system to minimize contact with harmful stimuli. Depending on its location, duration, and etiology, pain can be somatic or visceral, acute or chronic, nociceptive, neuropathic, or nociplastic. In addition, several unidimensional and multidimensional scales exist to assess pain intensity. Chapter 10 covers these topics and reviews pain physiology, the role of opioids in acute pain treatment, and the limitations of their use in chronic pain management. It also mentions opioid rotation (switching from one agent to another) as a strategy to improve pain control. The tables included in this chapter summarize information related to pain assessment scales, dosing, administration routes, half-lives, and metabolism of opioids frequently used and guidelines for equianalgesic dosing of different opioid agonists.

Patients with chronic and life-threatening illnesses should receive palliative care as part of human health rights. Chapter 11 reviews the pathophysiology of pain in terminal illness, the risk of drug-drug interactions in patients receiving several medications, and the advantages and limitations of using opioids in chronically ill patients. The specific situation of providing palliative care in children is also discussed, as is the importance of evaluating the risk of opioid misuse in all patients. The tables in this chapter provide guidelines on using specific opioids in people with advanced chronic kidney disease, impaired liver function, and the terminally ill.

Persons who use opioids are more prone to infections. Opioids act directly on the immune system (IS) by binding to specific classic opioid receptors, atypical MRGPRX2 receptors, and toll-like receptors (TLRs) expressed in immune cells. Opioids also indirectly affect the IS through the sympathetic nervous system and the hypothalamus-pituitary-adrenal axis. Chapter 12 presents an overview of the IS, the direct and indirect opioid actions on innate immune cells and lymphocytes and how genetic polymorphisms and aging can modulate them. Evidence exists that opioids play a significant role in cross-regulating the IS and the CNS. This chapter, which is supported by figures and tables, explains that the relevance of studying opioid-mediated IS-CNS communication lies in the possibility of modifying the immune response against pathogens, acute hyperinflammatory states (such as those seen in patients with COVID-19), and tumor growth.

Prolonged opioid use can lead to tolerance, hyperalgesia, dependence, and withdrawal. These adaptations occur due to changes in opioid receptor signaling, the mesocorticolimbic dopaminergic system, the noradrenergic locus coeruleus, and other areas associated with pain transmission. Upregulation of pronociceptive systems and opioid effects on TLRs also play a role. Chapter 13 reviews these processes and presents the scales that clinicians use to rate the severity of withdrawal in opioid users and newborns of mothers who used these substances during pregnancy. This chapter also describes animal models used to study opioid dependence liability (drug self-administration, drug discrimination, and conditioned-place preference tests) and presents some approaches to prevent the development of tolerance and dependence.

Medication-assisted treatment (MAT) is a complex biopsychosocial intervention aimed at diminishing opioid use and craving and improving the health and well-being of opioid users. This treatment recognizes the multiple adaptations of prolonged opioid use, combining opioid substitution therapy with behavioral and psychological therapies. Chapters 14 and 15 describe the process from enrolling and diagnosing the patients to implementing individualized care plans for opioid-assisted treatment (OAT), maintenance, and medical reviews, within the public UK health system. As detailed in Chap. 14, methadone and buprenorphine (alone or combined with naloxone) are the preferred medications for OAT. Methadone is an orally active opioid receptor agonist with a slow onset of action and prolonged duration of effects. Buprenorphine is a partial agonist which dissociates very slowly from opioid receptors. Both drugs prevent withdrawal and stabilize patients. This chapter includes suggestions for conducting the initial patient assessment and examples of methadone titration regimes and buprenorphine microdosing schedules. Chapter 15 covers detoxification and aftercare for patients that discontinue opioid substitution therapy, including the symptomatic relief in detoxification, the use of naltrexone to prevent relapse, and psychosocial interventions. It also reviews precautions needed while providing MAT to pregnant women, persons with mental health problems, or polydrug users.

Chapter 16 describes new trends in opioid use, particularly the increased availability of synthetic opioids as adulterants and new psychoactive substances (NPS), their precursors, and the changes in illicit opioid transactions in the digital era. Fentanyl and its analogs are responsible for many opioid overdoses worldwide. New non-fentanyl opioids detected in the NPS market include brorphine, isotonitazine, U-477700, analogs, and plant-derived compounds with opioid effects (mitragynine and salvinorin A). In addition, commercial transactions occur through the mail (regular or express), the Internet, and increasingly the darknet, where cryptocurrencies are the preferred form of payment. The final chapter of this book includes information on selected fentanyl and non-fentanyl analogs sold as NPS and reviews strategies to face the challenges posed by new trends in opioid markets.

Contents

About the Editor

Silvia L. Cruz, PhD, is Professor of Pharmacobiology at Cinvestav in Mexico City. She has previously served as graduate program coordinator and chair of the Department of Pharmacology at Cinvestav. She has worked as a full-time researcher and lecturer in the drug abuse field for the past 30 years, and she is author of the chapter on opioids in the Springer book *Neuroscience in the 21st Century.*

Part I
Social Aspects of Opioid Use

Chapter 1
A Brief History of Opioids and the Evolution of Concepts Associated with Substance Use Disorders

Silvia L. Cruz and Claudia Rafful

Abstract This chapter provides an overview of the history of opioids and people's different relationships with them across time and geography. It begins with the origins of the poppy flower, the recreational and medical use of opium in ancient cultures and Asia, and its pathway to European users. This chapter also reviews the different medicinal opium preparations (e.g., beverages, syrups, concoctions) used until the nineteenth century and the Opium Wars between England and China. Another section describes the discovery of morphine, the development of semisynthetic and synthetic opioids, the identification of endogenous opioid peptides, and the pharmacological characterization of opioid receptor subtypes. Finally, this chapter provides a brief description of the role of opioids in modern societies and reviews how basic concepts related to opioids and opioid use disorders (OUDs) have evolved through time.

Keywords History · Poppy flower · Synthetic opioids · Addiction · Substance use disorders

---

S. L. Cruz (✉)
Department of Pharmacobiology, Cinvestav, Mexico City, Mexico

Opioids Working Group, Global Studies Seminar, Faculty of Medicine, National Autonomous University of Mexico (UNAM), Mexico City, Mexico
e-mail: slcruzm@cinvestav.mx

C. Rafful
Opioids Working Group, Global Studies Seminar, Faculty of Medicine, National Autonomous University of Mexico (UNAM), Mexico City, Mexico

Faculty of Psychology, UNAM, Mexico City, Mexico

Center for Global Mental Health Research, National Institute of Psychiatry Ramón de la Fuente Muñiz (INPRM), Mexico City, Mexico

S. L. Cruz (ed.), *Opioids*, https://doi.org/10.1007/978-3-031-09936-6_1

1.1 Origins of the Poppy Flower and the Use of Opium in Ancient Civilizations

Opium (from the Greek word "*opos*" or "juice") is the oxidized milky sap of the immature capsules of *Papaver somniferum*. This poppy plant is native to Turkey and has a growth cycle lasting approximately 4 months. There is evidence of plant domestication by the end of the fourth millennium BC [1], and its use for psychoactive effects dates to the Bronze Age (1600–1200 BC) [2].

In Mesopotamia (now Iraq), the Sumerians cultivated the opium poppy plant and referred to it as "the plant of joy." More than a thousand years later, clay Babylon tablets refer to poppy juice as a medicinal substance. Opium is harvested by making shallow parallel incisions to the unripe poppy's seedpods with a sharp instrument and collecting the latex, which soon dries in contact with the air and turns into a brownish sticky paste. This gum is then boiled and filtered to get rid of impurities. The drying and molding process may have varied across Indian, Bengali, Turkish, Persian, and Egyptian civilizations, but the overall process has remained the same for over 2000 years [3].

Opium has several alkaloids, including morphine, its primary psychoactive compound. From the chemical point of view, alkaloids are bases (as opposed to acids) that contain a nitrogen atom and a characteristic bitter taste that may induce vomiting. To make it palatable, the whole poppy head was crushed and mixed with wine, honey, and water. Another preparation involved opium with nutmeg, cardamom, or mace mixed with saffron and ambergris [2]. Other compounds present in opium are codeine, thebaine, papaverine, and noscapine. Morphine, codeine, and thebaine have a similar structure, but the first two produce sedation, whereas thebaine has excitatory and toxic effects.

Box 1.1 What is the difference between opiates and opioids?
The term opiate originally applied to alkaloids derived from opium. After identifying morphine's chemical structure and extensive synthesis work, this term was extended to include all compounds with morphine-like structures, such as hydromorphine, heroin, oxycodone, and naloxone. Because one of the most characteristic effects of morphine is sleep, another way to refer to opioids is narcotic drugs. The more inclusive term opioid applies to opiates, endogenous peptides, and other compounds with morphine-like actions but different chemical structures.

The Assyrians expanded the knowledge of poppy cultivation and opium pharmacological properties to the civilizations they traded with, including the Egyptians and the Greeks. As a result, merchants as well as opium users were experts at assessing the quality, strength, and variety of the opium traded [3].

The Egyptian Ebers Papyrus, written around 1500 BC, included "a remedy to prevent excessive crying of children" with some grains of the poppy plant mixed with "excretions of flies found on the wall...." By 400 BC, the Greek physician Hippocrates made frequent references to opium's therapeutic properties and prescribed it to patients with insomnia. Later, Theophrastus of Eresus, from the third century BC, wrote a treatise on botany in which he listed opium as a bowel purge.

Dioscorides, a Greek physician-botanist who was born in the first century AD, visited the territories under Roman rule while serving under the Emperor Nero as a soldiers' physician. Being an avid scientist, he collected information from natural remedies wherever he travelled and wrote a medical compendium that was highly valued in the centuries to come. In his work *De Materia Medica*, Dioscorides referred to opium as an analgesic and a remedy against coughing and intestinal disorders [4]. He also described the method to recollect opium in a similar manner to how it is done today:

> Those who start to collect the liquor [latex], after the dew drops have dried, have to cut the little star on the head with a knife, so that it does not penetrate to the inside, and thus make oblique cuts on the surface of the heads in a straight line, collect the tears that come out of them with the finger and introduce them into a shell; again, not long after, it is necessary to review, because the liquor is coagulated, and the next day it is also so; you have to crush it in a mortar, shape it, and store it.

He also mentioned "meconium," a specific opium preparation known to Greeks, obtained by crushing, pressing, and grinding the whole poppy capsules in a mortar.

1.2 Expansion of Opioid Use in the Middle Ages

The Muslim physician Hussain Ibn Sîna, better known as Avicenna, wrote the *Canon of Medicine*, a comprehensive and influential work, in the first years of the eleventh century. In this work, Avicenna recommended opium to treat "diseases of teeth" and to "produce severe drunkenness to carry out painful treatment of an organ," among other medical indications.

Avicenna's description of opium contains the following information taken from an English translation made by Aziz and colleagues [5]:

> "Definition: Opium is the extract on the black poppy. Smelling the Egyptian variety induces sleep. The oral dose should not exceed two grains."
>
> "Selection: The type that should be used is resinous with a pungent odor."
>
> "Properties: Anesthetic and analgesic to all pains whether it is taken orally or applied locally. The oral dose is the size of a lentil."
>
> "Organs of the head: Induces sleep if taken rectally and produces analgesia if instilled in the ear in a mixture of rose-oil, myrrh, and saffron. It also relieves chronic headache and reduces comprehension and intelligence."
>
> "Organs of breathing: Relieves intractable cough."
>
> "Organs of nutrition: Causes the stomach to contract when it is in a relaxed state."
>
> "Waste organs: Stops diarrhea and cures intestinal ulcers and erosions."

The following citations, taken from Heydari and colleagues [6], exemplify the accuracy of Avicenna's knowledge of opium effects (Chap. 8):

"Opioids are drugs that prevent nerves to conduct painful sensory impulses."

"If all treatments fail, opioids can be used to treat severe cough."

"Opium may lead to difficulty breathing, which can lead to death."

"If you have no other option but to use opioids, closely monitor patient's pulse to avoid overdosing."

"If diarrhea leads to fainting and other treatments fail, you can use opioids to stop diarrhea."

The *Canon of Medicine* extended its influence on several countries until the sixteenth century through many translations and beautiful copies.

1.3 From Opium Dens in Asia to Europe

1.3.1 Opium Use in India

Arab traders introduced the opium poppy into China, Persia, India, North Africa, and Spain approximately in the eighth century, and into India in the ninth century [7]. During Muslim rule, opium was extensively cultivated, mainly for medicinal purposes. A customary greeting was "Take your opiate," and legal documents included the stamped inscription "Take a draught of opium." Opium was considered a household remedy and then transitioned to social and religious use. The Muslim Mughal dynasty, of Turkic origin, ruled India from the sixteenth to the eighteenth century. Under this dynasty, poppy cultivation became a state monopoly in 1524; cultivation grew extensively on the West coast, as did opium trade with China and the East.

In 1707, the monopoly was turned over to the British East India Company, which actively promoted drinks and drugs under the British Empire, and recreational opium use increased exponentially. Later, in 1773, General Warren Hastings, British Governor of India, inflicted strict control measures and restrained internal opium consumption because of its pernicious effects. The *British Pharmacopoeia* published the procedure of preparing opium for smoking in the early nineteenth century. It consisted in reducing the mass by about 50%, which doubled its concentration.

The sovereignty of the East India Company was transferred to the British Crown in 1858, and there were several attempts in the following decades to abolish the opium monopoly, which would have removed the role of drug merchants to the government while still raising revenues. Finally, in 1924, Gandhi passed a resolution in which he stated that opium policy was contrary to the moral welfare of the population.

In India, opium use was never as high as in China, mainly because it was eaten instead of smoked. However, in 1937 when the Indian National Congress came to power, an opium prohibition was enacted in some provinces. Later, the state banned the sale and use of opium in the districts where it was most prevalent. This policy also included media campaigns, public meetings, registration of "addicts," and free medical inpatient treatment.

All opium cultivation and manufacture became the government's responsibility with the establishment of the Narcotics Commission in 1920. As a result, nonmedical use decreased, while prices increased. In 1944, the United States expressed its concern about opium use in India, but the British government replied that all cultivation and trade followed established policies and trade agreements. The Indian government, noticing pressure, prohibited smoking opium, but this effort was halted toward the end of World War II. Three years later, with India's independence came the complete prohibition of opium production, except for medical and scientific purposes. In addition, more free treatment centers opened, and smuggling was punished. A decade later, opium sales were banned, and only registered "addicts" were allowed oral consumption on medical grounds. By 1970, oral consumption and smoking decreased. Most users were 50 years or older, international obligations (United Nations conventions; Chap. 4) restricted opium cultivation, and only licensed growers were allowed to produce and sell opium internally, as well as export it for medical purposes [3].

1.3.2 Opium Use in China

China had opium pills by the twelfth century. Three centuries later, China cultivated native opium for oral consumption and had extended its medical use. On the other hand, opium smoking began in the eighteenth century. Spanish ships brought tobacco to Europe, and the Portuguese introduced the smoking pipe to China [8]. Opium was initially smoked with tobacco, and this new route of administration triggered a significant social and health problem. Early Chinese pipes resembled hookahs, in which the opium pill was placed at the bottom to melt and evaporate. The user would breathe deeply and inhale the fumes through a tube [3]. With smoking, opium dens spread everywhere, and the number of users escalated.

Although the Portuguese, the Dutch, and the French were the first to trade opium with China, Britain became the dominant force when it entered this market. In 1729, the Chinese empire banned opium smoking, and, for 180 years, there was an unsuccessful effort to forbid its use and import, even when opium sales were a crime and merchants could receive the death penalty. Buyers, however, were not punished. In 1799, opium became an illicit commodity, but this measure increased trafficking in regions not underseen by the government. A year later, another imperial edict prohibited domestic cultivation and reiterated the prohibition against opium importation. By this time, opium use was associated with corruption and anti-government groups. Despite these facts, in 1821, opium revenue trade from India to China surpassed that from cotton, and there were approximately three million Chinese citizens/residents addicted to opium.

Opium grew in the East Indies, and the British companies exported it through Bengal to Canton despite imperial edicts. British traders argued that Chinese authorities had not implemented the mandate because they accepted opium if they received a bribe. Opium imports and use were so widespread that, in 1839, the Chinese emperor promulgated the most stringent law stipulating execution for users and sellers, including foreign importers and traders. This legal disposition was the beginning of the first war between China and the British Empire.

1.3.3 The Opium Wars

The Chinese fought against the British during the conflicts known as Opium Wars from 1839 to 1843 and from 1856 to 1860. Opium trade played a crucial role in the British Empire's economy, expansion, and capitalism. Historians have argued that the Opium Wars were not only related to opium but also to the British and overall Western attempt to expand free trade and impose capitalism into other cultures. However, opium, with its psychoactive effects, played a major role given the increased use among the Chinese population and the public concern regarding its economic consequences.

In 1839, Lin-Hse-Tsu (or Lin Zexu), a special commissioner under the Chinese emperor's orders, seized the illegal opium cargo from the ships stationed in Canton and burned it. Lin also wrote a letter to Queen Victoria asking her to halt the illicit opium trade:

> We find that your country is sixty or seventy thousand li [three li equal one mile] from China. Yet there are barbarian ships that strive to come here for trade for the purpose of making a great profit. The wealth of China is used to profit the barbarians. That is to say, the great profit made by barbarians is all taken from the rightful share of China. By what right do they then in return use the poisonous drug to injure the Chinese people? Even though the barbarians may not necessarily intend to do us harm, yet in coveting profit to an extreme, they have no regard for injuring others. Let us ask, where is your conscience? I have heard that the smoking of opium is very strictly forbidden by your country; that is because the harm caused by opium is clearly understood. Since it is not permitted to do harm to your own country, then even less should you let it be passed on to the harm of other countries - how much less to China! [9].

In response, Britain sent troops and warships, occupied Canton and other main coastal cities, and forced China to sign the Treaty of Nanking in 1842. By this treaty, China was forced to pay for the opium destroyed by Lin-Hse-Tsu, open new ports for international trading, and cede Hong Kong to the British. Despite this defeat, China refused to legalize opium.

The Second Opium War began in 1856. The supposed motive was the killing of a French missionary in China, but it was actually related to a cultural crash and a change in the Chinese government that tried to retake control over foreign presences. France, Russia, and the United States joined the war and defeated China. The Treaty of Tientsin (or Tianjin) stipulated new trading ports and unrestricted movement to Christian missionaries. China initially refused to accept it, but after several battles, it signed in 1860 the Peking Treaty, which incorporated the previous demands and opium legalization. Hence, from 1860 to 1906, the opium trade in China was legal and taxed, just like any other commodity. As a result, according to an official report, more than a quarter of adult Chinese males smoked opium at the beginning of the twentieth century [8].

In 1906, Britain ended the opium trade with China due to intense international and domestic opposition.

1.4 Opium and Morphine Use in the Western: Laudanum, Dover's Powder, and Soothing Syrups

There is evidence that a sponge soaked in a liquid with opium, mandragora, and hemlock was used in Europe by the late thirteenth century. Sniffing this "spongia somnifera" helped to control the pain, but the practice was dangerous and unpredictable, and it did not become prevalent. In 1541, the Swiss alchemist Philippus Aureolus Theophrastus Bombastus von Hohenheim, better known as Paracelsus, prepared a liquid form of opium that he called "laudanum," from the Latin word "*laudare*," meaning "to praise." In addition to opium, his recipe contained orange and lemon juices, some spices, powdered gold, crushed corals, and pearls. Paracelsus made pills of dried laudanum, considered opium the best of all the available remedies of his age, and successfully used it among his patients.

In 1660, the English physician Thomas Sydenham simplified laudanum's recipe using opium, cloves, cinnamon, saffron, and sherry wine. Opium dissolves better in alcohol than in juices, and this preparation became extremely popular, not only as a painkiller and an antidiarrheal drug but also as an intoxicating drink [8, 10]. Sydenham was well aware of opium's therapeutic effects:

> Among the remedies which it has pleased Almighty God to give to man to relieve his sufferings, none is so universal and so efficacious as opium.

A student of Sydenham, Thomas Dover, developed another famous opium formulation called Dover's powder. According to an old paper published in the 1846 edition of *The Lancet* [11]:

> Take Opium one ounce, Salt-Petre [potassium nitrate] and Tartar [potassium hydrogen tartrate] vitriolated [mixed with sulphuric acid or other sulfate] each four ounces, Ipecacuanha one ounce. Put the Salt-Petre and Tartar into a red-hot mortar, stirring them with a spoon until they have done flaming. Then powder them very fine; after that slice in your opium, grind them to a powder, and then mix the other powders with these. Dose from forty to sixty or seventy grains in a glass of white wine Posset going to bed, covering up warm and drinking a quart or three pints of the Posset-Drink while sweating.
>
> In later editions, he said that some apothecaries had desired their patients to make their wills and settle their affairs before they venture upon so large a dose as from 40 to 70 grains: "As monstrous as they may represent this, I can produce undeniable proofs where a patient of mine has taken no less a quantity than a hundred grains, and yet has appeared abroad the next day."

Undoubtedly, Dover's patients had developed tolerance to the respiratory depression produced by opium.

The oral use of opioids continued to grow in Europe and the United States through medicine remedies such as Mother Bailey's Quieting Syrup and Mrs. Winslow's Soothing Syrup. The latter was, by far, the most popular of these remedies. An advertisement published in the nineteenth-century newspapers reads:

> ADVICE TO MOTHERS! -Are you broken in your rest by a sick child suffering with the pain of cutting teeth! Go at once to the chemist and get a bottle of MRS WINSLOW'S SOOTHING SYRUP. It will relieve the poor sufferer immediately. It is perfectly harmless and pleasant to taste, it produces natural quiet sleep, by relieving the child from pain, and the little cherub awakes as bright as a bottom [12].

Jeremiah Curtis and Benjamin A. Perkins marketed this remedy used by a nurse (Curtis's mother-in-law) in 1845. "Mrs. Winslow's Syrup" had morphine and alcohol, but the labels were not required to list the ingredients in their pharmaceutical products. As a result, mothers treated their babies with this and other soothing syrups without knowing what was in them. By 1868, Curtis declared that his company was selling 1.5 million bottles of this remedy per year. Despite evidence of accidental death of babies receiving Mrs. Winslow's Syrup and its nickname "baby killer," this concoction continued to be sold as late as 1930.

In England, opium-containing remedies were so prevalent that Friedrich Engels wrote that children that took liquid opium became "pale, feeble, wilted, and usually died before completing the second year" ([13], cited in 8). The Harrop's Soothing Syrup was another preparation for infants sold in England. *The Lancet* published the following note in 1875 [12]:

> A CHILD has been poison'ed *(sic)* at Barnsley by taking Harrop's Soothing Syrup, which, on analysis by Mr. Palmer, of the firm of Worley & Co., M'trket- *(sic)* street, Manchester, was found to contain morphia! When will Government compel the publication of the composition of quack medicines, and appoint some responsible person to forbid the sale of such as are dangerous to life?

1.5 From Morphine Isolation to Opioid Use Restriction

1.5.1 The Isolation of Morphine

In the early 1800s, Friedrich Wilhelm Adam Sertürner, a young apprentice of pharmacology, mixed opium with different solvents, searching for the active compound responsible for opium's medicinal effects. He gave each extract with sugar to dogs and observed if they became sleepy. He also tried the extracts until he found the one that had opium effects. From it, Sertürner, aged 21, isolated an alkaline organic compound that he initially called "principium somniferum" and then "morphium" after Morpheus, the Greek god of dreams. Recognition of this historical finding occurred more than 10 years later, in 1817, when the renowned French chemist and physicist Joseph Louis Gay-Lussac read and commented on Sertürner's work. It was also Gay-Lussac who coined the term morphine that is still in use today [3, 10]. Morphine could be bought in Europe in the 1820s and one decade later in the United States.

It took more than a century to discover the chemical structure of morphine. Still, experimentation with the new isolated substance began immediately in an attempt to understand its properties, define its chemical composition, and produce new derivatives. Heinrich Emanuel Merck commercialized morphine in 1827 and codeine in 1836 after its isolation by Pierre-Jean Robiquet in 1832 [10].

Before morphine isolation, there were references to "confirmed opium-eaters" or to the "power of opium habit," indicating the occurrence of what are today called "substance use disorders" (SUDs). People referred to "morphinists" and "morphinism" to those who used excessive morphine and the condition derived from that behavior. Recognition of the problems that occurred with opium discontinuation, such as "unbearable pains" and significant discomfort, clearly existed. Still, it became more evident with pure morphine than with opium.

1.5.2 The Hypodermic Needle and Heroin Synthesis

The early 1850s brought the invention of the hypodermic needle and a significant change in the relationship between humans and drugs, including opioids [14]. New intramuscular, subcutaneous, and intravenous administration routes avoided the variability of gastrointestinal absorption, produced effects more rapidly, and required lower drug doses than the oral route. These characteristics allowed for better control of drug delivery, dosing, and analgesic effect, which was a significant therapeutic advantage, fully appreciated during conflicts such as the Crimean and Franco-Prussian wars in Europe and the Civil War in the United States.

The higher addiction risk of the intravenous route was neither anticipated nor recognized for a while. As a matter of fact, drug desire was often associated with craving in the sense of increased appetite for the substance. Physicians believed that the "hunger" component would disappear by surpassing the gastrointestinal tract. It took several years until they realized that intravenous morphine injection led to addiction more rapidly than oral administration. In the 1860s, morphine addiction became a significant problem associated with the Civil War in the United States, to the point that it was known as "the soldier's disease" [15]. Some authors consider the Civil War-related opioid use the first opioid epidemic in the Americas. However, it was only until the HIV epidemic that injection drug use became a significant public health concern. To this day, most harm reduction interventions (see Chap. 7) have their origins in preventing and controlling the HIV epidemic.

In 1874, Alder Wright conducted a milestone experiment in the history of opioids. He boiled morphine with acetic anhydride and obtained a molecule similar to morphine but attached to two acetyl groups. Although the structure of morphine was still unknown at that time, chemists already knew its chemical formula: $C_{17}H_{19}NO_3$. Like several other scientists, Wright wanted to discover an analgesic as good as or better than morphine but devoid of its adverse effects, including "morphinism." Instead, he synthesized diacetylmorphine ($C_{21}H_{23}NO_5$), later marketed by Bayer under the name of "heroin." This new drug promised to be better than morphine and addictive-free. At that time, preclinical investigations were virtually nonexistent. The only way to know if a substance was addictive was when users became dependent on that substance, which soon happened with heroin [3].

1.5.3 The Harrison Narcotics Tax Act

Colonial opium trading occurred from the seventeenth to the early twentieth century [16]. In the post-war years, the consolidation of the hegemonic role of the United States pushed drug prohibition, punishing opium trade and use.

In the United Kingdom, Asia, and the United States, opium use was widespread in the late nineteenth century, and addiction was already a concern. As a result, several countries signed the Hague International Opium Convention in 1912 to control opium production and commercialization. The signing of this treaty was propelled by problems associated with opium use in the Far East, specifically in China [15]. Two years later, with the support of the temperance movement, the United States enacted the Harrison Narcotics Tax Act, stating that this legislation was made:

> To provide for the registration of, with collectors of internal revenue, and to impose a special tax on all persons who produce, import, manufacture, compound, deal in, dispense, sell, distribute, or give away opium or coca leaves, their salts, derivatives, or preparations, and for other purposes.

This act required opium producers and sellers to register, pay taxes, and keep records of their drugs. Registered physicians could prescribe opioids, but heroin production and import were banned in 1924, effectively reducing access to opioids. Rigorous application of sanctions to sellers reduced the legal production and use of opioids for pain treatment, and the illegal heroin market increased. According to Austin, from 1925 to 1950, opioid use was considered a behavior that inevitably led to "dependence, death, and a criminal, immoral lifestyle" [17]. Stigma became firmly established in society, and it became effectively unthinkable to seek therapeutic options for opioid users.

The twentieth century introduced the continuous challenge of searching for a balance between meeting the needs of opioids for patients in pain, providing attention to people with opioid use disorders (OUDs), and preventing the spread of opioid misuse and its associated harms (Chap. 2).

After the introduction of the Harrison Narcotics Tax Act, the prohibition of opioids in the United States, and the adoption of restrictions on recreational opioid use by the League of Nations in 1931 [3], using opioids became a crime punished by prison. In 1935, a new initiative took form with the opening of the US Narcotic Farm in Lexington, Kentucky. Conceived as a hybrid federal prison/hospital, it aimed to rehabilitate inmates with addiction, using occupational therapy (work at the farm), psychoanalysis, and proper nutrition in a healthy environment. The farm also housed the Addiction Research Center (ARC) to study the effects of drugs on consenting inmates. People with opioid addiction could enter the facilities to seek treatment. This was a well-meaning approach to addiction when strict ethical guidelines for human studies did not exist. Animal studies were being performed simultaneously in the ARC to understand the effects of drugs in the organism. People seeking treatment turned themselves in, got clean, exited, relapsed, and returned. During this time, methadone was first used as medication-assisted therapy for inmates (see Chap. 14). Methadone was soon recognized as a molecule of interest because it is a good analgesic, produces less euphoria than morphine, has a slower onset, and has a longer duration of action. Moreover, cessation of methadone produces a less intense abstinence syndrome [18]. Despite these findings, clinical physicians realized that "curing addiction" was not an easy task, and the idea of rehabilitation progressively faded away.

On the other hand, the ARC received funding to test new synthesized and marketed drugs on inmates that voluntarily consent to participate. As a result, some prisoners received not only opioids but also barbiturates or LSD for some time and were then put into withdrawal to determine drug addiction liability and the intensity and duration of the abstinence syndrome. The Narcotic Farm closed in 1975. In terms of science, it was a big success because the ARC was the leading provider of science-based information on the behavioral effects of drugs in the world; however, as a rehabilitation center, it never worked [19].

1.6 Synthetic Opioids, Opioid Receptors, and the Discovery of Endogenous Opioid Peptides

1.6.1 Opioid Synthesis

Another milestone in opioid history is the elucidation of morphine's chemical structure by Robert Robinson in 1925, an accomplishment that led to him being awarded the Nobel Prize in Chemistry in 1947 [15].

In the decades after Robinson's work, the search for "the holy grail," i.e., effective analgesics with fewer adverse side effects, led to the active synthesis and screening of hundreds of morphine's derivatives. Preclinical investigation became common, and new substances underwent several animal tests to determine if they had actions similar to morphine but with fewer undesirable effects [20]. This work led to the synthesis of drugs that are still used for pain treatment (Chaps. 10 and 11) and other ailments (Chap. 8), drugs used to treat OUDs (Chaps. 14 and 15), and drugs that were discarded as failed analgesics and are now sold as "new psychoactive substances" (Chap. 16). Extensive research from 1950 onward also led to identifying opioid receptors and endogenous peptides with morphine-like effects and a better understanding of dependence and tolerance development (Chap. 13).

As previously mentioned, the characterization of the effects of semisynthetic and synthetic molecules required extensive human and animal testing. Progressively, methods to test opioids in simplified preparations became common. One of these was the isolated guinea-pig ileum, a segment of the small intestine that contracted in response to electrical stimulation and the subsequent acetylcholine release. This preparation allowed the test of opioid effects on many neurotransmitter systems because the ileum has a neuronal network between the longitudinal and transversal smooth muscle layers of the intestine (the myenteric plexus) considered a miniature nervous system [21]. Identification of agonist and antagonist actions was relatively easy with this preparation. Agonists like morphine inhibited acetylcholine release and muscle contraction, whereas antagonists reversed the effect of morphine. Interestingly, when morphine remained in touch with the intestine for prolonged periods, naloxone not only counteracted morphine's actions but also produced a muscle contraction that was later recognized as an abstinence response, indicative of the dependence liability of opioid agonists [22].

Meperidine was the first synthetic opioid drug with a very different chemical structure from morphine. Then came methadone, in 1946, also with an unusual chemical structure. Finally, in 1972, methadone became the first approved opioid agonist for OUD treatment (Chap. 14).

Nalorphine, or N-allyl-morphine, unlike meperidine and methadone, shares many structural features with morphine but has a very different pharmacological profile. Synthesized by Weijlard and Erikson in 1942, nalorphine has both agonist and antagonist properties. On the one hand, nalorphine produces analgesia when given by itself, but it antagonizes morphine-induced respiratory depression and sometimes produces dysphoric effects. Because of this, nalorphine and other compounds with similar actions are called mixed agonist-antagonist opioids [18, 23]. The adverse effects of nalorphine restricted its potential use as an analgesic drug, but further chemical modifications produced naloxone and naltrexone, two molecules devoid of agonist properties that effectively counteract the effects of morphine, heroin, and other morphine-like drugs.

1.6.2 Opioid Receptor Subtypes

The different pharmacological profiles of semisynthetic and synthetic opioids and the extreme potency of some drugs pointed to highly specific receptors. In 1976, Gilbert and Martin suggested the existence of three receptor subtypes, the μ (mu)-opioid receptor that mediated the effects of morphine and morphine-like drugs, the κ (kappa)-opioid receptor where drugs like ketocyclazocine and ethylketocyclazocine acted, and the σ (sigma)-receptors that mediated the effects of SKF 10.047.

Later, it was recognized that the σ-receptor has different pharmacology and functions and is not an opioid receptor [22].

In 1973, Candace Pert and Solomon Snyder [24] demonstrated stereospecific opiate binding sites in the nervous system. Stereospecificity, in this case, means that of the two isomers of each opioid (levorotatory and dextrorotatory), only the levorotatory can bind opium receptors. Pert and Snyder also provided a simple method to distinguish opioid receptor agonists from antagonists by adding sodium to in vitro preparations expressing opioid receptors. When sodium was present, antagonists increased their binding, and agonists decreased it [25]. This "sodium effect" not only allowed researchers to predict the effects of newly synthesized drugs but also advanced the understanding of how receptors behaved (Chap. 9). In 1976, Kosterlitz identified the δ (delta)-opioid receptors in the mouse *vas deferens*. However, identifying opioid receptor subtypes as biochemical entities occurred until the early 1990s [3, 26]. More recently, Maurice and Su recognized the nociceptin/orphanin FQ receptor as the newest member of the opioid receptor family [27].

1.6.3 Endogenous Opioid Peptides

The search for endogenous opioids was a logical consequence of previous work to understand why animals have receptors for a plant-based drug. In the early 1970s, Hughes and Kosterlitz in the United States and Terrenius and Wahlstrom in Sweden identified enkephalin, the first endogenous opioid peptide. Its name means "in the brain" because it was isolated from animal brain extracts. Other peptides and their precursors' identification quickly followed [22]. At about the same time, Pert and Snyder refined binding techniques, which help identify opioid receptor distribution throughout the nervous system. The mapping of opioid receptors was essential to understand the role of endogenous opioid peptides and exogenous opioids' effects in the body. Since then, it was possible to better understand the multiple roles of endogenous and exogenous opioids, their therapeutic possibilities, and the risks associated with opioid misuse [20] (see Table 1.1).

1.7 Opioids in the Context of Globalization

After three centuries of legal trade during colonialism, the prohibition of nonmedical opioid use by the United Nations in 1961 could not repress the opium trade [16]. Opium trade did not disappear, but it shifted from a public and legal income to a central, still public, but now illegal and secret income to fund counterinsurgency [28]. In 1969, heroin refineries opened in the Golden Triangle (Thailand, Laos, and Myanmar), intended to satisfy the opioid's demand by soldiers fighting in Vietnam and later by people in Europe and the Americas. In the same year, Iran absorbed the

Table 1.1 Timeline of opioid history and research

Date	Events
c. 3000 BC	Ancient cultures used and cultivated the opium poppy. The Sumerians referred to it as the "plant of joy"
c. 1500–1300 BC	From Egypt, the Ebers Papyrus included opium as a remedy "to prevent excessive crying of children." There were poppy fields and opium trade in Egypt by 1300 BC
c. 400 BC	Hippocrates used opium to treat insomnia
First century	Dioscorides wrote *De Materia Medica*, where he described how to harvest opium and its efficacy as a drug to treat pain, coughing, and diarrhea
Eighth century	Arab traders introduced the opium poppy to China, Persia, India, and North America
Eleventh century	Avicenna, in his *Canon of Medicine*, wrote that opium was "the most powerful of stupefacients" and warned of the risk of overdose
Twelfth century	China used medicinal opium extensively
Sixteenth century	**1524** Poppy cultivation was a state monopoly in India **1541** Paracelsus prepared "laudanum," a liquid with opium, citric juices, species, powdered gold, crushed pearls, and corals
Seventeenth century	Opium was smoked in China, Vietnam, and Taiwan **1680** Thomas Sydenham introduced a simplified version of laudanum with opium, wine, and herbs to treat numerous ailments. The drink rapidly became popular
Eighteenth century	**1733** Thomas Dover wrote *The Ancient Physician's Legacy to His Country, Being What He Has Collected Himself in Forty-Nine Years of Practice*, where he gave the formula for his famous Dover's powders with opium and ipecacuanha to treat "fevers"
Nineteenth century	Opium use was common and legal in Britain, America, and Western Europe **1803** Friedrich Sertürner isolated the active compound of opium and named it "morphium" (morphine), after Morpheus, the god of dreams **1827** E. Merck and Co. commercialized morphine Morphine-containing remedies were prevalent, in particular, Mrs. Wilson's Soothing Syrup for teething babies **1839–1841** The First Opium War. The British sent warships to China in response to China's decision to suppress opium traffic. In 1841, China was forced to pay an indemnity and to cede Hong Kong to Britain **1843** The hypodermic syringe was introduced and, with it, a new and more efficient route of drug administration **1856–1860** The Second Opium War. England and France defeated China, which was forced to legalize opium importation **1861–1865** The US Civil War. Morphine use was known as the soldiers' disease **1874** Charles Romley Alder Wright synthesized heroin **1898** The Bayer Company introduced heroin for medical use

(continued)

Table 1.1 (continued)

Date	Events
Twentieth century	Heroin addiction rose to alarming rates in the United States **1906** Britain halted opium sells to China **1914** The US Senate signed the Harrison Narcotics Tax Act controlling opium **1925** Robert Robinson identified morphine's chemical structure **1935–1975** Lexington's "Narcotic Farm" opened. It was a federal prison and a hospital run by physicians for persons with opioid addiction. In addition, it housed the Addiction Research Center to conduct studies on drug effects on inmates **The 1950s–1960s** Clinical and preclinical characterizations of different opiate compounds lead to proposing the existence of opioid receptors. In 1967, Billy Martin suggested the existence of more than one opiate receptor **1961** Signing of the Single Convention of Narcotic Drugs **1968** The International Narcotics Control Board was established **1972** Methadone, first synthesized for use as an analgesic in World War II, was approved by the Food and Drug Administration (FDA) to treat opiate addiction **1973** Opioid receptors were identified and characterized in binding assays **1975** Identification of endogenous opioid peptides **1992–1993** Cloning of delta-, mu-, and kappa-opioid receptors **1994** Cloning of the nociceptin/orphanin FQ receptor
Twenty-first century	The number of opioid prescriptions and opioid-related deaths increased in the United States **2002** The FDA approved buprenorphine for opioid addiction treatment **2010s** A growing number of synthetic opioids were sold as new psychoactive substances. There was extensive characterization of biased ligands, allosteric modulators of opioid receptors, single-nucleotide polymorphisms (SNPs), and opioid receptor dimers Opioid overdose deaths increased in the United States and Canada, involving heroin, fentanyl, and fentanyl analogs **2020** Nonmedical use of pharmaceutical opioids became prevalent in many countries, with almost 58 million people using opioids globally (prescription or illicit) This situation coexisted with the COVID-19 pandemic

most surplus of opium worldwide, and Turkey provided around 80% of the heroin used in the United States [16].

In the 1970s, President Richard Nixon's War on Drugs complicated the opium trade, intensified criminalization, and caused more deaths. One of the unintended consequences was the price increase and purity decrease of street opioids. The eradication of poppy cultivation in Asia, along with the international policy strategies of President Jimmy Carter, contributed to the steep reduction of opium traffic from Asia to the United States, followed by the increased production in Mexico and Colombia to cover unmet demand [16].

Opium trade from Asia to Europe diminished in the Cold War, and smuggling routes closed under the Chinese and Soviet Union communist regimes. However, in the years after the collapse of the Iron Curtain, new sources of supply and demand developed in former soviet countries [16].

According to Ciccarone [29], three independent waves of opium misuse can be identified since the 1990s. The first wave initiated when pharmaceutical companies developed long-acting opioid formulations, which were advertised to general practitioners and patients as safe analgesics devoid of addiction risk.

The second wave, intertwined with the first, consisted of a spillover effect of persons who developed OUDs from prescription opioids and transitioned to heroin use due to its lower cost and availability. This transitioning coincided with the reformulation of oxycodone to be abuse-deterrent.

The third wave began in 2013, driven by synthetic opioids, mainly fentanyl and its analogs (e.g., carfentanil, sufentanil). Synthetic opioids and their precursors are primarily manufactured in China and shipped to the Americas. Mexico, Canada, and the United States have also reported significant seizures.

In 2016, the government of British Columbia, Canada, declared a public health emergency related to the opioid overdose epidemic [30]. In 2018, prescription opioids, heroin, and synthetic opioids were involved in 70% of all fatal overdose events [31].

The transition from prescription opioids to heroin is far more frequent than contrariwise; in the United States, approximately 79.5% of persons with heroin use had previously used prescription opioids, whereas only 3.6% transitioned the other way around [32]. The opioid crisis is further explained in Chap. 5.

1.8 Evolution of Concepts Associated with Substance Use Disorders

The use of a particular substance, including opioids, was not considered problematic by the society until it was associated with contexts, behaviors, and specific populations, such as race, socioeconomic status, and criminality [33]. Although substance use-related behaviors have been criticized and seen as a moral failure throughout history, the scientific classification of substance use disorders (SUDs) only began in the early nineteenth century [7].

The American Psychiatric Association (APA) has published several editions of the *Diagnostic Statistical Manual of Psychiatric Disorders* (DSM). This manual is one of the two most important categorizations of psychiatric disorders internationally. The other, published by the World Health Organization, is the International Classification of Diseases [ICD].

The DSM-I and DSM-II included addiction as a manifestation of underlying psychopathology [7, 34]. Specifically, addiction was one of the four conditions of sociopathic personality disturbance, and it was considered a transient situational personality disorder [34]. The first two DSM editions were based on psychodynamic theory, while DSM-III and the following versions have been atheoretical, relying mainly on empirical data [34] (Table 1.2).

Table 1.2 Evolution of addiction concepts

Year	Document/meeting	Comments
1952	**DSM-I** (*Diagnostic and Statistical Manual of Mental Disorders*) Addiction is a symptom of a "sociopathic personality disorder"	Concepts are heavily influenced by psychoanalysis
1957	**WHO Expert Committee on Addiction-Producing Drugs** Drug addiction has "detrimental effects on the individual and on society." "Drug habituation" has "detrimental effects, if any, primarily on the individual"	A group of experts distinguishes between addiction and drug habituation
1967	**ICD-8** (International Classification of Diseases) For the first time, "substance use disorders" (SUDs) are included as a category independent of personality disorders	
1968	**DSM-II** Alcoholism and drug dependence are still specific categories of "personality disorders and certain other non-psychotic disorders"	Signs and symptoms are poorly defined, and stigma prevails
1973	**Symposium on Psychic Dependence** Psychic dependence is "a state where a person feels an overpowering desire, a compulsion or an irresistible drive to a) obtain sensations in mood which are experienced as pleasurable (positive reinforcements), or b) to avoid discomfort (negative reinforcements)"	
1980	**DSM-III** Substance abuse is characterized by a pattern of pathological drug use, impairment in social or occupational functioning associated with it, and a duration of at least 1 month. Substance dependence is more severe than abuse and occurs when there is evidence of "physiological dependence," such as tolerance or withdrawal	"Substance use disorders" (SUDs) is a category on its own. There is a distinction between dependence and abuse
1987	**DSM-III-R** Substance abuse is "hazard use or continued use despite social consequences." Substance dependence occurs if three of nine symptoms related to tolerance, withdrawal, compulsive use, and negative consequences are present for at least a year	Substance dependence includes physiological and behavioral symptoms and is still different from abuse
1992	**ICD-10** "Harmful drug use" is a pattern of use that causes damage to physical or psychological health. Drug dependence is defined by the presence of at least three of several criteria in the past year: tolerance, withdrawal, craving, difficulties controlling the onset, levels of drug use, increased time using the drug, reducing activities for using, and use despite negative consequences	There is a distinction between harmful drug abuse and drug dependence
1994	**DSM-IV** Abuse definition is similar to DSM-III, but neither tolerance nor withdrawal is necessary or sufficient for diagnosing dependence	This version makes a distinction between substance abuse, dependence, intoxication, and withdrawal

(continued)

Table 1.2 (continued)

Year	Document/meeting	Comments
2013	**DSM-5** "Substance-related and addictive disorders" include "opioid use disorder," defined as "a pattern of maladaptive opioid use that leads to significant impairment or distress." As with other SUDs, it can be mild, moderate, or severe	The term "addict" is no longer used
2019	**ICD-11** "Disorders due to substance use and addictive behaviors" include a spectrum of substance use: low-risk use, hazardous substance use, episode of harmful substance use, harmful patterns of substance use, or dependence	Addiction can occur even without substances (e.g., gambling)

In 1980, the DSM-III first included SUD as an independent psychiatric disorder and classified substances into narcotics, depressants, stimulants, and hallucinogens [35]. Moreover, the DSM-III distinguished between use, abuse, and dependence [34]. In the DSM-IV, published 14 years later, 11 drug classes appeared, including opiates [36]. The DSM-IV renamed the category "substance-related disorders" and distinguished substance use disorders (abuse and dependence) from substance-induced disorders (e.g., intoxication, withdrawal, substance-induced delirium) [34]. In this edition, tolerance and withdrawal were not necessary or sufficient for a SUD diagnosis, but only specifiers (with tolerance or withdrawal or both) [34]. In the context of the transition from the fourth to the fifth DSM edition, several statistical analyses and clinical studies showed no evidence to support the hierarchical distinction between abuse and dependence. For instance, in national representative samples of adolescents [37] and adults [38], OUD showed a statistically better fit when the categories of abuse and dependence were combined into a dimensional approach.

In the current edition (DSM-5), published in 2013, the category again changed to "substance-related and addictive disorders" [39]. This edition dropped the term "addiction" for the stigmatizing connotation it has had in academia and clinical settings and among the general population. Also, each substance-related disorder, such as OUD, can be mild, moderate, or severe. Finally, an OUD is defined in this edition as "a pattern of maladaptive opioid use that leads to significant impairment or distress" (see Chap. 13) [40].

The ICD-11 includes disorders due to substance use or addictive behaviors embedded in mental, behavioral, or neurodevelopmental disorders. The disorders due to substance use consider four hierarchical and mutually exclusive diagnoses: substance dependence, harmful pattern of substance use, episode of harmful use, and hazardous substance use [41, 42]. Additionally, the ICD-11 includes substance intoxication, substance withdrawal, and substance-induced mental disorders (Table 1.2).

As explained in this chapter, opioids have been used medically and for their psychotropic effects for several centuries. It is important to distinguish between physical dependence and OUDs [40]. Physical dependence may be present in most patients who are adequately prescribed opioids as pain medication and who may suffer withdrawal symptoms if opioid administration is abruptly stopped. However, that does not imply that they became "addicted." An OUD is the clinical assessment of significant impairment or distress [40]. About 10% of persons with long-term opioid prescription use may develop OUDs, which will have long-lasting effects after drug cessation [43].

1.9 Conclusions

The history of opioid use cannot be disentangled from the geopolitical context of its production. Politics plays a crucial role in who uses which opioids and how. It also determines what kind of treatment is available and who receives it. For centuries, opium has been a medicinal and psychotropic substance; it has accompanied civilizations, mainly as a legal commodity with a sociocultural value. Colonialism expanded opium trade, and scientific advances allowed for the development of more potent substances and efficient and effective routes of administration. Such advances should improve lives, not harm them.

In the following chapters, the authors present an overview of challenges posed by opioids as excellent medications, which can produce fatal overdoses when misused, the need for tailor-made regulations to control their use, the impact of OUDs in people's living conditions, and the structural interventions that have proven to reduce opioid-induced harms. The second part of this book reviews the main opioids, their properties, and chemical structures; opioids used in pain; opioid effects on the immune system; tolerance and dependence development and the mechanisms involved; medication-assisted treatment for OUDs; and opioids as new psychoactive substances.

References

1. Jesus A, Bonhomme V, Evin A, Ivorra S, Soteras R, Salavert A, et al. A morphometric approach to track opium poppy domestication. Sci Rep [Internet]. 2021;11(1):9778. Available from: https://doi.org/10.1038/s41598-021-88964-4
2. Heyman GM, Mims V. What addicts can teach us about addiction: a natural history approach. In: Heather N, Segal G, editors. Addiction and choice: rethinking the relationship [Internet]. Oxford University Press; 2016. p. 385–408. Available from: http://search.ebscohost.com/login.aspx?direct=true&db=psyh&AN=2016-47926-021&site=ehost-live.
3. Booth M. Opium: a history. Simon & Schuster Ltd; 1996. p. 359.
4. López Eire A, Cortés Gabaudan F. Hierbas y plantas: Adormidera. Dioscórides interactivo. Sobre los remedios medicinales-manuscrito de Salamanca, 2006.
5. Aziz E, Nathan B, McKeever J. Anesthetic and analgesic practices in Avicenna's Canon of Medicine. Am J Chin Med. 2000;28(1):147–51.
6. Heydari M, Hashempur MH, Zargaran A. Medicinal aspects of opium as described in Avicenna's Canon of Medicine. Acta Med Hist Adriat [Internet]. 2013 [cited 2021 Sep 15];11(1):101–12. Available from: https://pubmed.ncbi.nlm.nih.gov/23883087/

7. Robinson S, Adinoff B. The classification of substance use disorders: historical, contextual, and conceptual considerations. Behav Sci (Basel) [Internet]. 2016;6(3):1–23. Available from: http://www.mdpi.com/2076-328X/6/3/18
8. Fernandez H, Libby TA. Heroin: its history, pharmacology, and treatment, vol. 148. 2nd ed. Hazelden Publishing; 2011. p. 148–62.
9. University of Southern California. Lin Zexu (LinTse-hsu) writing to Britain's Queen Victoria to Protest the Opium Trade, 1839 | US-China Institute [Internet]. US-China Institute. 2021 [cited 2021 Sep 15]. Available from: https://china.usc.edu/lin-zexu-lintse-hsu-writing-britains-queen-victoria-protest-opium-trade-1839
10. Presley CC, Lindsley CW. DARK classics in chemical neuroscience: opium, a historical perspective. ACS Chem Neurosci. 2018;9(10):2503–18.
11. Osler W. A Contribution to medical history: Thomas Dover, M.B. (of Dover's Powder), physician and buccaneer. Lancet [Internet]. 1896;147(3775):4–7. Available from: https://linkinghub.elsevier.com/retrieve/pii/S0140673601606858
12. The Lancet Commission. Sanitary condition of pur public schools [Internet]. 1875. Available from: https://ia800708.us.archive.org/view_archive.php?archive=/22/items/crossref-pre-1909-scholarly-works/10.1016%252Fs0140-6736%252802%252930549-x.zip&file=10.1016%252Fs0140-6736%252802%252930644-5.pdf.
13. Engels F. The conditions of the working class in England, Oxford World's classics. Oxford University Press; 1844.
14. Lawrence G. The hypodermic syringe. Lancet [Internet]. 2002;359(9311):1074. Available from: https://linkinghub.elsevier.com/retrieve/pii/S0140673602081011.
15. Brook K, Bennett J, Desai SP. The chemical history of Morphine: an 8000-year journey, from resin to de-novo synthesis. J Anesth Hist [Internet]. 2017;3(2):50–55. Available from: https://doi.org/10.1016/j.janh.2017.02.001.
16. McCoy A. From free trade to prohibition: a critical history of the modern Asian opium trade. Fordham Urban Law J. 2000;28(1):307–49.
17. Austin GA. Perspectives on the history of psychoactive substance use [Internet]. Rockvile: National Institute on Drug Abuse; 1978. 280 p. Available from: https://books.google.com.mx/books?hl=es&lr=&id=UQE3rikfXSgC&oi=fnd&pg=PR5&dq=Perspectives+on+the+history+of+psychoactive+substance+use&ots=oS59dkrNbj&sig=kbFIvuZEX3ujqIfg_-WVChrpoKA.
18. Brownstein MJ. A brief history of opiates, opioid peptides, and opioid receptors. Proc Natl Acad Sci U S A [Internet]. 1993;90(12):5391–3. Available from: https://www.mendeley.com/viewer/?fileId=2748b363-305f-4311-6a3c-171b74a43332&documentId=6b4dc5ca-8403-3307-9265-ceb50db3e6ec
19. Campbell ND. "A new deal for the drug addict": The Addiction Research Center, Lexington, Kentucky. J Hist Behav Sci [Internet]. 2006;42(2):135–57. Available from: https://onlinelibrary.wiley.com/doi/10.1002/jhbs.20167
20. Corbett AD, Henderson G, McKnight AT, Paterson SJ. 75 years of opioid research: the exciting but vain quest for the Holy Grail. Br J Pharmacol [Internet]. 2006;147(SUPPL. 1):153–62. Available from: https://onlinelibrary.wiley.com/doi/10.1002/jhbs.20167
21. Kosterlitz HW, Waterfield AA. The effect of the interval between electrical stimuli on the acetylcholine output of the myenteric plexus-longitudinal muscle preparation of the guinea-pig ileum. Br J Pharmacol [Internet]. 1970;40(1):162P–3P. Available from: https://www.ncbi.nlm.nih.gov/pmc/articles/PMC1702700/
22. Cox BM. A concise review of concepts in opioid pharmacology up to the discovery of endogenous opioids. Mol Pharmacol [Internet]. 2020;98(4):392–400. Available from: http://molpharm.aspetjournals.org/lookup/doi/10.1124/mol.120.119420
23. Harris LS, May EL. Historical introduction and review of chemistry. Drug Alcohol Depend [Internet]. 1985;14(3–4):227–32. Available from: https://linkinghub.elsevier.com/retrieve/pii/0376871685900584
24. Pert CB, Pasternak G, Snyder SH. Opiate agonists and antagonists discriminated by receptor binding in brain. Science (80-). 1973;182(4119):1359–61.

25. Pert CB, Snyder SH. Opiate receptor binding of agonists and antagonists affected differentially by sodium. Mol Pharmacol. 1974;10(6):868–79.
26. Snyder SH, Pasternak GW. Historical review: opioid receptors. Trends Pharmacol Sci. 2003;24(4):198–205.
27. Maurice T, Su TP. The pharmacology of sigma-1 receptors. Pharmacol Ther [Internet]. 2009;124(2):195–206. Available from: https://doi.org/10.1016/j.pharmthera.2009.07.001.
28. McCoy AW. Coercion and its unintended consequences: a study of heroin trafficking in Southeast and South West Asia. Crime Law Soc Chang. 2000;33(3):191–224.
29. Ciccarone D. The triple wave epidemic: supply and demand drivers of the U.S. opioid overdose crisis. Int J Drug Policy. 2019;71:183–8.
30. Bardwell G, Boyd J, Tupper KW, Kerr T. "We don't got that kind of time, man. We're trying to get high!": exploring potential use of drug checking technologies among structurally vulnerable people who use drugs. Int J Drug Policy. 2019;71:125–32.
31. Wilson N, Kariisa M, Seth P, Smith H, Davis NL. Drug and opioid-involved overdose deaths — United States, 2017–2018. MMWR Morb Mortal Wkly Rep [Internet]. 2020;69(11):290–7. Available from: https://www.cdc.gov/nchs/data/nvsr/nvsr61/
32. Terry NR. Structural determinism amplifying the opioid crisis : It's the healthcare, stupid !*. Northeast Univ Law Rev [Internet]. 2018;11(1):315–71. Available from: https://www.samhsa.gov/data/sites/default/files/NSDUH-
33. Martin CS, Chung T, Langenbucher JW. Historical and cultural perspectives on substance use and substance use disorders. In: Sher KJ, editor. The Oxford handbook of substance use disorders, volume 1 [Internet]. Oxford University Press; 2014. p. 29–49. Available from: http://search.ebscohost.com/login.aspx?direct=true&db=psyh&AN=2016-26127-001&site=ehost-live.
34. Nathan PE, Conrad M, Skinstad AH. History of the concept of addiction. Ann Rev Clin Psychol [Internet]. 2016;12(1):29–51. Available from: http://www.annualreviews.org/doi/10.1146/annurev-clinpsy-021815-093546
35. Cotman CW, McGaugh JL. Behavioral Neuroscience. New York: Academic Press; 1980. p. 857.
36. American Psychiatric Association. Diagnostic and statistical manual of mental disorders. 4th ed. Washington, DC: American Psychiatric Association; 2000. p. 866.
37. Wu L-T, Ringwalt CL, Yang C, Reeve BB, Pan J-J, Blazer DG. Construct and differential item functioning in the assessment of prescription opioid use disorders among American adolescents. J Am Acad Child Adolesc Psychiatry [Internet]. 2009;48(5):563–72. Available from: https://linkinghub.elsevier.com/retrieve/pii/S0890856709600733
38. Wu L-T, Woody GE, Yang C, Pan J-J, Blazer DG. Abuse and dependence on prescription opioids in adults: a mixture categorical and dimensional approach to diagnostic classification. Psychol Med [Internet]. 2011;41(3):653–64. Available from: https://www.cambridge.org/core/product/identifier/S0033291710000954/type/journal_article
39. American Psychiatric Association. Diagnostic and statistical manual of mental disorders [Internet]. 5th ed. Washington, DC: American Psychiatric Publishing; 2013. p. 947. Available from: https://linkinghub.elsevier.com/retrieve/pii/B0126574103004578
40. Volkow ND, Blanco C. Medications for opioid use disorders: clinical and pharmacological considerations. J Clin Invest [Internet]. 2020;130(1):10–3. Available from: https://www.jci.org/articles/view/134708
41. Poznyak V, Reed GM, Medina-Mora ME. Aligning the ICD-11 classification of disorders due to substance use with global service needs. Epidemiol Psychiatr Sci. 2018;27(3):212–8.
42. World Health Organization. ICD-11 - ICD-11 for mortality and morbidity statistics [Internet]. International Statistical Classification of diseases and related health problems. 2019. Available from: https://icd.who.int/browse11/l-m/en#/http%3A%2F%2Fid.who.int%2Ficd%2Fentity%2F1602669465.
43. Kalivas PW, Volkow ND. The neural basis of addiction: a pathology of motivation and choice. Am J Psychiatry [Internet]. 2005;162(8):1403–1413. Available from: http://psychiatryonline.org/doi/abs/10.1176/appi.ajp.162.8.1403.

[illegible]

Chapter 2
The Two Sides of Opioid Use: Unmet Needs of Opioids for Pain Management and the Role of Opioids in Substance Use Disorders

María Elena Medina-Mora, Raúl Martín-del-Campo, Nayely V. Salazar-Trujillo, Hilda Dávila, Clara Fleiz, and Jorge Villatoro

Abstract This chapter describes the two sides of the opioid crisis: the unmet needs of opioids for pain management and the role of prescription and nonprescription opioids (heroin, fentanyl, and new synthetic opioids) in substance use disorders. It draws attention to the unfulfilled United Nations General Assembly (UNGASS) agreements to adopt a public health approach, the United Nations Sustainable Development Goals to strengthen prevention and treatment, the resolution of the World Health Assembly to develop palliative care, and the recommendation by the United National Academies to balance individual and societal needs. It describes illegal and regulated markets and their role in the opioid crisis, pain management and the disparities in the availability of medication between high- and low- and

M. E. Medina-Mora (✉)
National Autonomous University of Mexico (UNAM), Mexico City, Mexico

National Institute of Psychiatry, Ramón de la Fuente Muñiz (INPRM), Mexico City, Mexico

Opioids Working Group, Global Studies Seminar, Faculty of Medicine, UNAM, Mexico City, Mexico

R. Martín-del-Campo
National Institute of Psychiatry, Ramón de la Fuente Muñiz (INPRM), Mexico City, Mexico

Opioids Working Group, Global Studies Seminar, Faculty of Medicine, UNAM, Mexico City, Mexico

International Narcotics Board Control, United Nations, NewYork, NY, USA

N. V. Salazar-Trujillo · H. Dávila
Opioids Working Group, Global Studies Seminar, Faculty of Medicine, UNAM, Mexico City, Mexico

C. Fleiz · J. Villatoro
National Institute of Psychiatry, Ramón de la Fuente Muñiz (INPRM), Mexico City, Mexico

Opioids Working Group, Global Studies Seminar, Faculty of Medicine, UNAM, Mexico City, Mexico

S. L. Cruz (ed.), *Opioids*, https://doi.org/10.1007/978-3-031-09936-6_2

middle-income countries, the world opioid problem, the role of stigma in inadequate access to care and social pain, opioid-assisted treatment, and policy recommendations.

Keywords Opioids · Pain management · Nonmedical use · Stigma · Prevention · Treatment

2.1 Introduction

Nonmedical opioid use has spread around the world. The prevalence during 2018 was estimated at 62 million people globally (1.1% of the population), while 70% (0.5 million) of the total mortality rate attributable to drug use was related to opioids. Of these deaths, 30% were due to overdose. Furthermore, according to the United Nations Office on Drugs and Crime (UNODC), opioid use in the past decade doubled due to increased opioid use in Asia and Africa [1, 2]. Despite this, five billion people worldwide lack access to prescription drugs. At the same time, low- and middle-income countries (LMICs) have access to just 12.8% of the morphine produced for medical purposes, far less than their actual needs [3].

This chapter outlines opium use disorders (OUDs) due to heroin and, more recently, fentanyl and new synthetic substances that rely on non-regulated markets. It also describes how some people developed OUDs through the prescription of narcotics available on the regulated market, detailing the transitions between the two markets, on the one hand, and the unmet needs of people with OUDs and pain-related disorders, on the other hand.

2.1.1 Background

People with drug use disorders, particularly those who inject drugs, are discriminated against and have limited or no access to treatment and social services. These problems are still happening in 2022, 6 years after the 30th Special Session of the United Nations General Assembly (UNGASS) [4] that reflected the global interest in public health, rather than a criminal justice approach to address drug problems, and a renewed interest in prevention, treatment, and care and harm reduction interventions [5]. The aforementioned issues prevail 7 years after adopting the United Nations Sustainable Development Goals [6], which involved strengthening the prevention and treatment of substance abuse, including narcotic drugs.

There is an urgent need to advance the reduction of stigma and discrimination, as well as treatment gaps. Unfortunately, however, there are not enough actions to accompany the World Health Assembly resolution on "strengthening palliative care as a component of comprehensive care throughout life" [7, 8], urging national governments to take actions to develop palliative care and calling for its inclusion in the Sustainable Development Goals.

In 2017, the United National Academies of Sciences, Engineering, and Medicine of the United States [9] recommended developing improved methods for measuring pain and the effects of alternative treatment modalities, as well as implementing intensive surveillance of opioid-related harm within a balanced public health framework. These initiatives provide opportunities for a new, integrated policy to address public health challenges, the increasing magnitude of OUDs, and the insufficient availability of pain medication for those in need. International cooperation is essential to meet the goals adopted globally.

2.2 Drug Markets

Drug markets are part of today's world, and understanding their supply and demand forces is an essential component of an integrated policy. In the early 1970s, drugs could be bought at unregulated open markets on the streets. Potential retail markets included people who used drugs and sold them to meet their own needs, while places to buy drugs included bars, meeting places outside schools or parks, and other spots where young people congregated. Dealing houses were also open, and some people synthesized stimulants at homes in communities. Gangs organized and fought for control of different sites, and the increase in drug problems elicited an increased police response [10].

Developing new formulas for clients led some dealers to produce mixtures to satisfy customers' needs and test the new combinations themselves [11]. Since the early 2000s, sales tactics changed to be less visible. Using mobile phones and other technologies [12] as well as organized mail system became common methods to deliver orders. Also, the darknet and group organizations are used to supply nearby markets. These distribution systems coexist with middle-level markets that support the retail trade [10].

On the other hand, regulated markets for pain management pharmaceuticals have grown mainly because of the aging population and a higher prevalence of chronic diseases such as cancer, diabetic neuropathy, and osteoarthritis. This situation has contributed to the opioid epidemic with two coexisting public health challenges: reducing the burden of pain, increasingly regarded by physicians as a disease in itself, and controlling the rising costs associated with the use of opioid medicines [13].

2.2.1 Unregulated Markets

According to Babor et al. [14], two non-regulated markets warrant consideration: those for non-authorized drugs, such as heroin, and those of diverted psychopharmaceuticals. Both markets intersect in numerous ways. Their existence and characteristics shape drug use, its consequences, and policies, sometimes unintentionally. Black markets can harm producers, distributors, buyers, and the community in many ways, including corruption and violence. They also raise questions about prohibition as a solution when adverse consequences exceed the harm they are intended to reduce [14].

Only three areas produce opium worldwide: Afghanistan, Myanmar, and Mexico. Supplies for the illicit market originate mainly in these production sites. In contrast, the origin of synthetic opioids is far less geographically limited, with manufacturing concentrated in East and Southeast Asia and, to a lesser extent, North America and Europe. Some synthetic opioids are legally manufactured before being diverted to illegal channels, while some are illegally manufactured and subsequently trafficked. Drugs reach consuming countries through international trafficking and users through wholesale distribution and retail marketing chains between production and retail sale [1].

Heroin entered the market as a pain medicine in 1898, and by the 1910s, it had become a banned drug of abuse. As a result, its use decreased after 1931. Since then, illicit production has become the primary heroin source, gradually expanding and achieving a tenfold increase after 1970 [15].

Approaches for creating new non-regulated drug markets vary. Some have included market strategies that begin by smuggling drugs from producing areas to consuming countries. Sam Quinones [16] describes one such strategy to introduce black tar heroin into the United States. After arranging for regular supply from Mexico, providers hired poor young people in their hometowns who dreamed of going to the United States to make money. They were trained to look for regular clients by offering a reliable service; they also received substances and vehicles that would not attract attention, in which they could deliver quality products to stable users. When clients could not afford to buy more drugs, these young dealers gave them free doses to prevent users from experiencing withdrawal or changing suppliers. In addition, as these dealers did not use the drug themselves, they were not motivated to adulterate them with other products or alter the usual concentration of their products. With these strategies, they were able to introduce their products to new and unconquered markets where they did not interfere with well-established dealers in big cities.

By the end of the twentieth century, a shift in the opioid market from heroin use to the nonmedical use of pharmaceutical opioids became a problem in countries such as the United States [1, 17], where prescriptions were increasingly filled for chronic non-cancer pain management [18]. After introducing restrictions to reduce hazardous use, an increase in the use of heroin with higher purity occurred, and nowadays, there is a dual epidemic in many parts of the world [19].

The third wave of accidental exposure to opioids was observed when fentanyl was mixed with heroin and then cocaine, methamphetamines, and MDMA or used to produce counterfeit pharmaceuticals such as oxycodone or hydrocodone, which, in turn, led to an increase in opioid-related overdose deaths [1, 20, 21].

People who buy these substances in unregulated markets are often unaware of what they are using, a trend that poses new challenges for treatment [22].

New psychoactive substances have reached this unregulated market. Some sectors of the population that use these new substances are or were opium users. These substances include synthetic research opioids, including fentanyl; some are used in self-medication to cope with pain, stress, and anxiety and, sometimes, withdrawal and dependence [1, 21]. Dealers are unaware of the quality of the product they sell, as are buyers.

Consequences of non-regulated drug production and preparation for retail sale include low quality of products and the inability to sign contracts in the long chain between production and selling to users in the community, often creating political instability and undermining the power of states [14].

In the United States, the shift in the opioid market from heroin use (83% young men) to nonmedical use of pharmaceutical opioids (75% older urban men and women, mainly provided by pharmaceuticals) began in 1977 due to a more than 500% increase in prescriptions for chronic non-cancer pain management. The year 2006 saw a rise in the use of heroin with greater purity and the production of crush-proof pharmaceuticals that were less liable to be misused [17, 23]. Today there is a dual epidemic. In 2019, 9.3 million people self-reported nonmedical use of pharmaceutical opioids in the past year, while 745,000 people reported having obtained heroin in the expanding illicit market. There are estimates that, between 2011 and 2013, the United States consumed 68% of the world's prescribed analgesics. However, these data do not include homeless and institutionalized people. Opioids are estimated to be used by 0.1% of the population, yet account for most overdose deaths [1].

Overdose mortality, the leading cause of premature death among opioid-dependent people, may occur if opioids are combined with depressant drugs or after periods of abstinence due to treatment or incarceration, when tolerance diminishes, or when used in higher purity than usual. Mortality due to overdose in the United States is among the highest globally, where adult life expectancy has fallen in recent years. Moreover, mortality is estimated to have increased since the late 1970s because of the three forms of substance use described: heroin, prescription drugs, and heroin combined with fentanyl. Furthermore, deaths in each cohort have increased, occurring at an earlier age than before [1].

In the United States, overdose deaths due to prescription opioids (natural and synthetic, including methadone) ranked first in all overdose deaths from 2001 to 2015, when other synthetic opioids ranked first. Fentanyl, in particular, accounted for over 36,359 deaths in 2019 [24].

2.2.2 *Regulated Markets*

Regulated markets operate within market economies, with powerful industries, usually international companies, and resources to prevent drug misuse. Diverting from controlled to uncontrolled markets is a common channel for reaching users. Deviation of precursors and counterfeit production is another source. Other methods include doctor shopping, delivery by family members, and leakage of unused medications [14].

In the United States, the shift from illegal heroin to the prescribed market resulted from pharmaceutical actions [23, 25, 26]. Purdue Pharma invested heavily in marketing potent opioids, focusing mostly on prescribers, who received the products in their work setting. The company's marketing message misleadingly stated that patients under medical treatment would be unlikely to develop dependence [27, 28], which led to oxycodone overprescription. Purdue Pharma also lobbied in other countries such as Australia, Brazil, China, Colombia, Egypt, Mexico, the Philippines, Singapore, and South Korea, with prescriptions in Spain reaching record numbers [29].

It is estimated that 21% to 29% of patients who use prescribed opioids to treat chronic pain misuse them. Moreover, between 8% and 12% of people who use an opioid for chronic pain develop an OUD [30], between 4% and 6% of those who misuse prescription opioids go on to use heroin, and approximately 80% of people who use heroin had previously misused prescription opioids [31–33].

The likelihood of developing OUDs depends on many factors, including the length of opioid treatment prescribed for acute pain and how long people continue to take opioids, either as prescribed or inadequately [33, 34].

Another contributing factor to the opium crisis is the low availability of pain management pharmaceuticals and treatment alternatives [3], including medication-assisted treatment with opioid agonists such as methadone or buprenorphine to treat persons with OUDs in regulated markets or as part of free universal health packages. Under-regulated prescription markets, common in low-resource settings, encourage people to turn to heroin because it is easily accessible on the unregulated market.

There is an urgent need for universal coverage for persons in pain and with OUDs, including the availability of medications. Measures to prevent diversion and guidelines for prescription narcotics to prevent dependence are also required [3].

2.3 Opioids for Pain Management

2.3.1 *Pain Treatment Needs*

One of the pillars of palliative care is the availability of pain management medication. Despite this, the International Narcotics Control Board (INCB), a quasi-judicial United Nations body responsible for implementing the International

Conventions that mandate the adequate provision of drugs indispensable for medical purposes, has stated that:

> The differences in consumption levels among countries continue to be very significant. For example, in 2019, 80.4 per cent of the world's population, in low- and middle-income countries, consumed only 12.8 per cent of the total amount of morphine used for the management of pain and suffering, or 1 per cent of the total 379.2 tons manufactured. Although that is a slight improvement from 2014, when 80 per cent consumed only 9.5 per cent, the disparity in the consumption of narcotic drugs for palliative care continues to be a matter of concern [3].

Pain is one of the most devastating and feared complications for patients with cancer and other diseases that require treatment and palliative care to ensure quality of life. It is a source of suffering for patients and their families, and failure to manage it is inhumane. Studies have estimated that pain is a concern for over 10 million people worldwide with some form of cancer. Approximately one-third of cancer patients actively in treatment live with pain, as well as two-thirds of those with the disease in an advanced stage. Attending to the needs of cancer survivors is still a work in progress [35]. The International Association for Hospice and Palliative Care estimated that over 61 million people experience severe health-related suffering (SHS), and 80% of the burden of SHS occurs in LMICs where access is limited [36]. 2.5 million children die with SHS every year, 98% of them in LMICs [37, 38].

Understanding pain as a multidimensional construct and considering the etiology of pain and its intensity is essential to determining the benefits of safe opioid prescription for its clinical management (see Chap. 10).

Pain is a complex experience, modulated by the multiple dimensions of the human being (physiological, sensory, affective, cognitive, behavioral, and sociocultural) [39], including subjective perceptions, requiring a multimodal, personalized diagnostic-therapeutic approach [40].

In 1982, when the World Health Organization (WHO) declared pain a public health problem, it developed a method to assess the "analgesic ladder" [41], and despite constant debate and controversy, pain management has substantially improved since its inception [42].

The analgesic ladder was designed to treat cancer pain at a time when clinical evidence was limited. It was primarily related to the use of specific analgesics depending on three different levels of pain intensity and, as an educational tool, was soon highly appreciated. Over the years, it has undergone various modifications, including assessing other types of pain, such as postoperative pain and non-oncological pain. In addition, a fourth step has been added for intense pain treated with interventionist techniques. At the same time, medication needs for patient support in the context of palliative care have also been considered (see Chap. 10).

The WHO regards a country's medical opioid use as a good predictor of pain management. Opioids are considered the pharmacological treatment of choice in moderate to severe pain according to the analgesic ladder, yet their use remains controversial for chronic non-cancer pain (see Chap. 11), due to limited evidence regarding their efficacy, as well as potential adverse side effects in long-term treatment [18, 43–45].

The current opioid crisis has drawn renewed attention to opioid prescription, use of opioids for the chemical coping of stress and other emotions, and dependence. In the United States, dependence among patients with non-cancer pain has been estimated at 18% and less than 5% among those living with cancer [46, 47]. However, chemical coping and dependence in patients with advanced diseases require an early approach and referral to a mental health and palliative care team.

Children are especially vulnerable to pain under-treatment, even though morphine and other opioids are safe for treating moderate to severe pain under medical supervision. Adverse effects and dependence in adult palliative care are low when opioids are appropriately prescribed and used. There is no evidence that there is any difference with child palliative care [47].

Socioeconomic status constitutes a significant predictor of opioid dispensing rates in geographic areas and highlights the disparity in access to medications included in the WHO essential medicines list. There are many reasons for this disparity. From a structural perspective, large referral hospitals concentrate in the most prosperous cities and states. From an economic perspective, the added costs of storing opioids and protecting them from theft, as well as the limited affordability of these drugs in poorer areas, could explain why they are less likely to be available there. From the patients' and healthcare providers' perspectives, there may also be differences in cultural perceptions of pain and its treatment, which should be explored and considered when implementing public policies.

2.3.2 *Disparities in the Availability of Pain Medication*

The INCB publishes an annual analysis of the use of the main opioid analgesics (codeine, fentanyl, hydrocodone, hydromorphone, morphine, and oxycodone), expressed in defined daily doses for statistical purposes (S-DDD) [3]. In 2019, the main trends in the manufacturing, export, import, and use of these opioid analgesics showed that the highest consumption occurs in developed countries in Europe and North America. Countries reporting the highest average consumption of opioids for pain management in the period 2017–2019 were the United States (25,368 S-DDD), Germany (22,517 S-DDD), Austria (18,489 S-DDD), Belgium (15,487 S-DDD), and Canada (14,073 S-DDD).

Using the data provided by each country, the INCB has drawn attention to the fact that a regional analysis confirms the persistence of a global disparity in the use of opioid analgesics. The reported consumption of some countries in North America, Oceania, and Western Europe yielded regional averages of over 9,000 S-DDD (26,151 S-DDD for North America, 9,984 S-DDD for Oceania, and 9,098 S-DDD for Western Europe). Overall, North America remains the region with the highest consumption of opioids for pain management worldwide.

Other regions report an average consumption that is 10 or even 20 times below that of developed countries. Among these, South America reported the highest consumption in 2019 (603 S-DDD) as part of a general upward trend since the early

2000s, closely followed by Eastern Europe (601 S-DDD), which saw a significant increase in consumption from 269 S-DDD in 2018. This increase reflects the rising consumption in the Russian Federation, which almost doubled from 2018 (321 S-DDD) to 2019 (608 S-DDD). An overall upward trend in consumption has also been observed in West Asia from 2000 to 2020, albeit slightly decreasing from 536 S-DDD in 2018 to 479 S-DDD in 2019. The high average consumption in the region is driven by Israel (13,066 S-DDD in 2019) and, to a lesser extent, Turkey (606 S-DDD) [36].

In 2019, the INCB [3] also raised the issue of traditional, low-cost opioids such as opium and morphine replacement with synthetic opioids for medical purposes worldwide. The role of financial benefits for the private pharmacological industry, which may obtain higher profits from synthetic opioids, has also been discussed.

A comparison of individual substance use shows the prominence of fentanyl over the past two decades. However, after peaking in 2018 at 285,959 S-DDD, global usage of fentanyl decreased to 224,805 in 2019. Oxycodone use has also increased but to a lesser extent, and, since 2009, it has replaced morphine as the second most widely consumed opioid (after fentanyl), reaching an all-time high of 45,726 S-DDD in 2018 and decreasing to 42,592 S-DDD in 2019. Oxycodone use is highest in North America, Oceania, Western and Central Europe, and West Asia, although it is also consumed in other regions. Conversely, morphine use trends remained stable between 2004 (25,644 S-DDD) and 2019 (27,746 S-DDD). After decreasing steadily since 2014, hydrocodone use increased from 14,161 S-DDD in 2018 to 20,415 S-DDD in 2019, levels last seen in 2015. Hydromorphone use decreased from 11,834 in 2018 to 7,713 in 2019, its lowest level since 2008. The United States accounted for nearly all hydrocodone use worldwide (99.3%) [3].

The WHO and INCB consider opioid analgesic use between 100 and 200 S-DDD inadequate and highly inadequate below 100 S-DDD. In this context, the average levels of consumption reported in 2019 in East and Southeast Asia (207 S-DDD), Central America and the Caribbean (160 S-DDD), Africa (90 S-DDD), and South Asia (20 S-DDD) are of particular concern. INCB, therefore, warned the international community of the "urgent need to increase levels of opioid analgesic use in all countries, reporting inadequate and very inadequate S-DDD demands targeted to public policies and support from governments, civil society, the pharmaceutical industry, and the international community" [3].

2.4 The Global Opioid Problem

Opioid dependence is a chronic, relapsing condition that affects 40.5 million people, with periods of active use, abstinence, and relapse. Mortality is high, with 109,500 people dying annually from opioid overdose, 43% of whom are in the United States; other countries with high rates are Russia and Eastern Europe. Increased risk has been observed after periods of abstinence, which often happen after periods in jail or treatment. The risk is higher if people use alcohol, cocaine, amphetamines, or benzodiazepines [48].

Adolescents and young adults are particularly susceptible to the nonmedical use of prescription drugs, especially those between 12 and 15 years of age because they experience a shorter period of substance misuse before injecting other drugs [49]. People who use opium extra-medically often become dependent on multiple substances; between 8% and 12% of those who use opioids for chronic pain develop OUDs [31–33]. On the other hand, depression, anxiety, and posttraumatic stress disorders increase the risk of substance dependence. In addition, HIV/AIDS and hepatitis C viral (HCV) infections are significant risks for people that inject drugs; in South Asian countries, non-injecting routes are more common. The pooled, all-cause crude mortality rate has been estimated at 1.7% per year, ten times higher than expected for people the same age.

2.4.1 Stigma, Discrimination, and Human Rights

People with substance use disorders, particularly those who inject drugs, are subject to social stigma and exclusion, which constitute a barrier to treatment and care delivery and also exacerbate the disorder at the individual level. Underlying misconceptions about this group include wrongdoing, such as stealing to buy drugs and behaving aggressively. These behaviors are commonly elicited by withdrawal or extreme anxiety, agitation, or paranoia during intoxication. The misguided belief that willpower is enough to control use also plays an important role, making it difficult for family and other people to express empathy [50].

Healthcare workers are not immune to these beliefs, including the fear that those seeking help might be attempting to obtain drugs, which in turn increases the risk of denial of treatment and care; a relatively common belief is that these disorders are not a medical condition [50]. Unavailability of services, distance from places where often marginalized people who inject drugs gather, and the costs of transportation and medications, including methadone, are structural barriers to care [51]. People with substance use disorders can develop internalized stigma, which constitutes a barrier to seeking help. These factors often result in social isolation, reinforcing the vulnerability of this group and further hampering recovery [52, 53].

Persons with OUDs and their families, the general public, and care workers require education on drug effects. Ideally, this should include knowledge of the interaction between substances, people's genetic heritage, and life experiences. In addition, it is necessary to understand the role of the environment and mechanisms whereby different drugs alter the brain circuitry involved in reward, stress, and mood processing, as well as the biological and social need for external support care.

During the early stages of drug experimentation, when use is still voluntary, users seek pleasure from the effects of opioids; however, during the following stages, drug-seeking behavior is not motivated by volition or pleasure but by craving and the desire to diminish stress and pain. At these stages, users need external health and support; when rejection at health services occurs, the disorder is reinforced.

In the contextual arena, several lessons should be considered. Discrimination and aggressive environments can facilitate the development of self-stigma, leading to what is known as social pain. Evidence shows that brain areas that process physical pain also process social pain, fostering vulnerability and increasing the difficulty of recovering from substance use disorders [50].

According to the Universal Declaration of Human Rights, all people are born free and with equal dignity and rights. Although this applies to everyone, people who use drugs, particularly those who inject drugs, are seldom considered. This has an enormous impact on their well-being, as it increases vulnerability to overdose, transition to heavier use, blood-borne infections, mental disorders, and suicide. Coercive treatment with violence, a common practice, also violates human rights, as does the development of policies without consulting the voices of experts who have experienced an OUD [52, 53].

2.4.2 *Opioid Agonist Therapy (OAT)*

Methadone continues to be the most prescribed substance when OAT is available, followed by buprenorphine or buprenorphine combined with naloxone (see Chaps. 14 and 15). Long-acting subcutaneous and subdermal formulations of buprenorphine have also been available in some countries since 2020. The INCB 2020 Report [54] notes that there has been a steady increase in the use and manufacture of methadone between 2000 and 2019. Few countries concentrate on methadone use, and there are significant differences in global use patterns. The largest methadone-consuming country is the United States (24.8 tons, or 54.5% of global consumption), followed by Iran (5.4 tons, or 12%), Canada (1.5 tons, or 3.4%), the United Kingdom (1.5 tons, or 3.3%), Vietnam (1.3 tons, or 3%), Italy (1.3 tons, or 2.8%), France (1.2 tons, or 2.6%), and Germany (0.9 tons, or 1.9%). The INCB points out that the various consumption levels were mostly related to the presence or absence of people who inject drugs in those places.

Buprenorphine programs are also on the rise. From the estimates provided to the INCB by the countries themselves, it appears that since the late 1990s, global buprenorphine manufacture has increased, reaching a peak of 17.2 tons in 2018. The main countries importing buprenorphine in 2019 were, in descending order of the amount imported, the United States, France, Germany, the United Kingdom, Canada, Austria, Italy, and Belgium [45, 52].

Possible ways to reduce the increase in mortality overdose include the immediate use of naloxone and the promotion of agonist treatment with methadone or buprenorphine, the promotion of treatment-seeking, the reduction of stigma against drug users, and the education of public and health professionals on the benefits of maintenance treatment for the health and well-being of people who have developed a disorder [55, 56]. It would also be helpful to adopt the public health approach suggested by the UN/WHO Informal Scientific Research Network and to correct misconceptions that imprisonment and involuntary treatment with abstinence as the goal are effective, when, in fact, evidence shows that people in these situations are at greater risk of overdose mortality [5, 57].

Degenhardt and colleagues [48] used mathematical modeling to estimate OAT benefits in a range of settings. They concluded that OAT could be highly effective at reducing illicit opioid use and improving multiple health and social outcomes, including reduced overall mortality and key causes of death, such as overdose, suicide, and other injuries. Scaling up and retaining people in OAT, including in prisons, could avert an average of 7.7%, 14.5%, and 25.9% of deaths over the next 20 years. They also found that other pharmacological and non-pharmacological treatments and harm prevention have varying levels of evidence for effectiveness and patient acceptability. Despite robust evidence of the effectiveness of a range of interventions, the coverage is low even in high-income countries, the available treatment often lacks quality, and the criminalization of illicit opioid use and dependence causes social and economic harm.

2.5 Recommendations

Scientific evidence supports effective strategies to reduce the burden of the opioid crisis that include:

- Ensuring availability of pain management medication for those who need it, particularly in LMICs
- Balancing the two public health challenges of reducing pain and controlling the harm caused by opioid medication abuse
- Adopting alternative policy frameworks that protect human rights
- Decriminalizing drug use behavior and implementing measures to reduce drug-related harm at the population level, abolishing imprisonment for drug use, and adopting a public health model
- Embracing egalitarian policies with a humanitarian approach and the well-being of communities as the goal
- Fighting stigma and educating people with OUD and their families, the public, and care workers, including knowledge on the interaction between substances, people's genetic heritage, psychiatric and social factors, and the nature of the brain, chronic relapsing, and treatable disease
- Recognizing and taking actions to ensure the human rights of those affected, including access to voluntary treatment and welfare that includes the provision of housing, education, and job opportunities
- Investing in prevention and treatment (especially for those who inject drugs), harm reduction, and universally available pharmaceuticals for OUDs (methadone, buprenorphine)

- Ensuring availability of naloxone and strengthening opioid overdose care through naloxone programs in the community and training and equipping local emergency systems
- Strengthening care services for illnesses associated with drug use such as HIV/AIDS, HCV, tuberculosis, and mental health disorders
- Providing treatment for the health consequences of contaminants and new substances added to the traditional drugs offered on the illegal market
- Allowing access to justice for those whose human rights have been violated due to drug policies
- Investing in and enabling harm reduction and treatment programs in several settings, including prisons
- Reviewing indicators to evaluate the impact of drug policies and strengthening information systems to document progress on these issues
- Increasing funded research on the frequency of chemical compensation and opioid dependence in the context of pain and palliative care and introducing prescription drug monitoring programs at the local and global levels
- Supporting the training of health professionals from universities, disseminating opioid prescription guidelines for chronic pain management in palliative settings, applying screening tools, and evaluating risk factors to document the diversion of opioids from medical use or as chemical compensation

References

1. United Nations Office on Drugs and Crime: World Drug Report 2021. Booklet 3 [Internet]. (United Nations publication. Sales No. E.21.XI.8) Vienna; 2021 Jun [cited 2021 Aug 9]. Available from: https://www.unodc.org/unodc/data-and-analysis/wdr2021.html.
2. World Health Organization. Opioid overdoses [Internet]. 2021 Aug [cited 2021 Aug 15]. Available from: https://www.who.int/news-room/fact-sheets/detail/opioid-overdose.
3. International Narcotics Control Board. Report 2020 [internet]. United Nations Publication. Sales No.: E.21.XI.3; 2021. (E/INCB/2020/1) Viena; 2021 Mar. [cited 2021 Aug 12] Available from: https://www.incb.org/documents/Publications/AnnualReports/AR2020/Annual_Report/E_INCB_2020_1_eng.pdf.
4. United Nations Office on Drugs and Crime. Outcome document of the 2016 United Nations General Assembly Special Session of the World Drug Problem. [Internet]. UNGASS 2016, New York; 2016 Apr [cited 2021 Jul 29]. Available from: https://www.unodc.org/documents/postungass2016/outcome/V1603301-E.pdf.
5. Volkow ND, Poznyak V, Saxena S, Gerra G. UNODC-WHO Informal International Scientific Network. Drug use disorders: impact of a public health rather than a criminal justice approach. World Psychiatry. 2017;16(2):213–4. https://doi.org/10.1002/wps.20428.
6. United Nations. Sustainable Development Goals [Internet]. 2015 [cited 2021 Aug 12]. Available from: https://www.undp.org/sustainable-development-goals.
7. World Health Assembly. Strengthening of palliative care as a component of comprehensive care throughout the life course (No. WHA67. 19) [Internet]. 2014 May [cited 2021 Aug 15]. Available from: https://apps.who.int/iris/bitstream/handle/10665/162863/A67_R19-en.pdf?sequence=1&isAllowed=y.
8. Clark J, Barnes A, Gardiner C. Reframing Global Palliative Care Advocacy for the sustainable development goal era: a qualitative study of the views of International Palliative

Care Experts. J Pain Symptom Manag. 2018;56(3):363–70. https://doi.org/10.1016/j.jpainsymman.2018.06.009.
9. National Academies of Sciences, Engineering, and Medicine; Health and Medicine Division; Board on Health Sciences Policy; Committee on Pain Management and Regulatory Strategies to Address Prescription Opioid Abuse. In: Phillips JK, Ford MA, Bonnie RJ, editors. Pain management and the opioid epidemic: balancing societal and individual benefits and risks of prescription opioid use. Washington (D.C.): National Academies Press (U.S.); 2017.
10. May T, Hough M. Drug markets and distribution systems. Addict Res Theory. 2009;12(6):549–63.
11. Martínez MM, Ricardo J, Medina-Mora ME, Romero MP, Berenzon S. Los nuevos empresarios: Trayectoria del uso a la venta de drogas en contextos de fiesta. Salud Mental. 2012;35(6):475–81.
12. Johnson BD. Patterns of drug distribution: implications and issues. Subst Use Misuse. 2003;38(11–13):1789–806. https://doi.org/10.1081/JA-120024241.
13. Bonnie RJ, Schumacher MA, Clark JD, Kesselheim AS. Pain management and opioid regulation: continuing public health challenges. Am J Public Health. 2019;109(1):31–4. https://doi.org/10.2105/AJPH.2018.304881.
14. Babor T, Caulkins J, Fischer B, Foxcroft D, Humphreys K, Medina-Mora ME, et al. Drug policy and the public good. 2nd ed. Oxford, UK: Oxford University Press; 2018.
15. Hosztafi S. A heroin története [The history of heroin]. Acta Pharm Hung. 2001;71(2):233–42.
16. Quinones S. Dreamland: the true tale of America's opiate epidemic. 1st ed. New York: Bloomsbury Press; 2016.
17. Ciccarone D. The triple wave epidemic: supply and demand drivers of the U.S. opioid overdose crisis. Int J Drug Policy. 2019;71:183–8. https://doi.org/10.1016/j.drugpo.2019.01.010.
18. Volkow ND, Medina-Mora ME, Poznyak V, Saxena S, Gerra G, the UNODC-WHO Informal Science Network. Addressing the opioid crisis globally. World Psychiatry. 2019;18(2):231–2. https://doi.org/10.1002/wps.20633.
19. United Nations Office on Drugs and Crime: World Drug Report 2020 [Internet]. (United Nations publication. Sales No. E.20.XI.6) Vienna; 2020 Jun [cited 2021 Aug 15] Available from: https://www.casede.org/index.php/biblioteca-casede-2-0/crimen-organizado/narcotrafico-y-drogas/595-world-drug-report-2020/file.
20. Dudley S, Bonello E, Lopez-Aranda J, Moreno M, Clavel T, Kjestalt B, Restrepo J [Internet] 2019 [cited Aug 2021]. Mexico's role in the deadly rise fentanyl. Insight Crime & Wilson Center. February 2019. Available from: https://www.wilsoncenter.org/publication/mexicos-role-the-deadly-rise-fentanyl.
21. RAND Corporation. The future of Fentanyl and other synthetic opioids. [Internet] 2019 [cited Aug 2021]. Library of Congress Cataloging-in-Publication Data is available for this publication. ISBN: 978–1–9774-0338-4 Available from: https://www.rand.org/pubs/research_reports/RR3117.html.
22. Fleiz C, Arredondo J, Chavez A, Pacheco L, Segovia LA, Villatoro JA, et al. Fentanyl is used in Mexico's northern border: current challenges for drug health policies. Addiction. 2020;115(4):778–81. https://doi.org/10.1111/add.14934.
23. Van Zee A. The promotion and marketing of oxycontin: commercial triumph, public health tragedy. Am J Public Health. 2009;99(2):221–7. https://doi.org/10.2105/AJPH.2007.131714.
24. National Institute on Drug Abuse. Overdose Death Rates. [Internet]. 2021 Jan [cited 2021 Aug 11]. Available from: https://www.drugabuse.gov/drug-topics/trends-statistics/overdose-death-rates
25. Dasgupta N, Mandl KD, Brownstein JS. Breaking the news or fueling the epidemic? Temporal association between news media report volume and opioid-related mortality. PLoS One. 2009;4(11):e7758. https://doi.org/10.1371/journal.pone.0007758.
26. Zgierska A, Miller M, Rabago D. Patient satisfaction, prescription drug abuse, and potential unintended consequences. JAMA. 2012;307(13):1377–8. https://doi.org/10.1001/jama.2012.419.

27. Leung PTM, Macdonald EM, Stanbrook MB, Dhalla IA, Juurlink DN. A 1980 letter on the risk of opioid addiction. N Engl J Med. 2017;376(22):2194–5. https://doi.org/10.1056/NEJMc1700150.
28. Cruz SL, Pérez IP, Carranza CJ. El tratamiento del dolor y los opioides. Ciencia [Internet]. 2020 [cited 2021 Aug 18]; 71(2):52–61. Available from: https://www.amc.edu.mx/revistaciencia/images/revista/71_2/PDF/11_71_2_1228_Opioides-L.pdf.
29. Humphreys K. Avoiding globalization of the prescription opioid epidemic. Lancet. 2017;390(10093):437–9. https://doi.org/10.1016/S0140-6736(17)31918-9.
30. Vowles KE, McEntee ML, Julnes PS, Frohe T, Ney JP, Van der Goes DN. Rates of opioid misuse, abuse, and addiction in chronic pain: a systematic review and data synthesis. Pain. 2015;156(4):569–76. https://doi.org/10.1097/01.j.pain.0000460357.01998.f1.
31. Muhuri PK, Gfroerer JC, Davies MC. Associations of Nonmedical Pain Reliever Use and Initiation of Heroin Use in the United States. CBHSQ Data Review [Internet]. 2013 Aug;1:1–17. [cited 2021 Aug 18]. Available from: https://www.samhsa.gov/data/sites/default/files/DR006/DR006/nonmedical-pain-reliever-use-2013.htm.
32. Cicero TJ, Ellis MS, Surratt HL, Kurtz SP. The changing face of heroin use in the United States: a retrospective analysis of the past 50 years. JAMA Psychiat. 2014;71(7):821–6. https://doi.org/10.1001/jamapsychiatry.2014.366.
33. Carlson RG, Nahhas RW, Martins SS, Daniulaityte R. Predictors of transition to heroin use among initially non-opioid dependent illicit pharmaceutical opioid users: a natural history study. Drug Alcohol Depend 2016;160:127–34. https://doi.org/10.1016/j.drugalcdep.2015.12.026.
34. Volkow ND, McLellan AT. Opioid abuse in chronic pain--misconceptions and mitigation strategies. N Engl J Med. 2016;374(13):1253–63. https://doi.org/10.1056/NEJMra1507771.
35. International Association for the Study of Pain (IASP). Cancer Pain [Internet] Washington, D.C. 2021. [cited 2021 Aug 13]. Available from: https://www.iasp-pain.org/advocacy/global-year/cancer-pain/.
36. International Narcotics Control Board. Progress in ensuring adequate access to internationally controlled substances for medical and scientific purposes [internet]. United Nations Publication. Sales No.: E.19.XI.4; 2019. (E/INCB/2018/Supp.1) Vienna. 2019 Mar [cited 2021 Aug 12]. Available from: file:///C:/Users/Equipo21/Downloads/Supplement_E_ebook.pdf.
37. International Association for Hospice and Palliative Care (IAHPC). Global data platform to calculate SHS and palliative care need [Internet]. Houston: 2021. [cited 2021 Aug 13]. Available from: https://hospicecare.com/what-we-do/resources/global-data-platform-to-calculate-shs-and-palliative-care-need/.
38. Knaul FM, Farmer PE, Krakauer EL, De Lima L, Bhadelia A, Kwete XJ, et al. Alleviating the access abyss in palliative care and pain relief-an imperative of universal health coverage: the Lancet Commission report. Lancet. 2018;391(10128):1391–454. https://doi.org/10.1016/S0140-6736(17)32513-8.
39. McGuire DB. Comprehensive and multidimensional assessment and measurement of pain. J Pain Symptom Manag. 1992;7(5):312–9. https://doi.org/10.1016/0885-3924(92)90064-o.
40. Hui D, Bruera E. A personalized approach to assessing and managing pain in patients with cancer. J Clin Oncol. 2014;32(16):1640–6. https://doi.org/10.1200/JCO.2013.52.2508.
41. World Health Organization (WHO). Cancer pain relief. World Health Organization [Internet]. Geneva, 1986 [cited 2021 Aug 14] Available from: https://apps.who.int/iris/handle/10665/43944.
42. Breivik H, Collet B, Ventafridda V, Cohen R, Gallacher D. Survey of chronic pain in Europe: prevalence, impact on daily life and treatment. Eur J Pain. 2006;10(4):287–333. https://doi.org/10.1016/j.ejpain.2005.06.009.
43. Atluri S, Akbik H, Sudarshan G. Prevention of opioid abuse in chronic non-cancer pain: an algorithmic, evidence based approach. Pain Physician. 2012;15:ES177–89.
44. Trescot AM, Boswell MV, Atluri SL, Hansen HC, Deer TR, Abdi S, et al. Opioid guidelines in the management of chronic non-cancer pain. Pain Physician. 2006;9(1):1–39.

45. Sullivan MD, Von Korff M, Banta-Green C, Merrill JO, Saunders K. Problems and concerns of patients receiving chronic opioid therapy for chronic non-cancer pain. Pain. 2010;149(2):345–53. https://doi.org/10.1016/j.pain.2010.02.037.
46. Gómez MC, Zertuche T, Bruera E. Chemical compensation and opioid addiction: evidence, risk assessment and management in palliative care patients. Med Paliat [Internet]. 2018 [cited 2021 Aug 16]; 25(4):281–90. Available from: https://www.medicinapaliativa.es/(X(1)S(vd4mms1rvvhn2b2tlkk0du5m))/Documentos/ArticulosNew/REV.pdf
47. Kwon JH, Hui D, Bruera E. A pilot study to define chemical coping in cancer patients using the Delphi Method. J Palliat Med. 2015;18(8):703–6. https://doi.org/10.1089/jpm.2014.0446.
48. Degenhardt L, Grebely J, Stone J, Hickman M, Vickerman P, Marshall BD, et al. Global patterns of opioid use and dependence: population harms, interventions, and future action. Lancet. 2019;394(10208):1560–79. https://doi.org/10.1016/S0140-6736(19)32229-9.
49. Volkow ND. Prescription drug abuse. National Institute on Drug Abuse [Internet]. 2010 Sep [cited 2021 Aug 18]. Available from: https://archives.drugabuse.gov/testimonies/2010/prescription-drug-abuse.
50. Volkow ND. Stigma and the toll of addiction. N Engl J Med [Internet]. 2020 [cited 2021 Aug 18]; 382(14):1289–90. Available from: https://www.nejm.org/doi/full/10.1056/NEJMp1917360.
51. Fleiz C, Villatoro JA, Dominguez M, Bustos M, Medina-Mora ME. Opioid crisis along Mexico's Northern border: treatment needs Mexican opioid crisis. Arch Med Res. 2019;50(8):527–34. https://doi.org/10.1016/j.arcmed.2019.12.013.
52. United Nations Development Program, World Health Organization, United Nations Program on HIV/AIDS, United Nations Human Rights. International Guidelines on Human Rights and Drug Policy. [Internet]. 2019 [cited Aug 2021]. Germany March 2019. Available from: https://www.humanrights-drugpolicy.org/
53. Harm Reduction International. The Global State of Harm Reduction 2020 [Internet]. 2020 [cited Aug 2021]. 7th ed. London: Harm Reduction International. Available from: https://www.hri.global/global-state-of-harm-reduction-2020.
54. International Narcotics Control Board. Narcotic Drugs 2020 [Internet]. United Nations Publication. Sales No.: T.21.XI.2 (E/INCB/2020/2) Vienna, 2021 [cited 2021 Aug 10]. Available from: https://www.incb.org/documents/Narcotic-Drugs/Technical-Publications/2020/Narcotic_Drugs_Technical_publication_2020.pdf.
55. Jalal H, Buchanich JM, Sinclair DR, Roberts MS, Burke DS. Age and generational patterns of overdose death risk from opioids and other drugs. Nat Med. 2020;26(5):699–704. https://doi.org/10.1038/s41591-020-0855-y.
56. Hall W, Degenhardt L, Hickman M. Generational trends in U.S. opioid-overdose deaths. Nat Med. 2020;26(5):651–2. https://doi.org/10.1038/s41591-020-0875-7.
57. Rafful C, Medina-Mora ME, González P, Jenkins JH, Rangel MG, Strathdee SA, Davidson PJ. "Somebody is Gonna be hurt": involuntary drug treatment in Mexico. Med Anthropol. 2020;39(2):139–52. https://doi.org/10.1080/01459740.2019.1609470.

Chapter 3
Opioid Markets and Special Populations

Hilda Dávila and Alfredo Camhaji

Abstract Over the past two decades, opioid markets have become more dynamic and complex. The availability of drugs has affected consumption patterns and represents a significant problem in the world. This chapter presents information on the trends observed in the most widely used opioids in the world, organized by regions, drug groups, and affected populations. It analyzes the supply and demand of opioid markets, some changes that the COVID-19 pandemic has brought, and interventions and regulations needed to meet this challenge.

Keywords Epidemiology of opioid use · Market complexity · Diversification · Technological innovation · Changes in drug use patterns · Combinations of substances

3.1 Introduction

The growing supply of drugs is characterized by the increasing illicit production of opium and the diversification of available substances sold as new psychoactive substances (NPS; see Chap. 16). The variety and combination of recently available substances have dominated drug markets and posed challenges to both legal and illegal markets.

On the demand side, a growing number of people use, often unbeknownst to them, opioids laced with fentanyl, fentanyl analogs, stimulants, or other adulterants. In addition, synthetic opioids are becoming more prevalent among NPS, especially fentanyl analogs and other counterfeit, unregistered, or unlicensed opioids [1].

H. Dávila (✉) · A. Camhaji
Opioids Working Group, Global Studies Seminar, Faculty of Medicine, National Autonomous University of Mexico (UNAM), Mexico City, Mexico

S. L. Cruz (ed.), *Opioids*, https://doi.org/10.1007/978-3-031-09936-6_3

A parallel phenomenon is the combination of opioids with benzodiazepines prescribed in the legal market or with pain medications, cocaine, and other psychoactive substances that increase the risk of adverse effects. Combining opioids with benzodiazepines, in particular, enhances the risk of overdose [2]. On the supply side, there is an increase in the quantities of drugs seized, produced in illegal markets, or diverted from the legal market. This situation calls for effective and comprehensive interventions with a shared responsibility of all state and non-state actors involved.

3.1.1 Opioid Use

About 275 million people between the ages of 15 and 64 used a drug at least once in 2018, and just over 31 million suffer from drug use disorders [3]. This implies that, in that year, about 5.6% used some drug and 0.7% had a problematic use. In 2019, of the 275 million people, around 62 million used opioids (pharmaceutical or synthetic) for nonmedical reasons, representing 1.2% of the world's population [4].

Over the past 20 years, prolonged, inadequate, or unsupervised nonmedical use of opioid analgesics has increased considerably worldwide. The most commonly used opioids are heroin, fentanyl, morphine, and oxycodone.

Developing countries are and have been opioid producers. China and Thailand were among them, although not anymore. India is a producer for the legal market but has diversion problems. In addition to being countries with opioid use, Myanmar and Laos, Iran, Mexico, Colombia, and Guatemala have been transit countries in Latin America, West Asia, and Africa.

The use of synthetic opioids such as fentanyl and its analogs remains a red flag in both the United States and Canada. In recent years, these drugs have been the leading cause of overdose deaths due, in part, to the unpredictability of substances' potency and purity in illegal markets. The United Nations Office on Drugs and Crime (UNODC) reported that fentanyl and its analogs were involved in two-thirds of the 67,367 overdose deaths recorded in the United States in 2018 [1]. On the other hand, in West, Central, and North Africa, the nonmedical use of tramadol is a matter of concern (see Chap. 5).

Ease of production and low costs of synthetic opioids have been an attraction for traffickers who offer them on the black market [1].

The Global SMART Update 2019 [1] indicates that the nonmedical use of pharmaceutical opioids is a significant concern in West and North Africa, the Near and Middle East (tramadol), and North America (hydrocodone, oxycodone, codeine, tramadol, and fentanyl). There are also signs of increased nonmedical opioid use in Western and Central Europe (see Chap. 5), as reflected in the increasing proportion of admissions to treatment for such use.

In Eastern and Southeastern Europe, North America, and Asia, opioids have become the main drugs leading to users' treatment. Data from 2019 indicate that most are heroin users requiring medication-assisted treatment with opioid agonists (methadone or buprenorphine). The proportion of users with medication-assisted treatment for Western and Central Europe is 40% and 74% for Eastern and Southeastern Europe. The age range for users is over 30 years, and a quarter and a third of them, respectively, are first-time users [1].

3.1.2 Epidemiology of Opioid Use Disorders (OUDs)

Not everyone who uses non-prescription opioids develops dependence. A 1991 population survey in the United States suggested that about one in four people who used opioids (primarily heroin) became dependent at some point, while a UK study estimated that two in three people who used heroin were likely to develop OUDs [5]. The risk of developing opioid dependence increases when there is a history of substance use and dependence, psychiatric morbidities, or a social environment favorable to use or when opioids have been consumed by indiscriminate medical prescription (see Chap. 13).

The United Nations estimates that around half a million people died in 2019 from causes directly related to drug use, mainly from injectable opioids, including heroin and synthetics such as fentanyl. Of the 11 million people who injected drugs worldwide in 2019, 1.4 million were living with HIV/AIDS and 5.6 million with hepatitis C. Moreover, nearly 1.2 million lived with both diseases. The main risk of contagion stems from sharing contaminated injection material [4].

Opioid-associated morbidities account for 70% of the 18 million healthy life years lost to disability and premature death attributed to drug use disorders [4].

Deaths caused by drugs increased by 60% between 2000 and 2015. People over 50 years of age accounted for 27% of these deaths in 2000. This figure rose to 39% in 2015. About three-quarters of deaths associated with drug use disorders among those over 50 are from opioid use [6].

3.2 Vulnerability According to Age Groups and Gender

3.2.1 Drug Use Among Women

Women present specific drug use patterns both due to biological differences related to hormonal variations (alterations in the menstrual cycle, fertility, pregnancy, lactation, and menopause) and due to gender issues. In addition, women have different reasons for using drugs, including controlling their weight, fighting fatigue, managing pain, and dealing with violence and mental health problems on their own [7]. Economic, social, political, and cultural conditions determine drug use among women and how they are viewed and accepted; therefore, these factors change over time and vary in different countries.

According to the National Institute on Drug Abuse (NIDA), women use fewer drugs than men, but they may feel the effects more intensely, and dependence tends to develop faster than in men. Women may have difficulty seeking help with a drug use problem due to fear of legal and social consequences and stigma. They may also lack options to care for their children while they are in treatment. The use of opioids during pregnancy can harm not only pregnant women's health but also the newborn who may have withdrawal symptoms after birth, a disorder known as neonatal abstinence syndrome (see Chaps. 13, 14, and 15) [7].

Women and girls constitute one-fifth of the estimated number of people who inject drugs worldwide, and this proportion is higher in high-income countries. In addition, women who use drugs tend to start using substances later than men, but they escalate their use more quickly [8].

There is a strong link between sex work and drug use. People with substance use disorders may turn to sex work as a means of paying for drug use, while sex workers can use drugs to withstand the demands and nature of their work.

Over the past 35 years, there has been a substantial increase in the number of women arrested for drug-related offenses, and the prevalence of drug use is higher among women than men in the prison population [9]. Also, women who have experienced adversity in childhood may use drugs to self-medicate [8].

Women are more vulnerable than men to HIV, hepatitis, and other blood-borne infections. Numerous studies have shown that the female sex is an independent predictor of HIV or hepatitis C in people who inject drugs, particularly young women and those who have recently started injecting drugs. However, only one-fifth of people receiving treatment for drug abuse are women, as they must overcome significant systemic, structural, social, cultural, and personal obstacles to access it [8].

Compared to men, women are prescribed more opioids and anxiolytics, making them more likely to misuse those drugs. In addition, women with substance use disorders tend to report high levels of posttraumatic stress disorders (PTSD) and experiencing childhood adversities such as physical neglect, mistreatment, or sexual abuse. Finally, women's drug use is often related to their partner, while men are more likely to use drugs with male friends.

3.2.2 Young People

Data on chronic opioid use among youth and young adults are concentrated in the United States; therefore, more studies are needed to present conclusive claims. However, evidence indicates that mental health disorders are associated with an increased risk of chronic opioid use among adolescents and young adults [10].

A study conducted by NIDA concluded that nonmedical use of prescription medications (opioids, stimulants, and tranquilizers) increases the prevalence of drug use disorders among adolescents in a higher proportion than among young adults. For example, in the 12 months after the first misuse of prescription drugs, 11.2% of teens had a prescription opioid use disorder, compared to 6.9% of young adults. In addition, 13.9% of adolescents had a prescription stimulant use disorder, compared to 3.9% of young adults. Also, 11.2% of adolescents had a prescription tranquilizer use disorder, compared to 4.7% of young adults [11].

The health risks of heroin use are devastating and include overdose, severe addiction, hepatitis C, HIV, and other infections. In 2017, opioids were responsible for 47,600 (67.8%) of drug overdose deaths in the United States that occurred in people under the age of 25, according to data from the Centers for Disease Control and Prevention [12].

3.2.3 Older Adults

Drug use in older adults has increased at a faster rate than in younger adults. About 75% of deaths from drug use disorders among people age 50 and older are related to opioids [4]. Chronic health problems tend to arise or increase with aging; therefore, older adults often take more medications than other age groups. The combination of prescription drugs, over-the-counter medications, and dietary supplements is common, leading to a higher rate of exposure to potentially addictive drugs.

Some older adults may use drugs to cope with major life changes, such as retirement, grief, the loss of a loved one, deteriorating health, or a change in their housing situation. In addition, the use of opioid analgesics is more common in older adults who have other health problems such as advanced cancer or heart disease, among others.

As the body ages, its ability to eliminate substances diminishes, and it becomes more susceptible to the effects of drugs, including opioids. As we know little about the effects of drug use on the aging brain and effective treatment models, this is fertile ground for research.

People who lived through their teens when drugs were popular and widely available and those who began injecting heroin during the heroin epidemic of the 1980s and 1990s constitute a large cohort of opioid users who age and require specialized attention.

According to the 2020 report of the International Narcotics Control Board (INCB), health problems resulting from substance use in older adults increase the risk of death from illness, overdose, and suicide and decrease the median age at death. Furthermore, there is a higher risk of presenting premature degenerative disorders, cardiovascular disorders, liver disease, and chronic pain.

The older population of opioid users frequently has restricted physical functions, respiratory problems, diabetes, a higher risk of HIV infection, and hepatitis C. In addition, opioid use exacerbates other age-related diseases and increases the risk of falls, fractures, injuries, and traffic accidents due to impaired vehicle driving.

Older adults may also have more difficulties performing daily living activities, a higher incidence of mental health problems, increased risk of excessive sedation, overdose, confusion, and circulatory syncope [13]. If they are drug users, additional social problems, such as the associated stigma, can lead to a feeling of shame that prevents them from seeking help, so families and health workers may not detect the problem. Lack of attention, economic problems, unemployment, and loss of housing are frequent among older drug users. Also, social isolation, loneliness, exclusion, and limited contact with family may occur. People are more likely to be treated due to contact with the criminal justice system than by referral and access to the health system [13].

3.3 Consequences of Opioid Use

Drug use disorders are multifactorial. Socioeconomic inequalities, including poverty, limited access to education, lack of sanitation, healthcare, and employment, play a critical role in the risk of drug abuse and pave the way for vulnerable populations to become potential drug users. These factors can also be consequences of opioid use, self-perpetuating the cycle of poverty and drug misuse among vulnerable populations, who may have end working in the cultivation and trafficking of illicit crops.

Half of all people with psychiatric disorders may suffer from a substance use disorder at some point in their lives and vice versa [14]. Also, people with OUDs have higher risks of HIV, hepatitis C infections, and poor health in general.

3.3.1 Prenatal Exposure

Babies exposed to opioids during pregnancy may be born with neonatal abstinence syndrome (NAS). The main adverse effects associated with its consumption during pregnancy are growth restriction, preterm birth, ruptured membranes, low birth weight and height, infections and bleeding, respiratory complications, feeding difficulties, seizures, and problems affecting learning and behavior (see Chap. 13) [15].

3.3.2 Children and Adolescents

Children whose parents have prescription opioids at home face a much higher risk of overdose, with an estimated 9000 children's deaths from this cause alone. In addition, the number of children and adolescents hospitalized for opioid poisoning tripled between 1997 and 2012. While most overdose patients were adolescents, the largest overall increase in poisoning occurred among infants and preschoolers [16].

3.4 Challenges for the Treatment of OUDs

Access to controlled medicines for medical purposes, in particular for pain treatment, should be ensured as an essential component of the right to health in the context of a health-based drug control system. A better balance should be achieved in the availability of controlled pain medicines while avoiding the development of illicit markets for such products. Countries must address barriers to equitable access to pain management and palliative care medicines by reviewing policies, addressing challenges in the supply chain, supporting health workers, and raising awareness among the general public to increase access to controlled medicines while preventing diversion and nonmedical use (see Chap. 3).

Vulnerable populations need facilities where treatment is available and which offer programs that act on the possibility of facing recidivism (relapses) and low therapeutic adherence. To do this, therapeutic centers should have flexible schedules, weekend sessions, child care areas, and affiliation to cost-bearing health services.

Primary goals of treatment must include the early identification of other physical and mental disorders that may occur to people with drug dependence and providing them with accessible alternatives for comprehensive care and harm reduction.

Opioid users, in addition to being more vulnerable to the contagion and complications of COVID-19, may have a depressed immune system and need access to specialized treatment and other health services [17].

The gender perspective must be present in designing public policies and action programs that mitigate consumption, implement treatment programs with organized care systems, and respect human rights.

3.5 Supply

The UNODC states that opioid trafficking remains the most lucrative form of business for criminals, with an estimated annual value of $68 billion. The financial system launders 70% of illicit income, and less than 1% of laundered proceeds are intercepted or seized [18].

According to the Organization of American States (OAS) and the UNODC, the total value of controlled drug sales globally was $320 billion in 2003, a figure equivalent to 0.9% of the world gross domestic product (GDP). Retail drug markets in the Americas were estimated at $151 billion or about 47% of the global total. Dollar retail markets were about 44% of the global total in North America and 33% in Europe. Retail markets in South America, Central America, and the Caribbean were around 3% of the global total [19].

Innovations in retail distribution of substances include the sale of medicines and illicit drugs through web platforms. Internet service providers and courier companies might displace street sales, which are already losing ground to contactless methods such as online shopping and mail delivery, giving rise to a globalized market in which all substances, licit or illicit, are more available and accessible anywhere. As a result, drugs purchased on the dark web quadrupled between 2011 and 2017 and between mid-2017 and 2020 [20].

Seizures of pharmaceutical opioids peaked in 2014 and 2019. Fentanyl analogs remain the most seized substances in terms of dosages, with North America at the top. Heroin trafficking in Europe accounted for 27% of the global total in 2019 [20].

3.5.1 Legal and Illegal Markets

The functioning of legal and illegal opioid markets is a dynamic global phenomenon that substantially impacts humanity. On the one hand, opioids are potent and effective medications for treating pain; on the other hand, they are psychoactive drugs with high addiction liability and potentially fatal effects.

The legal opioid analgesic drug market has been affected by increasing life expectancy, aggressive marketing strategies to increase opioid prescription sales, medical malpractice, and the increased prevalence of pain-causing diseases such as arthritis, cancer, and back pain, even if opioids are not the best choice to treat these ailments (see Chap. 10). There has also been an increase in the illegal market due to the growing number of people affected with opioid use disorders (OUDs) and the diversion of prescription opioids sold as legal heroin replacements. Indeed, countries like the United States and Canada face an opioid epidemic that kills thousands of young people every year. Drug traffickers have exploited opioid production and distribution and have used regulation gaps to establish an uninterrupted and increasingly diversified supply of opioids [21].

3.5.2 Production and Traffic (Wholesaler Versus Retailer)

Availability of controlled substances for medical and scientific purposes should address barriers and differences in low- and middle-income countries. Lack of appropriate pain management which occurs primarily in developing countries is unacceptable. It is necessary to establish national legal systems and appropriate regulatory mechanisms with simplified procedures to ensure efficient distribution channels.

The indiscriminate use of prescription opioid analgesics increased in some countries, especially in North America, Oceania, and Western Europe between 2001–2003 and 2011–2013 [22]. The North American experience shows that about one-fifth of users who began using opioids for chronic pain treatment later switched to illicit opioids, such as heroin and fentanyl.

Regarding the illegal market, as mentioned above, the 2019 data reveal that 92% of worldwide opium production is concentrated in three countries: Afghanistan

(69%), Myanmar (14%), and Mexico (9%) [3]. The global supply of opium has stabilized since 2018; in 2020, it accounted to 7430 tons. Lack of economic growth and poverty have driven many communities into illegal drug cultivation.

According to the *World Drug Report 2019* [18], heroin produced in Afghanistan is trafficked through three main routes. The Balkan route accounts for nearly half of the morphine and heroin seized outside Afghanistan. It runs through Afghanistan, Iran, Turkey, and European countries. The southern route sends Afghan-produced opioids to South Asia and Africa via Pakistan. Finally, the northern route reaches the Russian Federation through Central Asia.

Opioid trafficking affects the quality of the drug sold because adulteration and drug reduction occur in all countries where the drug remains. The longer the pathways between drug production and use, the lower the purity of the drug.

Although heroin seizures were greater than those of pharmaceutical opioids for years, the diversification of substances on the market has changed this landscape [23]. The *World Drug Report 2021* reveals that "data from 2019 show that, for the third time in the last five years, the total amount of pharmaceutical opioids seized (228 tons) was greater than the total amount of heroin seized (93 tons). The pharmaceutical opioids seized in the largest quantities were codeine, followed by tramadol (an opioid that is not under international control), fentanyl, and methadone."

The largest quantities of pharmaceutical opioids seized in 2019 were reported by Bangladesh (mainly codeine), followed by Benin (mainly tramadol, which tends to be re-exported from there to other West African countries), India (mainly codeine), Malaysia (mainly codeine), and the United States (mainly fentanyl and fentanyl analogs).

It is worth noting that the legal status of many substances already on the market varies from country to country, which adds to the complexity of production and trafficking patterns [21]. In 1990, there were about 230 controlled psychoactive substances (cannabis, cocaine, opium, and heroin, among others). This situation changed in the 2020s, where the number of psychoactive substances under international control grew to 950 in 2019 (see Chap. 16).

In Latin America and the Caribbean, there is a market for psychoactive pharmaceuticals for nonmedical use, which is not limited to opioids. In addition, the Internet marked a shift in drug distribution networks. Online marketplaces were already up and running before the COVID-19 pandemic hit the world, and the rise of digital interconnectivity is key to understanding the rapid adaptation and innovation in the functioning of global drug supply chains. As a result, controlled and uncontrolled synthetic substances have become readily available products [1]. Selling opioids online reduces intermediaries, saves costs, and shortens supply chains because they are delivered directly to the buyers' door at the international shipping speed.

Fentanyl is so potent that the amount needed to produce the same effect as morphine is 100 times less. Carfentanil, a synthetic fentanyl analog, is 10,000 times more potent than morphine [20]. Therefore, thousands of doses can be mailed at a time, radically reducing transportation costs and changing the role of drug traffickers. Moreover, fentanyl and its analogs are entirely synthetic and can be produced in laboratories, making it easier to conceal illicit activities. Conversely, cultivating poppy seed plantations have more risks of being detected and can fail in case of droughts or labor shortages. This situation could cause prices to fall, increasing consumption and changing consumption patterns where heroin is expensive versus cheaper but more lethal synthetic opioids [1].

Estonia has had a long-standing fentanyl market since 2002, smuggled from Russia. The disruption in this market comes from law enforcement actions in 2017. A different situation occurred in Latvia, where users tended to omit fentanyl and went directly to consume fentanyl analogs. A preferred product in Sweden is a nasal spray bottle with fentanyl and fentanyl analogs which are almost exclusively sold on the Internet and coexist with the heroin market.

Mexico plays a key role in fentanyl trafficking because of its proximity to the United States, the largest market in the world, and because China, the largest producer of fentanyl and analogs globally, banned its production in 2019, which had an impact on strengthening the illegal market in Mexico [21].

3.6 Market Characteristics

Opioid markets tend to be highly competitive, fragmented, and heterogeneous, evolving rapidly to find new users [1]. The increased availability of new synthetic opioids that enables efficient production as well as the use of electronic systems for acquiring and distributing drugs defines the dynamism of opioid markets.

The emergence of a new generation of cheaper synthetic opioids for nonmedical purposes that are not subject to bans or immediate control has intensified opioid use [8]. Hundreds of new substances are detected yearly in illicit drug markets, indicating continuous innovation, and their presence is more common in Asia, Europe, and North America [1].

New psychoactive substances have a chemical structure or produce pharmacological effects similar to controlled drugs, but do not go through preclinical studies or clinical trials to know their effects on consumer behavior or the harm they cause. They characteristically have a very short life in the market (approximately 6 months; see Chap. 16) and are rapidly replaced by other synthetic analogs.

Between 2014 and 2015, the emergence of new synthetic opioids in the existing user population was a key characteristic of the expansion of drug markets. As a result, synthetic NPS increased from 166 in 2005–2009 to 950 in 2019 [18, 20].

A total of 294 psychotropic substances were under international control (see Chap. 4) by the end of 2020 [24]. By comparison, 1047 NPS had been identified by national authorities and forensic laboratories in 126 countries by the same date, three times the number of substances under international control. However, it is worth noting that many NPS emerge only for a short period before disappearing from the market [24].

Whether it is Internet-intensive contactless methods for purchasing or traditional street market operations, a number of actors involved in illegal drug business have well-organized structures and an immense economical and corruption power.

3.6.1 Combined Use of Opioids with Other Drugs

Polydrug use is becoming common, and combinations are constantly changing [16]. For example, fentanyl is frequently combined with cocaine or heroin. Also, adulterants such as levamisole, benzocaine, caffeine, and phenacetin are added to heroin, cocaine, methamphetamine, ketamine, and other drugs distributed on the illegal market [25]. In addition, drugs can contain other dangerous synthetic opioids without the consumer's knowledge.

The combination with other substances, the proliferation of synthetic drugs, and the decrease in purity bring an increased risk of overdose and death from acute intoxication. For example, fentanyl and its chemical analogs (such as carfentanil, acetyl fentanyl, butylfentanyl, and furanyl fentanyl), which are sold on the illicit market, have been linked to a 540% increase in overdose deaths in just 3 years [26]. Due to this increase in overdose deaths, it is very important to raise awareness among health professionals, communities, and users of its possible toxic and lethal effects.

As explained in Chap. 16, the term "new" does not necessarily refer to new inventions – several NPS were first synthesized decades ago – but to substances that have recently become available on the market. NPS vary in terms of the onset and duration of their effects, their potency, and the doses needed to produce the same desired effect [4].

These products can be especially dangerous when mixed with heroin or other prescription drugs. Therefore, many drug users whose tests found the presence of fentanyl and its analogs do not know that they took those substances (see Chap. 5) [27].

3.6.2 Indiscriminate Use of Prescription Opioid Analgesics

As previously mentioned, the use of prescription opioid analgesics has increased in several countries. In particular, the North American experience shows that about one-fifth of people who began using opioids for chronic pain treatment switched to illicit opioids, such as heroin and fentanyl.

In 2019, the United Nations Office on Drugs and Crime (UNODC) reported that nearly 62 million people worldwide were using opioids for nonmedical purposes [28]. North America accounted for 3.6% of the total and Europe for 0.8%. Estimates in Asia and Africa reveal that users have almost doubled in the last decade [1].

In the United States, from the 1990s to around 2015, OUDs spread into a pandemic, expressed in over-prescription of opioids by medical professionals due in part to aggressive marketing practices. In addition, pharmaceutical companies spread the idea among the medical community that opioid painkillers would not create addiction and offered various incentives to physicians who prescribed opioids. These practices led to a wide deviation in the use and misuse of opioids before it became clear that they could be highly addictive [11]. When patients could not pay for legal opioids, some turned to the illegal market to find another opioid, usually fentanyl or heroin, and opioid overdoses increased. Unethical practices by some pharmaceutical companies and a few physicians turned medicines into commodities and patients into mere consumers, a perverse system comparable to that of drug cartels [29].

3.7 Drugs, Crime, and Violence

The effects of illicit drugs, crime, and violence are highly detrimental to governments. They also harm local communities at the microsocial level because some of their members live amid illicit drug markets where crime and violence are often present, especially among vulnerable populations.

Substance use disorders have destabilizing consequences for societies, especially vulnerable populations. Lack of sustainable livelihoods may drive vulnerable populations to engage in illegal drug cultivation.

An INCB study concluded that most users are non-violent and often underage. In addition, economic-compulsive drug crimes, such as robbery and larceny, are more common than violent drug-induced assault.

Transnational organized crime is a global threat when linked to criminal groups and governments, which aggravate corruption, extortion, the development of illicit activities, and violence, destabilizing countries and hindering productive activities.

Violence and illicit drug-related activities are closely linked due, in part, to access to firearms and the corruption power of crime groups within societies. It is a vicious and self-perpetuating cycle that links poverty to drug problems [30].

3.8 Mexico-The United States: A Complex Bilateral Relationship

Since the 1940s, Mexico has been the leading supplier of heroin and other drugs to the United States. As Vanda Felbab-Brown points out in the Brookings Institute report "Fending off Fentanyl" [18], "Poppy cultivation in Mexico dates back to before World War II. During the war, Mexico supplied legal, medical opioids to the United States at Washington's request. When American demand for medical opioids declined after the war, poppy production shifted to heroin production and the supply of the illegal drug market in the United States. It flourished particularly in the 1970s, before declining due to poppy eradication campaigns between the United States and Mexico in Mexico and, more importantly, a widespread takeoff of cocaine consumption in the United States that mitigated the expansion of heroin consumption in the United States."

The extensive geographic border between Mexico and the United States has allowed drug trafficking to be a permanent and growing industry. The US market shifted from cocaine to heroin, and Mexico has remained the dominant source of drugs consumed in the United States. "In the case of Mexico, the US government has estimated a higher number of opium poppy crops in Mexico, with 32,000 ha [hectares] cultivated in 2016, 44,100 in 2017, and 41,800 in 2018. The cultivation of illicit crops is one of the most labor-intensive illicit economies, allowing those who sponsor it, in Mexico's case, organized crime groups—to obtain ample popular capital. As in other parts of the world, Mexico's poppy farmers are some of the poorest and most marginalized segments of Mexican society and often also members of indigenous groups" [18].

Mexico and China are the main suppliers for the Unites States of fentanyl and fentanyl analogs. Since April 2019, when China imposed controls on all fentanyl analogs and increased monitoring of its postal services to the United States, Mexican drug cartels have increased their market in the United States. However, networks created between Mexican criminal groups and the Chinese chemical precursor industry maintain routes to send precursor chemicals to Mexico where they are used in clandestine laboratories.

3.9 Demand

As indicated above, UNODC [1] estimates that nearly 62 million people use opioids for nonmedical purposes. North America accounts for 3.6% of the total and Europe for 0.8%. Estimates in Asia and Africa show that the number of users has almost doubled in the last decade. Therefore, opioid use has been more prevalent in developing countries and countries with economies in transition than in developed countries.

Stemming from population growth, Africa has the highest prevalence of drug use, with an estimated projection of 83 million people by 2030. However, the most prevalent opioids there (tramadol and codeine) are not as lethal as those consumed in other parts of the world, such as the United States. Population growth, widespread poverty, and lack of resources for drug prevention and treatment have an impact on the young population (25–34 years old) [24]. Another driver of drug abuse is urbanization, with which accessibility to drug markets is easier. Along with revenue, it increases the ability to buy medicines and also allows for expanding drug markets.

A school survey in Mexico shows that prevalence of nonmedical opioid use is 60% higher in urban areas than in rural areas. In some cases, such as Australia and the United States, the prevalence is higher in rural areas.

3.10 Impact of COVID-19 on Opioid Supply and Demand

The COVID-19 pandemic has affected drug use, trafficking, treatment, and service delivery. It has led to decreasing face-to-face medical appointments, a lack of sufficient capacity in hospitals, and inadequate supply of opioids for pain relief. It has also affected drug markets worldwide because of mobility restrictions, social distancing, and reduced commercial flights. Moreover, the availability of medicines decreased due to interruption of supply from manufacturing countries [31].

Geographically the impact was uneven, depending on the intensity of the lockdowns imposed. These lockdowns also had significant consequences on users and their ability to access illicit drugs [5]. However, as Angela Me, head of the UNODC Research and Analysis Branch, acknowledged, "...Traffickers have proven to be resilient and highly dynamic, and have adapted rapidly to the changes induced by COVID-19" [32].

Thus, drug markets adapted to changes over time through innovative and flexible, technology-intensive strategies that enabled full recovery by the end of 2020. For example, the COVID-19 pandemic increased shipments of drugs through sea routes via private ships [24].

Online drug advertising and sales strategies already in place intensified and allowed global drug supply chains to operate quickly. Contactless methods for purchasing medicines via both the Internet and the dark web adapted to changing circumstances efficiently and quickly. Internet-based distribution made drugs more accessible as new online shopping methods expanded in a way that knows no borders. Online selling of prescription or illegal drugs reduces intermediaries, saves costs, and shortens supply chains (see Chap. 16). In addition, drugs can be delivered directly to the buyer's door with the "speed of international shipping" [24].

The use of fentanyl and other alternatives to heroin increased among people who were already dependent on opioids. "A study in the US population diagnosed with or at risk for substance use disorders found that the frequency of cocaine, fentanyl, heroin, and methamphetamine use has increased after the COVID-19 outbreak" [24].

The future dynamics of the opioid market will be determined by new modes of production, supply, trafficking routes, and patterns of drug use with different geographical expressions that will require a multi-pronged approach to be effectively addressed.

3.11 Future Challenges: Interventions and the Future of Market Regulation

The complexity of opioid markets and the multifactorial conditions surrounding drug abuse require a set of measures that put people at the center and recognize the shared responsibility of international organizations, governments, and the private sector to address this challenge comprehensively.

Methadone and buprenorphine, the two opioid medications most commonly used to treat opioid use disorders, have become increasingly accessible over the past two decades. However, although the amount available for medical use has increased sixfold since 1999, from 557 million daily doses to 3317 million in 2019, this has not been the case in small communities, developing countries, and countries with economies in transition, where access to methadone is scarce or non-existent. Therefore, an immediate challenge is to provide the best available treatment for opioid use disorders (see Chaps. 14 and 15) as part of healthcare systems worldwide.

Opioid users may have trouble accessing treatment and other services despite their higher vulnerability to COVID-19 caused by depression of the immune system (see Chap. 12). Regardless of the present pandemic, people who use drugs should have the same rights to receive medical care as people who do not use drugs.

The UNGASS [33] concluded that international drug control conventions should build a system that guarantees access to controlled drugs for medical purposes and prevents abuse. Pain treatment is a key component of the right to health in the context of a health-based drug control system.

There is a need to improve communication, coordination, and cooperation between all actors involved. To achieve this, governments should:

- Follow communication protocols providing common ground for capacity building for prevention and supply reduction strategies.
- Develop collective and coordinated efforts with the public and private sectors to exchange relevant information to address supply reduction and trafficking.
- Enhance bilateral, regional, and international cooperation to build trust to effectively and permanently dismantle drug trafficking organizations.

It is essential to address social and economic needs of drug-using populations. Technological innovation and new modes of NPS production have brought an increased availability of drugs, which challenges drug misuse reduction efforts (see Chap. 16).

The economic impact of this growing, multimillion-dollar market and the capacity for adaptability of criminal organizations have led to confronting visions on whether reform of current drug policy is needed.

As the Global Commission on Drugs states, regulation must address the risks to our lives and communities. "The only responsible way is to control the illegal drug market. Therefore, governments must establish regulations and a new programming system, adapted to the hazard of each drug, and based on sound scientific assessments, and monitor and enforce these regulations" [32].

Ending the inconsistencies of the current programming system entails better coordinating the global governance of the international drug control regime. In addition, the WHO and interdisciplinary scientific research in developing evidence-based programming criteria should be more proactive in determining the scale of harms and benefits.

Although governments are responsible for establishing uniform and consistent procedures to regulate opioid supply and monitor its use, civil society organizations and communities, as well as users, must be actively involved in shaping a more flexible and balanced regulatory system.

Research should put individuals and communities at the center and be geared toward protecting human rights, public health, and sustainable development, bringing peace and security to all.

The COVID-19 pandemic brought new modes of drug consumption, production, and distribution. All actors involved in drug misuse prevention and treatment must recognize that they face a shared but differentiated responsibility to address drug misuse. COVID-19 has highlighted that governments have a key role in regulation, control, coordination, and communication that cannot be delegated to address a drug problem with local, national, regional, and global effects on our communities.

References

1. UNODC. Understanding the global opioid crisis [Internet]. Glob. SMART Updat. 2019.
2. CDC. Centers for Disease Control and Prevention, National Center for Injury Prevention and Control [Internet]. Other drugs. 2021.
3. UNODC. United Nations Office on Drugs and Crime (UNODC) [Internet]. World Drug Rep. 2021, Exec. Summ. Polices Implic. 2021.
4. UNODC. United Nations Office on Drugs and Crime (UNODC) [Internet]. World Drug Rep. 2018, Summ. Policy Find. Implic. 2018.
5. Degenhardt L, Grebely J, Stone J, Hickman M, Vickerman P, Marshall BDL, et al. Global patterns of opioid use and dependence: harms to populations, interventions, and future action. Lancet (London, England). 2019;394:1560–79. https://doi.org/10.1016/S0140-6736(19)32229-9.
6. Felbab-Brown V. "Fending off fentanyl and hunting down heroin Controlling opioid supply from Mexico", paper series "The opioid crisis in America: Domestic and international dimensions. [Internet] Brookings Inst 2020.
7. International Narcotics Control Board (INCB). Report 2016 Press material. Unite Nations informations Serv. 2017.
8. International Narcotics Control Board (INCB). Report 2019 Press material. Unite Nations informations Serv. 2020.
9. Herring T. Since you asked: what role does drug enforcement play in the rising incarceration of women? [Internet]. Prison Policy Initiat. 2020.
10. Martins SS, Segura LE, Santaella-Tenorio J, Perlmutter A, Fenton MC, Cerdá M, et al. Prescription opioid use disorder and heroin use among 12-34 year-olds in the United States from 2002 to 2014. Addict Behav. 2017;65:236–41. https://doi.org/10.1016/j.addbeh.2016.08.033.
11. National Institute on Drug Abuse (NIDA). Opioid overdose crisis [Internet]. Natl. Institutes Heal. (NIH), U.S. Dep. Heal. Hum. Serv. 2021.

12. Wilson N, Kariisa M, Seth P, Smith H, Davis NL. Drug and opioid-involved overdose deaths — United States, 2017–2018. MMWR Morb Mortal Wkly Rep. 2020;69:290–7.
13. International Narcotics Control Board (INCB). Report 2020 Press material. Unite Nations informations Serv. 2021.
14. National Institute on Drug Abuse. Comorbidity: substance use disorders and other mental illnesses. 2018;1–3.
15. Zapata Diaz JP, Rendón Fonnegra J, Berrouet MC. Síndrome de abstinencia neonatal. Pediatria (Santiago). 2017;50 https://doi.org/10.14295/pediatr.v50i2.60.
16. American Academy of Pediatrics (AAP). The opioid epidemic: how to protect your family [Internet]. Comm. Subst. Use Prev. 2020.
17. World Health Organization (WHO). Opioid overdose [Internet]. Press Cent. 2020.
18. UNODC. World drug report 2019, Global overview of drug demand and supply 2019.
19. Insulza JM. The OAS Drug Report, The drug problem in the Americas: consensus on the report entrusted to the OAS. Organ. Am. States. 2014.
20. UNODC. World Drug Report, 2021: The effects of the pandemic increase the risks of drugs, as young people underestimate the dangers of cannabis. United Nations Publ. 2021.
21. UNODC. World Drug Rep 2020, Cross-cutting issues: evolving trends and new challenges. United Nations Publ. 2020.
22. Nelson R. Global opioid use continues to be "Feast or Famine" [Internet]. Medscape. 2016.
23. Pardo B, Taylor J, Caulkins JP, Kilmer B, Reuter P, Stein BD. The future of fentanyl and other synthetic opioids. RAND Corporation; 2019.
24. UNODC. World Drug Report, United Nations publication. Sales, editor. 2021;E.21.XI.8.
25. NIDA. Fentanyl DrugFacts. Natl Inst Drug Abus 2021;1–8.
26. Forecast S. La crisis de los opiáceos en los EE. UU. [Internet]. First Gov. Acc. nationwide drug deaths. 2016.
27. Gibney A. The crime of the century [Internet]. HBO documentary; 2021.
28. UNODC. World Drug Report 2018, Global overview: drug demand [Internet]. United Nations Publ. 2021.
29. International Narcotics Control Board (INCB). Drugs, crime and violence: the microlevel impact. 2003;1–11. https://doi.org/10.4324/9781315252063-20.
30. NIDA. Substance use in women – drug facts. Natl Inst Drug Abus. 2020;
31. Dudley S, Bonello D, Lopez-Aranda J, Moreno M, Clavel T, Kjelstad B, et al. Mexico's role in the deadly rise of fentanyl. Wilson Cent. Woodrow Wilson International Center for Scholars - The Mexico Institute; 2019.
32. Global Commission on Drug Policy (CGPD). Classification of Psychoactive Substances: when science was left behind [Internet]. Reports 2019.
33. UNODC. Outcome document of the 2016 United Nations General Assembly special session on the world drug problem. Thirtieth Spec Sess Gen Assem New York. 2016;19–21.

Chapter 4
Toward a New Drug Policy Agreement

Alfredo Camhaji, Mariana Racotta, Raúl Martín-del-Campo, and José Franco

Abstract This chapter is divided into five sections. The first one describes the beginnings and evolution of the international drug control system. The second one delineates the institutional architecture in which this system operates within the framework of the United Nations organization. The third section refers to the Special Sessions of the United Nations General Assembly (UNGASS) on the World Drug Problem, emphasizing the one of 2016 where integration of human rights to the international drug control initiatives was encouraged. The fourth section addresses the present-day limitations and challenges to respond to the production and trafficking of illegal drugs. Finally, the fifth section, called "Toward a New Consensus," considers that the penalization and restriction policies for medical and scientific purposes have failed and proposes a tailor-made drug policy based on the United Nations 2030 Agenda and human rights.

Keywords International drug control · Human rights · Opioids · Policy

A. Camhaji (✉)
Opioids Working Group, Global Studies Seminar, Faculty of Medicine, National University Autonomous of Mexico (UNAM), Mexico City, Mexico

M. Racotta
Faculty of Higher Studies (FES) Acatlán, UNAM, Mexico City, Mexico

R. Martín-del-Campo
Opioids Working Group, Global Studies Seminar, Faculty of Medicine, National University Autonomous of Mexico (UNAM), Mexico City, Mexico

International Narcotics Control Board, United Nations, Vienna, Austria

National Institute of Psychiatry Ramón de la Fuente Muñiz (INPRFM), Mexico City, Mexico

J. Franco
Opioids Working Group, Global Studies Seminar, Faculty of Medicine, National University Autonomous of Mexico (UNAM), Mexico City, Mexico

Institute of Astronomy, UNAM, Mexico City, Mexico

S. L. Cruz (ed.), *Opioids*, https://doi.org/10.1007/978-3-031-09936-6_4

4.1 International Drug Control: Origins and Characteristics

The current international drug control regime is based on the following United Nations treaties:

- The Single Convention on Narcotic Drugs, 1961 (amended in 1972)
- The Convention on Psychotropic Drugs, 1971
- The Convention Against Illicit Traffic in Narcotic Drugs and Psychotropic Substances, 1988

This global drug control system aims to prevent drugs from harming individuals and society's lives and protect the physical and mental health of humankind and its well-being. Drug control treaties have focused on creating a drug-free world through policy and law enforcement, including criminal sanctions. However, this strict regime has failed to meet its objectives and has resulted in human rights violations. This chapter describes the consequences of the current drug policy master plan focusing on human rights and proposes a new policy approach based on the 2030 Agenda of the United Nations.

The drug control policy system dates back to the early twentieth century, with the creation of the League of Nations. During this time, opium was used as a medicine in various countries and consumed for its psychoactive effects in China and much of Southeast Asia. Thus, the international treaties that first addressed drug control referred to opium and its derivatives, seeking to restrict their use in nonmedical settings [1–4].

Various substances with therapeutic and addictive potential, such as morphine, codeine, cocaine, and diacetylmorphine, were discovered or synthesized during the nineteenth century (see Chaps. 1 and 8). In tandem, many cases of people with harmful consumption patterns emerged. As a result, regulating the use of these substances became a topic of debate and a significant diplomatic issue at the beginning of the twentieth century [1, 5].

4.1.1 First Steps in the Establishment of an International Drug Control System

With the participation of 13 countries, the Shanghai Opium Commission in 1909 was the first effort to establish an international drug control system focused on opium. Even before the Convention, some countries started reducing their opium imports and sales. The most important agreement was that China offered to eradicate opium cultivation and gradually reduce the number of opium dens, while Britain agreed to stop sending opium to China. The United States and China sought a generalized ban on the opium trade, while the European powers favored looser control. At the end of the meeting, the Commission adopted several non-binding recommendations, including that it was undesirable to export opium to countries that did not legally admit it, in a clear message to the situation imposed by Britain to China for several years (see Chap. 1) [1, 6]. Soon after, opium became the epicenter of a highly sophisticated initiative in the multinational arena: the international drug control regime. The first steps toward reaching a consensus on this issue came with the International Opium Convention, held at The Hague in 1912. This was the first in a series of global gatherings that would seek to limit and regulate the production, trade, and consumption of several substances exclusively for medical and scientific purposes (Table 4.1) [1, 3, 4].

4.1.2 The League of Nations in the International Drug Control Regime

After World War I, many more nations became involved in the international drug control regime, assuming policies to control and prohibit drug use and trafficking (Table 4.1). The Treaty of Versailles in 1919 laid the groundwork for creating the League of Nations, which would assume a lead role in the fight against the drug regime. From then on, treaties of a regulatory and non-prohibitionist nature were negotiated among interested countries. Although data and statistics on consumption were shared, participating parties were not asked to limit or eliminate substance use. This led to a continued increase in drug distribution and use; some pharmaceutical companies moved to countries that were not part of the control system, traffickers became more sophisticated, and drug users sought alternative supply sources [5]. To deal with this situation, the League of Nations convened a conference in Geneva in 1931 where participants adopted the "Convention for Limiting the Manufacture and Regulating the Distribution of Narcotics." That same year, an agreement concerning the Suppression of Opium Smoking in Bangkok was also approved [1, 3, 4]. Although, by 1934–1935, the legal manufacture of opiates and cocaine had fallen to the level of legitimate demand, the emergence of illicit activities continued, and smuggling of opium to produce heroin increased [5].

4.1.3 International Drug Control After the Second World War

After the Second World War, and through the Lake Success Protocol of 1946, the signatory governments of drug treaties transferred the international drug control from the League of Nations to the newly created United Nations, thus initiating a new chapter in drug regulation (Table 4.2).

Table 4.1 First multilateral meetings and treaties in the international drug control system

Date and place	Resultant document	Goals	Entry into force
02/1909 Shanghai, China	Report of the International Opium Commission	First discussion on imposing international control of opium commerce	None. Only nine general recommendations were issued
01/1912 The Hague, The Netherlands	International Opium Convention	To establish the first international treaty on drug control, including morphine, diacetylmorphine (heroin), and cocaine. To reduce the use of opium use. Each country should enact effective laws regarding drug production and distribution	06/1919
02/1925 Geneva, Switzerland *First one negotiated within the League of Nations	Agreement concerning the Suppression of the Manufacture of, Internal Trade in, and Use of, Prepared Opium	Focused on opium-producing nations. A production and distribution system with a State monopoly was adopted as a control mechanism	07/1926
02/1925 Geneva, Switzerland	International Opium Convention	To institutionalize the international control system and to extend the scope of control to cannabis. A system of import certificates and export authorizations was established. To control the supply of drugs, without including provisions on demand reduction nor criminal sanctions for users. It established the Permanent Central Opium Board (PCOB). States remained free to manufacture and use controlled substances, provided that they accurately reported to the PCOB	09/1928
07/1931 Geneva, Switzerland	Convention for Limiting the Manufacture and Regulating the Distribution of Narcotic Drugs	To establish a mandatory report system. Producer countries had to report on areas cultivated with coca and poppy to limit production to medical and scientific needs. A scheduling system for the classification of substances was established, and the Drug Monitoring Body was created	

(continued)

Table 4.1 (continued)

Date and place	Resultant document	Goals	Entry into force
11/1931 Bangkok, Thailand	Agreement concerning the Suppression of Opium Smoking in the Far East	It referred to a limited scope for action, emphasizing the need to limit and control opium poppy cultivation in the Far East through international action	04/1937
06/1936 Geneva, Switzerland	Convention for the Suppression of the Illicit Traffic in Dangerous Drugs	To make all trafficking-related activities an international crime subject to criminal sanctions To establish forms of cooperation in the international criminal sphere, such as the extradition of traffickers. Countries were called upon to create special police forces for the optimal implementation of the convention	10/1939

4.1.3.1 The Current Drug Control Regime: Three Existing Treaties

The expansion of drug use was accompanied by increased stigmatization of opioid users. At the same time, the existence of multiple international legal agreements on narcotics became a problem, as their provisions were complex, and several States had not signed and ratified all of them. As a result, 77 States adopted the Single Convention on Narcotic Drugs in 1961 (Table 4.3).

The International Narcotics Control Board (INCB), established in 1968, sought to simplify and rationalize procedures to increase the efficiency of drug control activities [1, 3, 4]. The INCB compiled estimates of needs and statistics on the use of opium, coca, and cannabis by governments [7]. The Single Convention classified psychoactive drugs into four lists (Schedules I, II, III, and IV) according to their therapeutic potential and risks (Table 4.4). A substance would be more or less controlled depending on the schedule.

The United States was dissatisfied with the results of the 1961 conference because of the INCB's lack of embargo powers against States not complying with their obligations. Thus, in the context of President Nixon's punitive positioning, Washington organized a Plenipotentiary Conference at Geneva in 1972 to amend the Single Convention [5]. The Conference was attended by representatives of 97 States and resulted in signing a protocol that amended the Single Convention on Narcotic Drugs.

The previous year, the Vienna Convention on Psychotropic Substances was held, an event in which synthetic substances were still grouped into four lists (Table 4.4). However, the new treaty aimed to distinguish between "narcotic drugs," controlled by the 1961 Convention, and "psychotropic substances" [1, 3, 4, 8]. Furthermore, to be considered in the 1971 schedules, a substance needed to have the ability to produce dependence and to stimulate or depress the central nervous system, resulting in hallucinations or alterations in motor function,

Table 4.2 First drug control treaties within the framework of the United Nations

Date and place	Resultant document	Goals	Entry into force
12/1946 New York, United States	Protocol amending the Agreements, Conventions and Protocols on Narcotic Drugs concluded at The Hague on 23 January 1912, at Geneva on 11 February 1925 and 19 February 1925 and 13 July 1931, at Bangkok on 27 November 1931, and at Geneva on 26 June 1936	To transfer the functions and apparatus of the drug control from the League of Nations to the newly created United Nations (UN)	1948
11/1948 Paris, France	Protocol Bringing under International Drugs Outside the Scope of the Convention of 13 July 1931 for Limiting the Manufacture and Regulating the Distribution of Narcotic Drugs, as amended by the Protocol signed at Lake Success, New York, on 11 December 1946	To authorize the World Health Organization to place under full international control any new drugs, including synthetic drugs, which could not be subject to such control through the application of the relevant provisions of the 1931 Convention and which are considered to be addictive or to become an addictive drug	12/1949
06/1953 New York, United States	Protocol for Limiting and Regulating the Cultivation of the Poppy Plant, the Production of, International and Wholesale Trade in, and use of Opium	To limit and regulate opium poppy cultivation and production. Only seven States would be entitled to produce opium – Bulgaria, Greece, India, Iran, Turkey, the USSR, and the former Yugoslavia – for medical and scientific purposes, with a State monopoly and receiving international inspections	1963

thinking, behavior, perception, or mood [6]. Thus, the definition of narcotic drugs and psychotropic substances was established following the already agreed schedules of both treaties, without a specific definition being given to each group of substances [9, 10].

While the repressive action on the production centers of drugs of natural origin was growing with the creation of the US Convention against illicit traffic of narcotic drugs and psychotropic substances, approved by the Conference in 1988, the so-called designer drugs, entirely synthetic, emerged on the market (Chap. 16) [5]. The traditional emphasis on control culminated in a new illicit trafficking treaty, and the agreement addressed many of the issues included in the 1936 treaty [1, 3, 4]. This Convention included two tables listing precursors and common chemical reagents used to synthesize psychoactive substances. Both precursors and reagents were also brought under different levels of control.

Table 4.3 Existing treaties of the drug control regime

Date and place	Resultant document	Goals	Entry into force
03/1961 New York, United Sates	Single Convention on Narcotic Drugs of 1961	To replace all existing conventions and protocols except for the 1936 Convention To extended the control to opium poppy, coca bush, and cannabis. They were subjected to the same controls as extracted alkaloids and concentrates such as morphine and cocaine To require parties to impose criminal penalties for illicit cultivation and prohibit all traditional uses. It also required them to submit estimates of needs and statistics on trade, production, existence, and consumption of narcotic drugs It incorporated four schedules (see Table 4.4)	12/1964
02/1971 Vienna, Austria	Convention on Psychotropic Substances	To extend the control to other substances, especially to those of synthetic origin such as amphetamine-type stimulants, hallucinogens (LSD), ecstasy, sedatives, anxiolytics, analgesics, and antidepressants A scheduling system was also established (see Table 4.4)	08/1976
03/1972 Geneva, Switzerland	Protocol amending the Single Convention on Narcotic Drugs of 1961	To strengthen the control system, expanding the role of the INCB in the control of illicit trafficking To include provisions authorizing States to use treatment, education, and rehabilitation measures, as well as criminal sanctions To fine-tune the existing provisions relating to the forecasting system and the collection and reporting of data	08/1975
12/1988 Vienna, Austria	Convention Against Illicit Traffic in Narcotic Drugs and Psychotropic Substances	To criminalize illicit trafficking in chemical precursors, laundering of assets, and international trafficking. The offense of conspiracy was recognized, and States were urged to enact legislation allowing for the confiscation of ill-gotten assets To require States to enact appropriate measures for the eradication of illicit cultivation, and as to offer mutual legal assistance, share information, and cooperate in law enforcement efforts It incorporated two tables for precursors and chemicals frequently used in the illicit manufacture of narcotic drugs and psychotropic	11/1990

Table 4.4 Schedules of the Single Convention on Narcotic Drugs of 1961 [2]

	Schedule I	Schedule II	Schedule III	Schedule IV
Single Convention on Narcotic Drugs of 1961	Highly addictive or likely to be abused substances and precursors that could become equally addictive and likely to be abused. Examples: cannabis, opium, heroin, methadone, oxycodone	Substances that are less addictive and less susceptible to abuse than those in Schedule I. Examples: codeine, dextropropoxyphene	Preparations intended for legitimate medical use and which are unlikely to be abused. Examples: <2.5% codeine, <0.1% cocaine	Certain drugs listed in Schedule I that are considered to be particularly harmful because of their addictive properties and their little or no therapeutic value. Example: heroin
Schedules of the 1971 Convention on Psychotropic Substances	Drugs presenting a high risk of abuse, posing a severe threat to public health, with little or no therapeutic value. Examples: LSD, MDMA, cathinone	Drugs presenting a risk of abuse, posing a severe threat to public health, with low or moderate therapeutic value. Examples: dronabinol, amphetamines	Drugs presenting a risk of abuse, posing a serious threat to public health, which are of moderate or high therapeutic value. Examples: barbiturates, buprenorphine	Drugs that present a risk of abuse, posing a serious threat to public health, with a high therapeutic value. Examples: tranquilizers such as diazepam

4.2 Drug Control Agencies of the United Nations System

The first international agencies in charge of the drug control system appeared within the framework of the League of Nations. Since the signing of the Protocol of 1946, the United Nations, with 193 Member States under its belt, has been responsible for the international drug control regime, including implementing treaties through different commissions, bodies, and boards.

4.2.1 The Commission on Narcotic Drugs (CND) of the Economic and Social Council

The Economic and Social Council (ECOSOC or the Council) has the power to carry out studies and reports on economic, social, and environmental issues and make recommendations on such matters to the General Assembly, the members of the United Nations, and the specialized agencies concerned.

The Commission on Narcotic Drugs (CND or the Commission) is the political body responsible for discussing and formulating international drug control policy guidelines. It was established as a subsidiary commission of ECOSOC in 1946 to assist the Council in monitoring the implementation of drug control treaties. The CND comprises 53 Member States, elected by ECOSOC, which subscribe to the Single Convention of 1961. In addition, the Commission discusses the annual reports of the United Nations Office on Drugs and Crime (UNODC) and the INCB. Its agenda is structured in two distinct segments: a normative one for the performance of normative and treaty-based oversight functions and a series of operational sessions to exercise the role of governing body of UNODC [1, 9].

Per the 1961 and 1971 conventions, the CND currently decides, based on World Health Organization (WHO) recommendations, whether to add or delete substances included in the narcotics and psychotropics lists. This Commission can also move substances from one schedule to another. In addition, per the 1988 Convention, the CND is responsible for establishing which chemical precursors should be under international control, based on INCB recommendations [3, 4].

The CND has five subsidiary bodies with the Heads of National Drug Law Enforcement Agencies for Europe, Latin America and the Caribbean, Asia and the Pacific, and Africa and the sub-commission for the Near and Middle East. They meet two times each year as the governing body of the US drug program: first in March when drug policy resolutions are discussed and adopted and then in December, to consider budgetary and administrative issues. During negotiations, governments can be represented by officials from the Ministries of Foreign Affairs and the Interior, Health, Justice, and Defense or representatives of their diplomatic missions in Vienna [8]. In addition, representatives of non-governmental organizations and civil society may also participate in the meetings as observers [1, 9].

Decisions within the CND are made under the "Vienna Consensus," whereby all resolutions and declarations must be adopted unanimously. However, due to the difficulty of reaching unanimity, diplomatic formulations and final decisions become ambiguous, which has made it easy for the Member States to block resolutions or decisions [11].

4.2.2 The International Narcotics Control Board (INCB)

The INCB is an independent and quasi-judicial control body responsible for implementing the international drug control treaties and monitoring compliance. It was established in 1968 by the Single Convention of 1961 and is composed of 13 members elected by ECOSOC: 3 with medicine or pharmacology backgrounds, chosen from a list presented by the WHO, and 10 elected from a list of persons provided by governments. The Board members serve in a personal capacity and not on behalf of their states or governments.

The functions of the INCB are:

- To ensure, in cooperation with governments, the adequate provision of controlled substances for medical and scientific purposes while safeguarding that there are no diversions to illicit channels
- To monitor government control of chemicals used in the illicit manufacture of drugs
- To identify the shortcomings of national and international control systems regarding illicit trafficking, production, and consumption and to rectify them
- To assess which chemicals are being used in the illicit manufacture of drugs and determine whether they should be subject to international control
- To administer a forecasting system on each country's narcotic drugs and psychotropic needs, monitoring licit drug activities to help governments balance supply and demand
- To promote measures for governments to prevent the diversion of chemicals used in the illicit production of narcotic drugs and psychotropic substances

The INCB is in permanent contact with governments to assist them in complying with their obligations under the international drug control treaties. In the event of apparent violations of the treaties, or when governments have difficulties applying them, the INCB is called to propose appropriate measures or to assist in overcoming such challenges. When remedial measures are not implemented or followed, the INCB may call the matter to the attention of the CND and ECOSOC. The Board can recommend third parties stop importing or exporting drugs to or from a defaulting country as a last resort. Each year, the INCB submits its annual report to ECOSOC through CND, with a global analysis regarding production, consumption, and trafficking of controlled substances, including the progress and setbacks in the control area [3, 4].

4.2.3 The United Nations Office on Drugs and Crime (UNODC)

The UNODC (or the Office) is the United Nations agency mandated to assist the Member States in their fight against drug trafficking, crime, and terrorism. It was established in 1997 through the merger of the United Nations International Drug Control Program with the Centre for International Crime Prevention and is the main body for drug control through the effective implementation of treaties [9].

Based in Vienna, the UNODC focuses on research and analytical, normative, and technical work. It operates 52 offices, covering more than 150 countries, and has liaison offices in New York and Brussels. It is responsible for compiling data, producing statistical analyses, and providing technical support to governments in implementing the existing international treaties on illicit drugs, organized crime, and terrorism. The UNODC also assists in creating local legislation that is in line with these treaties. On the demand side, the UNODC works to prevent and treat drug use disorders. In addition, it generates annual crop studies of the major drug-producing countries and works with these countries to establish sustainable alternative livelihoods for farmers and others involved in drug production.

The UNODC acts on behalf of the US Secretary-General when exercising its functions under international treaties and resolutions of US bodies. It also provides substantive services to the General Assembly, ECOSOC, and the committees and conferences dealing with drug control issues [1, 9]. Furthermore, the UNODC publishes the annual World Drug Report, one of the most comprehensive sources of information on illicit drug trends, and performs secretariat functions for several UN commissions, including the CND and INCB.

4.2.4 The World Health Organization (WHO)

The WHO is the United Nations agency specialized in public health. The UN drug control treaties of 1961 and 1971 mandate the WHO to conduct detailed analyses of substances proposed for (re)classification on control lists within the international drug control regime.

The WHO Expert Committee on Drug Dependence (ECDD) is an independent group of experts that assesses the medical properties of any given substance regarding its dependence liability. Depending on the outcome, the Committee can recommend placing it within the control regime, transferring it from one list to another, removing it from a list, or keeping it under surveillance for lack of evidence of actual misuse or dependence [1, 9]. With this information, the Director-General of WHO presents recommendations to the Secretary-General of the United Nations and the CND based on the best available scientific, medical, and public health evidence.

4.3 The United Nations General Assembly on the Drug Problem and Human Rights

The United Nations General Assembly is the main body for formulating public policies. It has 193 Member States, all with equitable representation. In addition, the General Assembly may convene Special Sessions (UNGASS) regarding drug control in response to requests from the Member States on specific topics [1, 9, 12].

4.3.1 The United Nations Decade Against Drug Abuse: 1991–2000

The United Nations General Assembly convened the first Special Session to address drug abuse in 1990 at the request of Colombia. This occurred within the context of the intensifying of the "war on drugs" and the ongoing dissolution of the Soviet

Union. There was hope that the fall of the Berlin Wall and the end of the ideological confrontation between countries would allow for improved international collaboration, especially in terms of the United States' anti-drug policy toward Latin America. During the first UNGASS, a Global Program of Action was approved, and the creation of the United Nations International Drug Control Program (UNDCP) was announced [12, 13].

4.3.2 *The Second UNGASS: 1998*

Seven years later, Member States met in Vienna and agreed on a new Political Declaration and Plan of Action on international cooperation for a comprehensive and balanced strategy to control the world drug problem. The INCB believed that demand reduction was a role that each country should perform independently, although international support was required in some cases. It also considered that "demand reduction programs should be developed at the national and local levels, according to the actual situation of drug abuse and taking into account the cultural, political, economic and legal environment" [13]. Despite dissenting voices, the drug control approach was ratified. In addition, for the first time, issues such as demand reduction, money laundering, chemical precursors, harm reduction, synthetic drugs, and higher investment in alternative development were raised to be debated in future assemblies.

Mexico stressed the need to review narcotic drug classification according to WHO criteria "to reduce the illicit drug market," thus suggesting that the control of certain substances should be abandoned [13]. Also, harm reduction measures, a consequence of the AIDS epidemic worldwide, highlighted the need to take action to reduce the risk of the HIV spreading through the injection of intravenous drugs [9, 12, 14].

4.3.3 *The Third UNGASS: 2016*

In October 2012, the governments of Colombia, Guatemala, and Mexico, driven by the increasing violence of the war on drugs, issued a joint statement requesting a new UNGASS on the pressing problem of drug policies, particularly arms trafficking and money laundering. The objective, according to Felipe Calderón, then president of Mexico, was "to review the current international strategy and above all, define better solutions from a human rights, prevention and public health perspective that put people's well-being at the center, bring a new approach that did not criminalize consumers, but give them opportunities and alternatives" [15].

The Assembly took place on 19–21 April 2016, bringing together the Member States, United Nations agencies, and civil society delegates. Attendees recognized a growing divergence in the world drug policy landscape among the Member States and a lack of coherence in the United Nations drug control architecture within its agencies. WHO, UNAIDS, and UNDP discussed decriminalization and harm reduction in their written and oral statements. The Office of the United Nations High Commissioner for Human Rights (OHCHR) and various special rapporteurs also drew attention to the continuing violations of these rights in the context of drug control. Several countries and many non-governmental organizations (NGOs) asked the Secretary-General to improve the functioning and coherence of the drug control architecture, inconsistencies in the treaties, and harmonization with the Sustainable Development Goals (SDGs) and human rights. The 2016 UNGASS resolutions were the following:

- To dismantle criminal organizations, it is necessary to intensify cooperation and extend information exchange and related actions.
- Better coordination and cooperation are needed among the specialized agencies within the United Nations system to address all aspects of the world drug problem.
- Public policies and actions derived from international drug policy must be aligned with the 2030 Agenda Sustainable Development Goals (SDGs) (Fig. 4.1).
- Social harm related to the illicit drug market must be addressed in communities controlled by organized crime. Likewise, the prevention of violence, exclusion, and weakening of the social fabric should be handled with social and recreational work alternatives.
- It is necessary to approach drug control from a human rights perspective; only then a more comprehensive, balanced, and development-promoting response can be offered.
- Punitive approaches must be modified to place people, their rights, and dignity at the center of all efforts instead of substances and judicial processes.
- Drug use must be treated as a public health problem since it threatens people's development. Drug addiction needs prevention mechanisms and comprehensive therapeutic solutions without criminalizing users.
- It is necessary to consider proportional penalties and alternatives to incarceration from a gender perspective because disproportionate sanctions punish vulnerable women and children and generate vicious circles of marginalization and criminality.
- Joint efforts should be implemented to prevent consumption in children and young people globally.
- Availability and better access to controlled substances for medical and scientific purposes must be ensured, avoiding diversion, misuse, and trafficking.
- Control should be done in terms of groups, not individual substances.

4.3.3.1 Meeting Progress

The 2016 UNGASS was instrumental in expanding the scope of global drug policy debates beyond the usual three pillars: demand reduction, supply reduction, and international cooperation. It also embraced other areas, such as health (including harm reduction and access to controlled medicines), development, human rights,

Fig. 4.1 Public policies and actions derived from international drug policy should be aligned with the 2030 Agenda for Sustainable Development Goals adopted by the United Nations in 2015

and other new challenges. The UNGASS provided a much-needed platform to bring some long-unresolved issues to the negotiating table and received constructive inputs from United Nations agencies, Member States, and civil society organizations. In addition, a more global approach to the drug issue and the 2030 Agenda for Sustainable Development was adopted, albeit rhetorically. The 2016 UNGASS created the necessary conditions so that human rights, the SDGs, and a consultative group of experts could be part of the agenda to improve the functioning and coherence of the global drug control system. It included the most ambitious provisions on human rights ever to be adopted in a United Nations resolution on drug control. Moreover, access to controlled drugs received significant attention for the first time, and some advances were made to specific references to naloxone and overdose prevention, "drug-assisted therapy programs," and "programs related to injection equipment" [14].

4.3.3.2 Limitations of 2016 UNGASS

Despite the meeting's progress, the UNGASS avoided grassroots structural problems, such as arms trafficking and money laundering, and stifled a debate that questioned the current architecture of the United Nations drug control system. Unfortunately, the final document did not explicitly refer to "harm reduction" but rather to reduce the consequences of consumption. Also, the adoption of the United Nations 2030 Agenda did not include specific recommendations for practical measures regarding decriminalization, death penalty abolition, harm reduction, and respect for Indigenous rights. Even though there has been an increase in the number of countries offering harm reduction interventions, only 1% of people who inject drugs live in countries with needle and syringe exchange programs and opioid substitution therapies [9, 12, 14, 16].

4.3.4 Human Rights as a Key Element in Drug Control

Before 2008, the International Drug Control System did not consider human rights. When the objectives of the international drug control regime clash with those of the international human rights regime, human rights obligations should prevail [17].

In that same year, the United Nations Commission on Narcotic Drugs (CND) adopted a resolution calling for the drug control system to work more closely with the human rights system of the United Nations [9, 10, 18]. The resolution was approved after all references to the death penalty and the declaration on the rights of Indigenous peoples were removed from the text [17]. This event opened the door for human rights safeguards to appear in CND resolutions and for the organizations that make up the SIDCD to include human rights issues, albeit partially.

In 2010, the United Nations released the Report of the Special Rapporteur, which focused on the human right to have the highest possible standard of physical and mental health. As a result, the UNODC received a clear mandate to review and assess its human rights responsibilities. In 2010, a report was prepared for the CND, and in 2012, staff received a guidance note regarding the human rights implications of their work [17, 18].

Two years later, the INCB began to defend the idea that drug policies should respect human rights and respond to the complexity of drug production, trafficking, and consumption phenomena. Currently, the INCB states that the Executive Director recognized “the unintended consequences of the international narcotics control system, among which is the phenomenon known as ‘policy displacement’” (emphasis on law enforcement and insufficient attention to public health) and the marginalization of people who use drugs [3, 4]. Issues of significant concern include violations of the rights to life, health, equality, and non-discrimination, the rights of Indigenous peoples, and children’s rights. The prohibition of torture and other forms of ill-treatment as well as the prohibition of arbitrary detention should also be enforced [10, 19, 20].

In September 2015, OHCHR presented a study on the impact of the world drug problem on the exercise of human rights [21]. The study addressed recommendations based on five categories:

Protect the right to health The threat of criminal sanctions and punitive practices leads drug users to avoid access to services because they fear being denied care, increasing their health risk. Therefore, medication-assisted therapy should be available in harm reduction programs, providing the necessary drugs for users (e.g., methadone, buprenorphine), especially those in prisons and other detention centers (Chaps. 14 and 15). In addition, better access to essential controlled drugs should be guaranteed, especially in developing countries, and investment in public health programs should increase.

Rights related to criminal justice It is necessary to protect the right to life of people convicted of drug-related crimes. This includes avoiding the death penalty and considering decriminalizing consumption and possession of drugs for personal use. In addition, every person must have the right to a fair trial under international standards. It is also necessary to prohibit arbitrary arrest and detention, torture, and other forms of mistreatment of persons detained or accused of drug-related crimes. It is important to consider closing compulsory detention centers and, when necessary, exert only proportional force. Extrajudicial executions must be the object of a prompt, independent, and effective investigation so that the perpetrators of the crimes are brought to justice, considering the severe consequences that a conviction for a drug-related crime can have on a person’s life. Alternatives to prosecution and incarceration of people for minor, non-violent drug-related offenses should be considered.

Prohibition of discrimination Ethnic minorities and women who own or use drugs or are “micro-distributors” must be protected against all forms of discrimination through laws to face the uneven impact of drug policies. To eliminate discrimination, law enforcement, health personnel, and social service workers who interact with drug users should receive training.

Child and Indigenous people's rights Children's rights must be protected by focusing on drug use prevention while providing them with information in a child-friendly, age-appropriate way. Children should not be subject to criminal prosecution; rather, the focus should be on health education and treatment, including harm reduction and social reintegration programs. Indigenous peoples have the right to preserve their traditional, cultural, and religious practices. In cases where drug use is part of these practices, the right to consume them for such strictly defined purposes must be protected in principle, subject to the limitations provided in human rights law.

The Convention that includes human rights In 2014, the INCB began to defend the idea that drug policies should respect human rights and respond to the complexity of production, trafficking, and consumption phenomena; they should also have a gender perspective: protect women, children, youth, and older adults. In addition, policies on drug users must have a public health approach based on prevention, treatment, rehabilitation, and re-socialization [3, 4].

4.4 International Regulations: A Hard to Change Straitjacket

As previously mentioned, the international drug control regime is defined as the set of principles, norms, rules, and decision-making processes to which practically all States and other international actors are bound to comply with the obligations derived from their treaties, develop coordinated actions, and adjust their behaviors to the agreed environment for drug control. If unable to comply, Member States are subject to sanctions.

Countries have a certain degree of flexibility to manage controlled drug use, but not their production. In addition, drugs must be controlled and limited to medical and scientific research, an aspect in which countries may have more restrictive national standards than those established in the conventions. In terms of international relations, a legal regime such as drug control is complicated to modify since changes require the agreement of all participating countries [9, 22–24].

4.4.1 *Main Obstacles for Content Conventions Modification*

4.4.1.1 Existing Regulations for Convention Revisions

There are two ways of revising conventions: modification and amendment. Modification occurs when a substance is eliminated or relocated to a different list within the 1961, 1971, and 1988 Conventions. Amendments entail a formal alteration of the treaty provisions, essentially the text of some of its articles, which would affect all the signatory parties. These Conventions require that the WHO or a State party propose a change based on relevant scientific information to relocate substances on the control lists. According to the 1961 Convention, a proposed change can be approved if most of the 53 members of the Commission on Narcotic Drugs agree to it. Two-thirds are required, according to the 1971 Convention [1, 9, 25]. Any party can propose an amendment to the treaties and communicate it to the general secretary. If none rejects the proposal in a given period, it is considered accepted and takes effect in the corresponding period. However, if one of the States rejects the proposal, it is passed to ECOSOC, which organizes a conference to address the issue. Because of this, getting a large enough majority of States to modify an article is unlikely [3, 4, 16, 22, 24].

4.4.1.2 Obligations by Treaties' Signatory Countries

The conventions establish six general obligations for the Member States:

1. First, establish and maintain an institutional infrastructure to manage the control of substances at the national and international levels.
2. Classify in their internal legislation each narcotic and psychotropic substance and each chemical precursor, ensuring the minimum control required by the conventions and implementing regulation for its trade for lawful purposes.
3. Prevent drug use, providing treatment and rehabilitation measures for users with substance use disorders.
4. Classify conduct related to illicit drug trafficking as severe crimes in national criminal law, establishing sanctions proportional to their severity.
5. Classify the acquisition or cultivation of drugs for personal consumption as a crime. Treatment and rehabilitation may be considered as alternative measures to conviction or punishment.
6. Participate in international cooperation in criminal matters with the rest of the parties in serious cases related to illicit drug trafficking and money laundering [26].

The signatory countries cannot change international commitments through their national legislation [3, 4].

4.4.1.3 Lack of Consensus

Two groups exist within the International Drug Control System (SICD). One is in favor of the control and criminalization of drugs. It includes the United States, Sweden, Japan, China, most Arab countries, the republics of the old Soviet Union, and those of Eastern Europe and others (with Buddhist or Confucian influences). On the other hand, the minority bloc comprises countries that express criticism of international drug policies and seek greater flexibility and tolerance.

Moral positions have always had a significant influence on the SICD. For example, the Preamble to the 1961 Convention characterizes drug use as "evil" [27]. Therefore, if drug use and dependence are criminal manifestations, the only logical solution is prohibition. However, the drug-producing countries questioned this position in the 1990s, claiming the so-called consumer countries' responsibility. As a result, the Special Session of 1998 adopted the principle of shared responsibility on the international agenda [3, 4]. Likewise, during the 1990s, European approaches to reduce the negative consequences of drug use were discussed.

4.4.2 *Flexibility Margins Within the System*

As a larger number of States adhered to the conventions, drug control policies were standardized globally so that all countries could control the same substances and prohibit and sanction similar activities related to their consumption, possession, production, and trafficking. However, the application of international standards is up to national authorities.

Complying with international drug control treaties may result in assistance, foreign aid, or trade relations, whereas failure to do so may dispense with these incentives. For this reason, some States prefer to take advantage of the existing flexibility margins and the different interpretations of the provisions contained in the conventions to implement policies according to their circumstances [1, 9].

Western European nations, Australia, and New Zealand have implemented harm reduction programs related to intravenous opioid injections. These programs include (a) distribution and exchange of needles and syringes for people who inject drugs or are dependent on heroin; (b) chemical analysis of the drugs that users take to health centers in order to prevent overdose or intoxication; (c) use of methadone to replace heroin; (d) house arrest for mothers with children; (e) confiscation of drugs and the deportation of "mules" instead of imprisonment in a foreign country, and (f) controlled provision of heroin to addicts through their doctors [3, 4, 28, 29]. In other cases, state policies have introduced drug consumption rooms – as happened in the 1980s in Germany, Switzerland, Spain, and the Netherlands and in the early 2000s in Canada and Australia.

According to the European Monitoring Centre for Drugs and Drug Addiction (EMCDDA), decriminalization involves removing specific conducts or activities from the sphere of criminal law. The prohibition remains, but the criminal sanctions for consumption disappear or are non-criminal sanctions. Thus, decriminalization implies a decrease in criminal sanctions, fundamentally avoiding the deprivation of liberty, while sanctions such as fines are maintained, and the crime is noted in the offender's record. Legalization is a different case because it involves the elimination of all activities related to a specific substance from the scope of criminal law and establishing regulation of its market by the State, including cultivation, production, distribution, and availability in the context of a specific jurisdiction, or the establishment of a free and legal market for the substance [27].

4.4.3 The Capacity of the Current International Drug Control Regime to Respond to Current Challenges

The regulatory framework for drug control built over the last 60 years is a complex system that faces significant challenges. Moreover, the world drug panorama is in constant flux, and the conditions in countries differ and will continue to do so in the coming years [3, 4]. Problems related to drugs derived from opium have not only failed to be solved through the current strategy but have even increased. In particular:

- According to the UNODC, between 2009 and 2018, opium poppy cultivation increased by 125%. There are also more synthetic or semi-synthetic opioids on the market (Chap. 3), and consumption of opioids increased by 16%, reaching 35 million people.
- The global prevalence of HIV among people who inject drugs has remained stable at 11.8%, as has the prevalence of hepatitis C at 51.9% and tuberculosis at 8%. However, the accumulated number of drug-related deaths (1999–2019) reached almost 500,000, with injectables being the most frequent [29–31].
- More resources go to law enforcement than to the detriment of measures focused on public health.
- When a substance is prohibited, users tend to shift to other drugs with similar psychoactive effects that are easier to obtain. Between 2009 and 2017, more than 950 new psychoactive substances appeared on the world drug market (see Chap. 16).
- The use of prescription opioids has significantly increased, causing numerous fatalities.
- Drugs are commonly laced with pharmacologically active adulterants to maximize economic benefits and enhance drug effects, including the more dangerous ones. For example, fentanyl and its analogs are frequent drug adulterants responsible for fatal overdoses.
- The drug production and distribution chain generates violence that manifests differently in various countries [30–32].

Money laundering associated with drug trafficking has not been controlled. According to data from the International Drug Policy Consortium (IDPC), "every year an amount that ranges between 800 million and 2 trillion United States dollars (USD) is laundered on a global scale, which represents between 2% and 5% of global GDP." It is estimated that 25% of the total income of transnational organized crime derives from drug sales and that the world drug market reaches a volume between 426,000 and 652,000 million USD. Of these, more than half of the gross profits generated are channeled toward money laundering, while the money confiscated remains at below 1% [12, 33].

4.4.4 *Limits of International Regulations*

4.4.4.1 Criminalization of Drug Production and Trafficking

The third UNGASS [14] made clear that the three conventions that make up the regime require that countries comply with minimum restrictive measures, so it is not indispensable to implement policies commonly associated with the "war on drugs" (see Chap. 1). Secondly, the UNGASS emphasized that sentences for drug-related crimes should be proportional to the crimes committed. Although the conventions do not prohibit the death penalty, countries are urged not to implement it for drug offenses. INCB, for its part, notes that policies related to the militarization of law enforcement agencies, lack of respect for human rights, excessive recourse to incarceration, denial of appropriate medical treatment, and inhumane approaches are not in line with the principles of treaties and political declarations [18].

With the so-called war on drugs launched by Richard Nixon in the 1970s, drug control became a matter of national security, linking drugs with violence related to international illicit trafficking and organized crime [1, 18]. This new vision would soon take shape in an international treaty, the 1988 Convention on Illicit Traffic. As already noted, its Preamble refers not only to dangers to public health but also to the "links that exist between illicit trafficking and other organized criminal activities related to it, which undermine legal economies and threaten stability, security and States sovereignty."

When drug abuse becomes a security problem, ordinary public policy and accountability channels, typical of democracies, are often abandoned, and violations of human rights related to gender equality, environmental protection, and socioeconomic development can occur. In addition, the disappearance and displacement of millions of persons increases [27].

Due to the growing prevalence of illicit trafficking and organized crime, other international conventions have been adopted, such as the United Nations Convention Against Transnational Organized Crime in Palermo in 2000 or the 2003 United Nations Convention Against Corruption. In addition, regional integration groups have echoed this "new" problem. For example, the 2003 European Security Strategy identified organized crime involved in drug trafficking as one of the main threats to the Union's security and linked drug revenues with terrorism and the weakening of state structures in various world regions.

Viewing drugs as a threat to national security, coupled with an increasingly punitive approach, encouraged policymaking that disengaged from public health objectives and instead focused on combating this situation through military force and criminal law enforcement [27].

The transition from a system based on rehabilitation to one establishing harsh criminal penalties resulted in mass incarceration for drug offenses, even for the possession of small amounts of controlled substances. It is estimated that a fifth of the world's inmates has been arrested for drug-related crimes and, for the most part, only for drug use [34].

The fundamental principle of international regulations is to limit controlled drugs to "medical and scientific research." However, these concepts are not defined in the body of the conventions and allow for various interpretations. According to INCB, some governments interpret them in such a way that calls into question their adherence to the objectives of the treaties.

In addition, as pointed out by Francisco E. Thoumi, translation problems occur in the different versions of the conventions. For example, the English version (the language in which the Convention was negotiated) and the Russian, Chinese, and Arabic translations refer to "the health and welfare of humankind." However, the Spanish and French versions refer to "the physical and moral health of humanity" ("Les parties, soucieuses de la santé physique et morale de l'humanité"). The 1971 Convention on Psychotropic Substances presents the same difference between the English version and the Spanish and French versions. The terms "well-being" and "moral" are not synonymous. If the word "moral" is accepted, the conventions would implicitly establish that it is possible to have a unique morality in the world, something that is impossible to sustain due to ideological-political and religious differences regarding this term. Finally, if the word "welfare" is chosen, the policies would be much more flexible, although there would also be differences in their interpretation [35].

The conventions have left the formulation of policies in the hands of doctors, public health experts, and State agents, taking it out of the realm of politics. This approach worked reasonably well for a few decades because drug addiction was marginal on the political agenda of most countries. Still, it underestimated the costs generated by an increase in drug use.

However, illegal drug production, trafficking, and consumption derive from complex multidimensional situations that cannot be addressed with simplified initiatives. The solution to addiction and associated illicit markets, as Thoumi suggests, cannot be formulated only from the perspective of doctors, economists, sociologists, and police. Unfortunately, even though the issue's complexity is generally acknowledged, the vast majority of the architects and scholars of drug policy analyze these issues from a perspective limited by the theoretical models developed within their academic discipline.

Medical epidemiology and the social and human sciences show that drug abuse varies substantially between societies. In some cases, genetics plays a role. However, social factors such as broken families, school failure, poverty, and social exclusion, among other factors, can also contribute to an increase in the prevalence of addiction. Conversely, other factors such as stable families, some religious beliefs, and school success can protect individuals and decrease the risk of drug addiction.

Drug abuse and trafficking illustrate societies' structural and institutional vulnerabilities, but they are not the roots of all social problems. Today's world faces many issues, including extreme poverty, inequality, greed, lack of cohesion and social trust, corruption, human and arms trafficking, high homicide rates, fraud, economic crisis, and wars. Current evidence indicates that social ills are interrelated but not necessarily in a causal manner; each problem contributes to the others because it increases their propensity to develop.

A multiple and flexible approach is required to understand different realities and maintain unity in diversity. The physical and social structure of a country determines its risks of allowing the development of solid criminal organizations that, in turn, produce illicit drugs. Thus, the concentration of drug production and trafficking and the high rates of violence are explained by external factors such as growing international demand and internal factors. Why are societies so vulnerable to illegal activities? Although they cannot be considered causal agents, some risk factors, such as poverty and inequality, can derive from economic crisis and corruption. The key factor is that illegal activities can generate a social support network; this means that there are groups for whom the unlawful activity is legitimate [35].

In today's world, many organizations with political interests frequently challenge the State's control: large transnational corporations (mainly chemical and pharmaceutical), non-governmental organizations, financial institutions, religions, citizen associations, criminal organizations, subversive groups, and others. As a result, traditional policies have not been very effective and are sometimes rather difficult to implement.

4.5 Toward a New Consensus on Drug Policy

The use, trafficking, and production of opioids are global social problems. Therefore, it is necessary to think globally and act locally, assuming a shared responsibility that requires intense and effective international cooperation, fully complying with the Universal Declaration of Human Rights principles.

The first conventions regulated the consumption of opioids that the pharmaceutical companies promoted to increase their profits. Unfortunately, today some pharmaceutical companies once again play a leading role in expanding opioid misuse and addiction, mainly in the United States, where the sales of their products far exceed pain relief needs. This has caused an ongoing epidemic that killed about half a million people between 1999 and 2019 (see Chap. 5).

This epidemic did not happen suddenly or by accident; one of the most significant factors was that some pharmaceutical industries falsely stated that the use of prescribed opioids was unlikely to cause addiction. Also, the variety of associated drugs has increased enormously. Their production and marketing have become more sophisticated, and the market players are different, making problems much more challenging to address with traditional measures (see Chap. 16). This has given rise to various opinion groups that demand a change in international drug control. However, the alternatives must be considered with caution because, aside from overdose risks, opioids are responsible for 50% of the years lost due to disability and premature death caused by drug use [36].

Drug policies aim to change behaviors. They face a complex problem that requires a strategy with a multiple, comprehensive, and balanced view, based on scientific evidence, and a public health approach with full respect for human rights that puts people, not substances, at the center. It is necessary to generate trust and empathy, promote dialogue and social agreement among all interested parties, and take into account all those affected in the formulation and implementation of policies: growers, producers, consumers, suppliers, prisoners for drug crimes, public officials, and other actors involved in criminal networks [35].

The United Nations entities and the Member States should also adopt an approach to drug control based on the right to health, ensuring respect and protection of the rights of people who use drugs. This does not imply ceasing to make political and judicial efforts to enforce the laws. However, drug dependence must be addressed with prevention mechanisms and comprehensive therapeutic solutions, not criminalizing consumers.

The current challenge is how to adjust public policies and actions derived from international drug policy with the United Nations 2030 Agenda for sustainable development by combining social, economic, and environmental aspects with a gender perspective, under the umbrella of human rights and the rule of law.

The next decade of world drug policy must be aligned with the 2030 Agenda, which considers inclusion and measures against environmental deterioration. There is a demand for deep transformations to which we have been unable to respond. The State must regain presence to address health and education challenges, combat hunger and poverty, and thus mitigate the pain and suffering of millions of people affected by current drug policies. This historical and universal agreement calls for a holistic approach to counter the most pressing problems humanity faces and the courage, as the United Nations Secretary-General has put it, to "consider all options."

Countries must recognize internal risk factors and the need to carry out structural reforms in their societies to implement the SDGs, thus expanding the role of the State in securing universal rights. Guaranteeing political, civil, social, economic, and cultural rights to prevent crime has brought positive results. Since 1995, the European Union has been adopting security policies that include measures to prevent organized crime in high-risk poor social areas. These programs have reduced the rates of juvenile offenses, organized crimes, and criminal relapse to levels between 55% and 70% below those observed in the United States [37].

It is urgent to rethink the general objectives of global drug policy and the indicators and measurement systems used to evaluate progress, considering the SDGs. This would make it possible to reduce drug-related harm to health, improve access to medical care, and defend fundamental human rights. Also, it would help toward achieving gender equality, reduce poverty in growing and trafficking areas, improve citizen security, reduce corruption, and strengthen and diversify international cooperation in the fight against organized crime.

Readjusting drug policy with the SDGs and human rights obligations requires a new international multilateralism of criminal, social, fiscal, and legal policies within the United Nations bodies. It is necessary to consider firearms trafficking, money laundering, corruption, and tax evasion when creating an agenda for sustainable development and the fight against drugs.

The 2030 Agenda for Sustainable Development represented a watershed moment in global efforts to prevent, combat, and eradicate the illicit arms trade, as specified in SDG No. 16: Peace, Justice, and Strong Institutions. Firearms trafficking affects almost all countries, compromises human security, and is at the center of law enforcement and order maintenance efforts. Firearms strengthen many forms of crime, such as gang intimidation, human trafficking, and terrorism related to the illegal drug trade [36]. According to the United Nations estimates, money laundering reaches up to 2.7% of the world's gross domestic product (GDP) each year. This crime, along with corruption and tax evasion, constitutes a fraud that deprives governments of resources that would help in responding to the emergencies that plague the planet. The adoption of international standards of financial integrity would make it possible to raise funds to alleviate contingencies and promote sustainable development. It is necessary to reduce illicit financial and arms flows, strengthen the recovery and return of stolen assets, and combat all forms of organized crime.

The 2030 Agenda and the SDGs seek an alternative path based on a new model of development that puts the well-being of people and the preservation of the environment at the center. According to ECLAC estimates, two-thirds of the SDG targets will be unattainable without substantial changes to the development model [38].

To lower stagnation, States must generate pacts to implement macroeconomic, industrial, social, and environmental policies to build new forms of collaboration between the public-private and social sectors, rethinking the relationship between science, technology, and the productive national system. This could help orient long-term budgets and recover a sense of public interest.

The results of the agreements and programs derived from the international drug control regime are modest. These international conventions helped lay the foundations to control the flow of substances that could harm communities. However, today's world constitutes a mosaic of realities that vary significantly from one country to another and even between communities. The problems generated by different substances can also be very different, even in neighboring countries.

The international drug conventions will be 61, 51, and 34 years old in 2022. They were good instruments for a global consensus, but the world has changed since they were written. Furthermore, due to the nature of binding treaties, their modification or updating is an almost impossible process. Perhaps what is needed is to retake what can be derived from these treaties but allow drug policies tailored to the problems of each country. These policies should:

- Strengthen a focus on health, prioritizing prevention, treatment, and harm reduction.
- Have a people- rather than a substance-centered perspective to allow the policies and those who apply them always to observe human rights.
- Give space to local decision-making, allowing States to focus specifically on their drug problems over compliance with international provisions.

In addition, international agencies should advise States that policies that contravene human rights should not be implemented, such as criminalizing drug use, the death penalty, extrajudicial executions, and compulsory coerced treatment. Carrying out these changes may lead to building a new regime that somehow disrupts the power structures strongly permeated by legal and illegal opioid trafficking, acknowledging nations' identities. We are in a transitionary period, in which the old fades, but the new is yet to be born, that is, when the elite and its "model" lose their legitimacy, but alternative discourses have not yet managed to generate sufficient credibility.

References

1. Chatterjee SK. Legal aspects of international drug control. 1st ed. The Hague: Martinus Nijhoff Publishers; 1981. 587 p.
2. Hallam C. The international drug control regime and access to controlled medicines. Int Drug Policy Consort [Internet]. 2014;(26):1–16. Available from: https://www.tni.org/files/download/dlr26-e.pdf.
3. International Narcotics Control Board (INCB). Celebrating 60 years of the single convention on narcotic drugs of 1961 "… a generally acceptable international convention …" and 50 Years of the Convention on Psychotropic Substances of 1971 "… an international convention is necessary …" [Internet]. Vienna: United Nations; 2021. 31 p. Available from: https://www.incb.org/documents/Publications/AnnualReports/AR2020/Supplement/00_AR2020_supp_full_document.pdf.
4. International Narcotics Control Board (INCB). Report of the International Narcotics Control Board for 2020 [Internet]. United Nations; 2021. 142 p. (Report of the International Narcotics Control Board). Available from: https://www.un-ilibrary.org/content/books/9789210056380.
5. McAllister WB. Drug diplomacy in the twentieth century: an international history. 1st ed. London/New York: Routledge; 2000. 368 p.
6. McAllister WB. Conflicts of interest in the international drug control system. J Policy Hist [Internet]. 1991;3(4):143–66. Available from: https://www.cambridge.org/core/journals/journal-of-policy-history/article/abs/conflicts-of-interest-in-the-international-drug-control-system/039C134034433B19D72C874470F6D09F.
7. Bewley-Taylor D, Jelsma M. Fifty years of the 1961 single convention on narcotic drugs: a reinterpretation [Internet]. Series on legislative reform of drug policies. Amsterdam; 2011. Available from: https://www.tni.org/en/publication/fifty-years-of-the-1961-single-convention-on-narcotic-drugs-a-reinterpretation.
8. Jelsma M, Armenta A. The UN drug control conventions [Internet]. Ideas into movement. 2015. Available from: https://www.tni.org/files/publication-downloads/primer_unconventions_24102015.pdf.
9. Babor TF, Caulkins JP, Fischer B, Foxcroft DR, Humphreys K, Medina-Mora ME, et al. Drug policy and the public good [Internet]. 2nd ed. Oxford University Press; 2018. 352 p. Available from: https://oxford.universitypressscholarship.com/view/10.1093/oso/9780198818014.001.0001/oso-9780198818014.
10. Global Commission on Drug Policy (GCDC). Classification of psychoactive substances: when science was left behind [Internet]. 2019. Available from: http://www.globalcommissionondrugs.org/wp-content/uploads/2019/06/2019Report_EN_web.pdf.
11. International Drug Policy Consortium (IDPC). The 2009 Commission on Narcotic Drugs and its high level segment-report of proceedings [Internet]. SSRN J. 2009. Available from: https://papers.ssrn.com/sol3/papers.cfm?abstract_id=1910557.
12. Jelsma M. UNGASS 2016: watershed event or wasted opportunity? [Internet]. Transnational Institute; 2016. Available from: https://www.tni.org/en/node/22954?content_language=en.
13. Jelsma M. Drugs in the UN system: the unwritten history of the 1998 United Nations General Assembly Special Session on drugs. Int J Drug Policy. 2003;14(2):181–95.

14. UNGASS. Special session of the United Nations General Assembly on the world drug problem [Internet]. UNODC United Nations Office on Drugs and Crime; 2016. Available from: https://www.unodc.org/ungass2016/en/documentation.html.
15. Peña E. Speech by the President of Mexico, Enrique Peña Nieto, during the Special Session of the General Assembly on the world drug problem (UNGASS 2016) [Internet]. 2016. Available from: https://www.gob.mx/cms/uploads/attachment/file/210802/7._PEUM-_UNGASS_2016.pdf.
16. Global Commission on Drug Policy (GCDP). Regulation. The responsible control of drugs [Internet]. idpc International Drug Policy Consortium; 2018. Available from: https://idpc.net/publications/2018/09/regulation-the-responsible-control-of-drugs.
17. Jensema E. Human rights and drug policy [Internet]. Transnational Institute; 2018. Available from: https://www.tni.org/en/briefing/human-rights-and-drug-policy.
18. Office of the United Nations High Commissioner for Human Rights (OHCHR). Building Bridges between the drug policy and human rights regimes of the United Nations [Internet]. idpc International Drug Policy Consortium; 2021. Available from: https://idpc.net/alerts/2021/02/building-bridges-between-the-drug-policy-and-human-rights-regimes-of-the-united-nations.
19. International Harm Reduction Association (IHRA). Global state of harm reduction 2010. Key issues for broadening the response [Internet]. London; 2010. Available from: http://www.ihra.net/files/2010/06/29/GlobalState2010_Web.pdf.
20. Pompidou Group. Policy paper providing guidance to policy makers for developing coherent policies for licit and illicit drugs [Internet]. 2011. Available from: https://rm.coe.int/CoERMPublicCommonSearchServices/DisplayDCTMContent?documentId=09000016806f40ab.
21. United Nations Human Rights Council (UNHRC). The impact of the world drug problem on the enjoyment of human rights [Internet]. Office of the United Nations High Commissioner for Human Rights. Available from: https://www.ohchr.org/EN/HRBodies/HRC/Pages/WorldDrugProblem.aspx.
22. Global Commission on Drug Policy (GCDP). Taking control: pathways to drug policies that work [Internet]. 2014. Available from: https://www.globalcommissionondrugs.org/wp-content/uploads/2016/03/GCDP_2014_taking-control_EN.pdf.
23. Hall W. The future of the international drug control system and national drug prohibitions. Addiction. 2018;113(7):1210–23.
24. Room R, Reuter P. How well do international drug conventions protect public health? Lancet [Internet]. 2012;379(9810):84–91. Available from: https://doi.org/10.1016/S0140-6736(11)61423-2.
25. Collins J. Rethinking' flexibilities' in the international drug control system—potential, precedents and models for reforms. Int J Drug Policy. 2018;60:107–14.
26. United Nations Office on Drugs and Crime (UNODC). UNODC. Annual report 2014 [Internet]. Vienna; 2014. Available from: https://www.unodc.org/documents/AnnualReport2014/Annual_Report_2014_WEB.pdf.
27. Sanchez Avilés C. El régimen internacional frente al “problema” mundial de las drogas. Evolución, políticas alternativas y resistencia al cambio [Internet]. Trans-border Institute. University of San Diego; 2012. Available from: https://justiceinmexico.org/wp-content/uploads/2010/07/2012_Sánchez-Avilés_Regimen-Internacional-Drogas.pdf.
28. Harm Reduction International (HRI). The death penalty for drug offences: global overview 2019 [Internet]. 2020. Available from: https://www.hri.global/death-penalty-2019.
29. United Nations Programme on HIV/AIDS (UNADIS). Health, rights and drugs – harm reduction, decriminalization and zero discrimination for people who use drugs [Internet]. 2019. Available from: https://www.unaids.org/en/resources/documents/2019/JC2954_UNAIDS_drugs_report_2019.
30. United Nations Office on Drugs and Crime (UNODC). World drug report 2019 [Internet]. Vienna; 2019. Available from: https://wdr.unodc.org/wdr2019/.

31. Bretteville-Jensen AL, Costa C, Kattau T, Mikulic S, Trigueiros F, Papamalis F, et al. Public expenditure on supply reduction policies [Internet]. 2017. Available from: https://rm.coe.int/public-expenditure-on-supply-reduction-policies/168075b9dd.
32. Hales G, du Mont S, Desroches C, Redgrave H. What is driving serious violence: drugs. "The marketplace is more volatile and violent than it has ever been." A report by Crest Advisory [Internet]. 2020. Available from: https://b9cf6cd4-6aad-4419-a368-724e7d1352b9.usrfiles.com/ugd/b9cf6c_5aad01b7fa2743d8b5508fb85faf2aa4.pdf.
33. International Drug Policy Consortium (IDPC). Progress report 2019–2020 [Internet]. 2020. Available from: http://fileserver.idpc.net/library/Progress_Report_2019_Final.pdf.
34. International Drug Policy Consortium (IDPC). Taking stock: a decade of drug policy – a civil society shadow report [Internet]. IDPC International Drug Policy Consortium; 2018. p. 132. Available from: https://idpc.net/publications/2018/10/taking-stock-a-decade-of-drug-policy-a-civil-society-shadow-report.
35. Thoumi FE. Reexamen de las bases " médicas y científicas " en la interpretación de las convenciones internacionales de drogas : ¿ será que el " régimen " está desnudo ? [Internet]. 2016. Available from: https://library.fes.de/opus4/frontdoor/index/index/docId/44428.
36. United Nations Office on Drugs and Crime (UNODC). World drug report 2020 [Internet]. Vienna; 2020. Available from: https://wdr.unodc.org/wdr2020/index2020.html.
37. Usi E. Buscaglia: "En México hay un pacto de impunidad" [Internet]. DW; 2014. Available from: https://www.dw.com/es/buscaglia-en-méxico-hay-un-pacto-de-impunidad/a-18001753.
38. Economic Commission for Latin America and the Caribbean (ECLAC). Building forward better. Action to strengthen the 2030 Agenda for Sustainable Development. Fourth report on regional progress and challenges in relation to the 2030 Agenda for Sustainable Development in Latin America and the Caribbean [Internet]. 2021. Available from: https://repositorio.cepal.org/bitstream/handle/11362/46696/1/S2100124_en.pdf.

Chapter 5
The Opioid Crises

Silvia L. Cruz and Raúl Martín-del-Campo

Abstract In the first two decades of the twenty-first century, over-prescription of opioid analgesics, diversion of opioid medications, opioid use disorders, and overdose deaths increased to unprecedented levels in the United States, causing a devastating crisis. Canada suffered a similar problem but to a lesser extent. These crises were triggered by aggressive marketing strategies promoting prescription opioids to treat chronic non-cancer pain and using pharmaceutical presentations without deterrents, easy to crush and misuse. Such practices ultimately favored an increase in heroin use. More recently, fentanyl and other synthetic opioids have become prevalent in North America, the European Union, and the United Kingdom and are sold through the dark web. In Africa and Asia, the high-risk use of tramadol currently affects several populations. This chapter gives an overview of opioid misuse in several countries and summarizes some evidence-based interventions and public health responses to the opioid crises.

Keywords Opioid crisis · United States · Canada · Africa · Asia

S. L. Cruz (✉)
Department of Pharmacobiology, Center for Research and Advanced Studies (Cinvestav), Mexico City, Mexico

Opioids Working Group, Global Studies Seminar, Faculty of Medicine, National Autonomous University of Mexico, Mexico City, Mexico
e-mail: slcruzm@cinvestav.mx

R. Martín-del-Campo
Opioids Working Group, Global Studies Seminar, Faculty of Medicine, National Autonomous University of Mexico, Mexico City, Mexico

National Institute of Psychiatry Ramón de la Fuente Muñiz (INPRFM), Mexico City, Mexico

International Narcotics Board Control, United Nations, Vienna, Austria

S. L. Cruz (ed.), *Opioids*, https://doi.org/10.1007/978-3-031-09936-6_5

5.1 Introduction

By the end of the nineteenth century, physicians and pharmacists prescribed morphine- and opium-based medications for all kinds of ailments, from menstrual discomfort, anxiety, "hysteria," dysentery, epilepsy, coughing, and intermittent fevers to soothing babies during teething. Opioids were central in treating soldiers from the American Civil War (1861–1865), the Franco-Prussian War (1870–1871), and the Boers wars (1880–1881, 1899–1902). At that time, opium and opiates (morphine, codeine, and heroin) were as readily available and frequently prescribed as aspirin (Chap. 1). Unfortunately, this practice resulted in a significant increase in opioid use disorders (OUDs) that led to strict restrictions to opioid use a few years later. The end of the twentieth century also faced a devastating opioid epidemic that continues to this day.

The opioid crisis is multifaceted. While some countries have insufficient access to controlled opioids for pain management, others have an unprecedented number of people affected with OUDs and overdose fatalities (Chap. 2). Global inequities are such that 20% of the world's population uses 90% of all licit morphine to treat pain in countries like the United States, Canada, Australia, and Germany. In contrast, two-thirds of the world, mainly low- and middle-income countries, have little to no access to this drug for medical use [1].

According to the *World Drug Report*, 2022 the number of past-year opioid users worldwide (15–64 years old) increased from over 31 million in 2010 to almost 61 million in 2020. This figure includes nonmedical use of prescription opioids and represents 1.2% of the global population in that age group. In addition, opioids caused 77% of the drug-related deaths in 2020 and contributed to almost 13 million disability-adjusted life years (DALYs, i.e., years of healthy life lost due to disability and premature mortality) [2].

Although the United States and Canada have been the most affected countries, other areas of the world, such as parts of Africa and Asia, have also experienced a surge in opioid misuse. Learning from these experiences could help prevent the occurrence of new opioid epidemics.

5.2 Opioid Misuse in the United States

5.2.1 The United States' Opioid Crisis in Figures

- Opioid analgesic prescriptions peaked in 2012 with a rate of 81.3 prescriptions per 100 inhabitants [3].
- Neonatal abstinence syndrome (see Chap. 13) rose fivefold from 2000 to 2012 [4].
- The estimated societal cost of prescription opioid misuse was approximately 56 billion dollars in 2007, compared to 504 billion in 2015 [5].

- One percent of opioid prescribers were responsible for 27% of all written prescriptions in 2017 [6].
- Data from 2018 indicate that approximately ten million people (12 years or older) misused opioids that year [7].
- Of the 71,000 drug overdose deaths in 2019 in the United States, 70% (approximately 50,000) involved an opioid [8].
- Approximately half a million people died from opioid overdoses in the United States from 1999 to 2019 [2], more than the recognized American casualties in the Second World War.
- According to the Centers for Disease Control and Prevention (CDC), nearly two million Americans currently have an OUD involving heroin or prescription opioids.
- Opioids caused an average of 186 deaths per day in the United States in 2020, i.e., one life was lost almost every 8 minutes [9].
- In 2021, there were more than 100,000 drug-related deaths, most of them caused by opioids [10].

5.2.2 Factors Enabling the United States' Opioid Crisis

Multiple factors have contributed to the current opioid crisis, including poverty, income inequality, lack of or insufficient health services, access to treatment, unemployment, and social isolation [11]. Addressing these factors requires well-orchestrated structural interventions and public policies. Other aspects that have played a significant role in the current American crisis are opioid availability, over-prescription, bad medical practices, and inadequate pain management, as discussed in this section.

5.2.2.1 Opioids' Availability

The United States has shown a high prevalence of illegal drug use for more than a century. In the 1940s and 1970s, heroin use increased along with a smuggling network that trafficked the drug from Asia to America to satisfy the increasing opioid demand by users, including World War II, and Vietnam War veterans. Soon, Colombian and Mexican providers established different commerce routes to that country.

Since the early 1980s, the supply of a dark sticky gum, known as black tar heroin, increased. This potent and difficult-to-adulterate heroin from Xalisco (Nayarit, Mexico) conquered new markets in Ohio and other States, becoming an alternative to the multi-cut powdered heroin available in big cities. In addition, black tar providers developed new distribution practices with home delivery service by young dealers who drove modest cars and offered to meet at inconspicuous spots (such as parking lots) to minimize the risks for opioid users that did not want to attend drug distribution points.

In the late 1990s, the prescription of opioids, especially extended-release oxycodone, increased exponentially due to marketing campaigns conducted by some pharmaceutical companies under the pretense that prolonged opioid use rarely resulted in addiction. More recently, fentanyl analogs and other synthetic opioids have emerged as new psychoactive substances that are easy to produce, transport, sell, and acquire, gaining wide acceptance among drug sellers. Therefore, dozens of opioids are now available through traditional markets and the dark web (Chap. 16).

5.2.2.2 Opioid Over-Prescription and Unethical Medical Practices

In a one-paragraph letter to the Editor published in 1980 in the *New England Journal of Medicine*, the authors Porter and Jick commented that after examining the files of 11,882 *hospitalized* patients "who received at least one narcotic preparation, there were only four cases of reasonably well-documented addiction in patients who had no history of addiction." The authors concluded that "despite widespread use of narcotic drugs in hospitals, the development of addiction is rare in medical patients with no history of addiction" [12].

This letter neither specified the opioid doses used nor the length of hospitalization. However, it did say that patients were hospitalized, meaning that they received opioids under medical supervision and in controlled conditions.

Few could anticipate the influence that this brief text would have in the decades to come. As pointed out by Leung and coworkers in another letter published in 2017 in the same journal, Porter and Jick's letter received 608 citations, while others published at about the same time in the same journal received a median of 11. Unfortunately, 80.8% of those citations falsely claimed that the risk of developing addiction with repeated opioid administration was minimal and failed to mention that the patients described in the letter were hospitalized [13].

References to Porter and Jick's "paper" were used in marketing promotional pieces as the scientific "support" to prescribe opioids for chronic non-cancer pain or moderate to severe acute pain in outpatients. Purdue Pharma, the company that manufactured and distributed MS Contin® and OxyContin® (slow-release morphine sulfate and oxycodone, respectively), trained sales representatives to assure physicians that opioids were practically devoid of addiction liability. Unfortunately, neither the sales representatives nor the medical practitioners checked the validity of this assertion. As a result, an unprecedented increase in opioid prescriptions dispensed by pharmacies in the United States occurred, changing from 76 million in 1991 to 219 million in 2011 [14]. The use of oxycodone, in particular, grew fivefold from 1999 to 2011. At the same time, opioid-associated deaths became increasingly common [15].

Although most physicians acted ethically, some took advantage of the situation and established pain clinics that provided prefilled opioid prescriptions to anyone paying for them, without conducting physical examinations or confirming a diagnosis that justified opioid use. These clinics, known as "pill mills," proliferated throughout the country and were recognizable by the long queues of clients waiting to receive a prescription for highly concentrated OxyContin® tablets, benzodiazepines, or both [16]. The more potent extended-release oxycodone tablets had 160 mg, when the usual initial dose for this opioid is between 5 and 15 mg. It is difficult to understand why such a concentrated pharmaceutical formulation without abuse deterrents was marketed. It did not take long until users realized that the pills could be crushed, sniffed, chewed, or dissolved in water and injected to produce a psychoactive effect comparable to that of the purest heroin available.

5.2.2.3 Inadequate Pain Management

The International Association for the Study of Pain (IASP) defines pain as "an unpleasant sensory and emotional experience associated with, or resembling that associated with, actual or potential tissue damage." Data from 2016 reported that approximately 9 to 12 million people suffer from chronic pain in the United States. Pain, especially chronic pain, can have a devastating impact on the patients' quality of life, and they deserve the best and safest treatments available.

To reduce the consequences of insufficient pain treatment, in 1996, the American Pain Society (APS) proposed to consider pain as the fifth vital sign and to assess it as frequently as the other four (body temperature, blood pressure, heart rate, and respiratory rate). APS recommended using unidimensional scales for pain evaluation and providing analgesics when necessary. Unfortunately, this measure was implemented while pharmaceutical campaigns were promoting opioid consumption, which caused indiscriminate use of highly concentrated opioids for acute and chronic non-cancer pain [17]. When the consequences of such practices became evident, various medical associations distanced themselves from the "pain as the fifth vital sign" promotion. Today, it is accepted that patients must receive the best available treatment based on the cause and type of pain. Still, evidence supporting opioid use to treat chronic pain is limited, and there is sound evidence that prolonged use can lead to OUDs, tolerance, and overdoses [18] (see Chap. 10).

5.2.3 Three Waves of Opioid Overdose Deaths in the United States

As mentioned in previous chapters, the Centers for Disease Control and Prevention (CDC) of the United States recognize three waves in the opioid crisis, characterized by different drugs and evolution patterns.

The first wave began in the late 1990s and was marked by proliferation of opioid prescriptions, particularly slow-release oxycodone, increased OUDs, and overdoses. The second wave began in 2010 and involved heroin as the cause of many overdose deaths. In the late 2010s, Purdue Pharma redesigned OxyContin® to deter abuse by covering each tablet with a coating that was impossible to crush or melt before ingestion. This change in the formula, as well as new FDA restrictions on prescribing opioids, had the unintended consequence of increasing heroin use. A significant proportion of opioid-treated patients transitioned from pharmaceutical opioids to heroin after requiring higher and more frequent doses and having difficulties getting multiple prescriptions. As a result of heroin injection, the number of patients with HIV/AIDS and hepatitis C increased. The third wave began in 2013 with an increase in overdose deaths caused by synthetic opioids. Today, heroin laced with fentanyl or fentanyl analogs is responsible for many fatal opioid overdoses [19]. Unbeknownst to users, drug dealers often add fentanyl to heroin or traffic this drug combination as medical opioid pills (mainly fake oxycodone). Data from 2017 show that deaths from fentanyl and fentanyl-like drugs reached more than 28,000 in 2017. On the other hand, deaths from heroin have stabilized, albeit at very high levels [9].

5.3 Opioid Misuse in Canada

5.3.1 Canada's Opioid Crisis in Figures

- Twenty percent of adults used prescription opioids in 2010. The same year, 15% of high school students reported nonmedical use of opioids in Ontario [20].
- In 2017, an estimated 12% of the Canadian population aged 15 and over (3.5 million people) used opioid pain relievers.
- Among Canadians who used opioid analgesics in 2017, about 3% used them for nonmedical purposes.
- Seventeen hospitalizations occurred per day due to opioid intoxication in 2017 [21].
- One out of eight people received opioid prescriptions in 2018.
- The opioid medications more frequently used were codeine (76%), hydromorphone or morphine (28%), oxycodone (20%), and fentanyl (5%).
- From January 2016 to June 2020, more than 17,500 opioid overdose deaths occurred, most of them involving fentanyl [22].

5.3.2 Differences and Commonalities with the Opioid Epidemic in the United States

There are some similarities in the factors enabling the opioid crisis in Canada and the United States. In particular, pharmaceutical opioids, diverted from legal channels and obtained from illicit sources, drove the opioid crises in both countries.

Also, at about the same time, the American and Canadian National Pain Treatment Guidelines recommended using relatively high doses of opioids for non-cancer chronic pain. Moreover, the addition of fentanyl and other potent synthetic opioids to heroin resulted in a growing number of overdose deaths.

A major difference between these two countries is that the number of people affected in the United States was much higher than in Canada (Table 5.1), but it must be taken into account that the United States' population is approximately ten times larger than Canada's. Another difference was the crises' evolution. There was a substantial increase in opioid prescriptions in all Canadian provinces from 1999 to 2011–2012, but they later declined to levels similar of those in 2005 due to two central interventions. The first one was the delisting of OxyContin® from all provincial formularies. By doing this, the cost of this drug was no longer covered by

Table 5.1 Opioid overdoses in the United States and Canada

Country	Period	Drug overdose deaths	Opioid overdose deaths	Comments
USA [54, 55]	From 1999 to 2021	841,000	500,000	247,000 of these deaths involved prescription opioids A third of all opioid-related deaths involved heroin
	2019	70,630	49,860	70.6% of all drug overdose deaths were related to opioids 72.9% of opioid-involved overdose deaths involved synthetic opioids
	2020	93,331	69,710	Increasing in drug overdose deaths up 29.4% over the previous year
Canada [21, 23, 24] Data per 100,000 population	From January 2016 to December 2019		15,393	The highest number of deaths occurred in 2018 In 2016, about 55% (1424) of all opioid-related deaths involved fentanyl-related opioids (e.g., fentanyl, carfentanil, furanyl fentanyl). This number rose to 78% in the first 9 months of 2019
	From 2016 to 2021		22,828	3658 deaths were reported in 2019 (10 deaths per day); 6265 deaths were reported in 2020 (17 deaths per day)
	January to March 2020		1073	12 deaths per day
	January to March 2021		1772	20 deaths per day 65% increase compared to the same period 2020 87% of fatalities involved fentanyl
	April 2019 to March 2020		3691	Pre-pandemic data
	April 2020 to March 2021		6946	Pandemic period 88% increase in opioid overdoses

insurance plans, which led to a sharp decrease in its use. In addition, several provinces expanded prescription monitoring programs to restrict opioid availability. The second general intervention was the introduction of new chronic pain treatment guidelines with a more cautious approach, recommending opioid use mostly as a last resort for severe pain management [20].

These interventions were breakpoints that reduced opioid misuse, but the balance achieved is still fragile. For example, the fatal overdose rate fell from 11.8 deaths per 100,000 inhabitants in 2018 to 10.1/100,000 in 2019. However, in September 2020, the overdose death rate rebounded to 16/100,000 residents [23]. Moreover, the COVID-19 pandemic worsened this scenario because there was a 96% increase in fatal opioid overdoses in the period comprising from April 2020 to March 2021 than in the year before (April 2019–March 2020) [24].

5.4 High-Risk Opioid Use in Africa and Asia

The opioid crisis in Africa and Asia is less visible than in America, but it affects very vulnerable countries. The increased use of tramadol in this region poses particular risks to communities with insufficient health services and structural interventions to mitigate opioid misuse consequences. Some experts attribute the nonmedical use of tramadol to gaps in regulating this narcotic and an underestimation of its adverse effects. Other specialists consider that classifying tramadol to limit its use would leave low- and middle-income countries with very few options for pain management.

5.4.1 Tramadol

Tramadol is a centrally acting opioid analgesic used for moderate to moderately severe pain management. The pharmaceutical company Grünenthal synthesized it in the 1960s, providing evidence of its safety when used at therapeutic doses, and therefore, it is not included in the list of substances controlled by the International Narcotics Control Board (INCB). However, recent data have shown that tramadol has significant addiction liability and other adverse effects when taken at doses higher than those recommended by the manufacturer [25–27].

Tramadol is not only an opioid analgesic, but also a weak inhibitor of serotonin and norepinephrine reuptake. Its active metabolite, O-desmethyltramadol (ODMT), is up to six times more potent than the parent drug. Therefore, for tramadol to be effective, it must be metabolized by the liver; this makes it more active when administered orally than when injected directly into the bloodstream.

CYP2D6, a highly polymorphic enzyme, is responsible for converting tramadol to ODMT. According to the enzymatic activity, individuals can be poor metabolizers, extensive metabolizers with regular enzymatic activity, or ultra-rapid metabolizers. The latter group has two copies of the most efficient enzyme variant and produces more ODMT. Therefore, there are significant differences in tramadol's metabolism and effects among individuals. For some, it may not be effective as an analgesic, while ultra-rapid metabolizers may experience toxic effects [28]. In most people, however, high doses produce stimulant effects due to increased norepinephrine levels

5.4.2 Opioid Use in Africa

Although heroin use is still prevalent in this continent, the nonmedical use of tramadol has become a significant problem in 22 out of 57 countries, particularly in West, Central, and North Africa [29]. To put the situation into perspective, between 2010 and 2019, the estimated number of opioid users quadrupled, and the past-year prevalence of opioid use tripled. Despite these data, however, there is practically no information available about overdose mortality rates in Africa [25].

Recent estimates indicate that 4.6 million people in Nigeria, or 6.0% of the male population and 3.3% of the female population aged 15–64 years, used opioids in 2018. Tramadol was the most commonly used opioid, followed by codeine and morphine. In particular, 2.3% of the population reported the nonmedical use of cough syrups containing codeine [30].

In Egypt, about 3% of the adult population misused tramadol in 2016. In the same year, 1.4% of high school students reported past-year misuse of tramadol.

In South Africa, treatment admissions for opioid use disorders increased from 16.1% in 2012 to 20% in 2017 [22, 31]. India and to a lesser extent China are the sources of tramadol counterfeit pills of 120 mg, 225 mg, 250 mg, and 500 mg tramadol trafficked to Africa. These doses are higher than the 50 or 100 mg recommended for analgesia and available in commercial tablets.

Users report that tramadol improves mood and energy, allowing them to endure arduous work hours. They also claim that this opioid improves sexual performance and is useful for men with premature ejaculation. Other users take tramadol to self-medicate for depression or anxiety and consider it a mood stabilizer. In some regions, farmers give tramadol to their cattle as an energy booster to make them work excessive hours [32]. There are also testimonies of employees who began taking tramadol at their employers' advice in order to be able to work longer hours [33].

5.4.3 Opioid Use in Some Asian Countries

The *World Drug Report 2022* indicates that more than half of opioid users worldwide live in Asia [2]. In India, opioid use prevalence is three times the global average. A recent report from the Government of India indicates that almost 29 million people (approximately 2.1% of the general population) use opioids, mainly heroin (1.14%), prescription opioids (0.96%), and opium (0.52%) [34].

The number of opioid users varies significantly among Indian states, with Uttar Pradesh, Punjab, and Haryana having more people requiring treatment for OUDs. The province of Punjab is considered the epicenter of the tramadol crisis in India. There, drug factories produce counterfeit pills and ship them worldwide. In 2019, the Indian government installed methadone treatment centers to prevent people dependent on tramadol from migrating to heroin. Thousands of patients have attended these services since [35]. Paradoxically, India has a severe pain management problem due to restricted legal access to morphine. This is not surprising, considering that 80.4% of the world's population, mainly in low- and middle-income countries, consumes 12.8% of the morphine used for pain management [27].

In Pakistan, Afghanistan, and Iran, heroin is still the most commonly misused opioid, but tramadol use is rising [36]. A systematic review calculated that the past-year prevalence of tramadol use among the general population in Iran ranged from 4.1% to 5.9% in males and was approximately 0.7% in females [37]. Another study conducted in urban cities in Iran estimated the number of tramadol users to be over 200,000, which corresponds to approximately 0.7% of the population aged 15–49 years [38].

In 2017, Italian authorities reported a seizure of $75 million worth of tramadol bound to Libya from India. The supposed intended buyers were members of the Islamic State terrorist group Boko Haram. The United Nations speculates that the tramadol trade has a direct role in destabilizing the Asian region [27].

A study conducted in 2015 with data from patients recruited from the National Rehabilitation Centre in Abu Dhabi showed that 67.2% of opioid users had consumed tramadol [39]. In addition, the United Arab Emirates reported that the most frequently used opioids from 2013 to 2018 were codeine, heroin, morphine, and tramadol [2].

5.4.4 The Response to the High-Risk Tramadol Use in Africa and Asia

Governments from India, Egypt, and Ukraine have recognized that the dangers of tramadol misuse may be more significant than previously believed and have adjusted their local laws to restrict its trade [27, 30, 34]. In addition, the United Kingdom and the United States in 2014 and Denmark in 2017 also implemented stricter regulations for tramadol. For example, tramadol is now a Schedule IV substance in the United States [40, 41].

The World Health Organization (WHO) is the international body recommending how drugs should be regulated depending on their medicinal properties and harm potential. Within the WHO, the dilemma posed by tramadol has been widely discussed. On the one hand, it is the only non-controlled opioid available in countries with low economic resources where pain treatment is scarce. On the other hand, tramadol's dependence liability is higher than initially thought [42]. In 2000, after reviewing the available evidence, the WHO's Expert Committee on Drugs and Drug Addiction concluded not to bring tramadol under international scrutiny. Since then, the Committee has reviewed the drug numerous times, recommending to keep it under local surveillance but without international regulation [27, 42]. Grünenthal, the German company that developed and first marketed the drug, also supports the status quo of tramadol regulation, arguing that counterfeit illicit pills are responsible for the current opioid crisis.

Regulation would not necessarily restrict illicit trade and could have unintended consequences, as was the case with the transition from oxycodone to heroin [35]. After the Food and Drug Administration of the United States restricted oxycodone prescription, many doctors began to prescribe tramadol for postoperative pain management because it was considered safer. However, studies have found that patients using tramadol for pain management were just as likely to switch to long-term use as those who used stronger opioids [19].

The importance of not controlling tramadol internationally is that it is the only opioid available in countries with severe deficiencies, including access to morphine, and philanthropic medical organizations rely heavily on its availability in war zones and natural disasters [42].

5.5 Opioid Misuse in Other Countries

In 2019, the United Nations Office on Drugs and Crime (UNODC) Global Synthetics Monitoring: Analyses, Reporting and Trends (SMART) Programme published a report on the "global epidemic opioid crisis." This report presented an update on opioid misuse around the world, urging for a coordinated response to reduce overdoses while providing adequate access to opioids for medical purposes to low- and middle-income countries.

The UNODC report and other publications indicate that there are warning signs of increasing opioid misuse in other world regions. For example, the prevalence of opioid use in Europe in 2018 was estimated at 0.7% among the population aged 15–64 years in 2019, that is, 3.6 million people, of which 1.3 million were high-risk users. Although heroin continues to be the most widely used opioid in the region, more than 30 fentanyl analogs have been detected in the European Union since 2012. In 2018, 8300 overdose deaths were reported in the European Union, primarily related to opioids, compared with 6000 reported in 2013. Overdose deaths in Germany and the United Kingdom accounted for half of the fatalities in the European region, and the median age of those who died continued to rise, reaching 42 years in 2018. This reflects the aging of a large proportion of opioid users.

Other opioids such as methadone, buprenorphine, fentanyl, codeine, morphine, tramadol, and oxycodone are also misused in the European Union. Sixteen percent of the patients who entered treatment for substance use disorders in 2019 did so because of synthetic opioid use.

The trend in opioid-related deaths in Northern Ireland, where the drug-related death rate is high (10 per 100,000 population) and has more than doubled over the last decade, is an example of diversification and spread of the different opioids on the market [43].

Despite recent reductions in opium production, Mexico is still an important supplier of illegal opium and heroin for other countries, mainly the United States [44]. Seizures of locally synthesized fentanyl or of fentanyl coming from China indicate the increasing availability of potent synthetic narcotic drugs. However, the past-year use prevalence of opioids in Mexico is lower than the international average (0.1% vs. 0.4%, respectively). Recent studies indicate that opioid overdoses have increased in the northern border States and that most heroin samples are laced with fentanyl [45]. This situation could lead to a crisis in the country because naloxone is rarely available and opioid consumers use a highly concentrated salt solution, trying to counteract opioid intoxication, a method with very limited efficacy [46].

5.6 What Have We Learned from the Opioid Crises?

5.6.1 Sequence of Events that Led to the Increase in Opioid Misuse

The sequence of events that led to an unprecedented increase in opioid misuse around the world is:

1. There was an uncontrolled increase in pharmacological opioid prescriptions for all kinds of pain without the observance of guidelines or monitoring mechanisms.
2. A large number of patients developed iatrogenic addiction, evidenced by using opioid drugs in higher quantities and more frequently than prescribed. To satisfy their increased opioid use, patients searched for new strategies to obtain prescriptions and medications, either through acquaintances or websites.
3. Fatal overdoses associated with pharmaceutical opioids increased.
4. Overdue regulations were implemented to prevent nonmedical opioid use without increasing medication-assisted programs for opioid-dependent patients.
5. A significant proportion of patients with iatrogenic addiction who could not get enough pharmaceutical opioids to prevent withdrawal symptoms migrated to heroin.
6. Deaths from heroin overdoses increased.
7. Drug cartels laced heroin with fentanyl, which led to an increase in overdose deaths that the State could no longer prevent through regulations pertaining to the pharmaceutical companies.
8. New types of fentanyl emerged in the illegal market, resulting in more opioid-related fatalities.

9. Due to the profits obtained with heroin laced with fentanyl, the addition of fentanyl and fentanyl analogs to other drugs, mainly cocaine and methamphetamine, became a common practice.
10. When the trade in fentanyl and its precursors became regulated, the cartels rescued failed opioid analgesics previously discarded for human use and other types of potent experimental synthetic opioids to be sold as new psychoactive substances (see Chap. 16).

Recognition of this sequence of events could help identify risk factors in other countries in the coming years and help policymakers intervene before the consequences of new epidemics of opioid misuse become widespread [47].

5.6.2 Measures to Approach the Opioid Crises

According to Blanco et al. [48], an integrated public health approach to the opioid crises "should seek to understand who is affected, where they are affected; the trajectories, pathways, and consequences of opioid use and misuse, and how these trajectories are changing." Therefore, controlling the opioid crises requires the active involvement of multiple actors and structural interventions, some of which are detailed in Chap. 7 and summarized below.

5.6.2.1 Harm Reduction Programs

Syringe exchange programs significantly reduce HIV and hepatitis C transmission among persons who inject drugs. In addition, syringe exchange program facilities can motivate users to initiate treatment. Establishing supervised injection facilities, providing take-home naloxone kits, and establishing drug-checking points to determine adulterants in drugs are also effective interventions to reduce overdose deaths (Chap. 7).

The WHO recommends providing naloxone to anyone who could witness an overdose, i.e., friends, families, and health and service providers, such as paramedics, police officers, and firefighters. Naloxone is a safe opioid antidote with minimal adverse effects. It displaces heroin, morphine, fentanyl, and other opioid analgesics from their receptors, thus reversing respiratory depression. In addition, because death does not occur immediately after opioid injection, there is a window during which people witnessing an overdose can intervene. As emphasized by the WHO guidelines on how to manage opioid overdoses:

> Death in opioid-overdose can be averted by emergency basic life support resuscitation and/or the timely administration of an opioid antagonist such as naloxone.

After successful resuscitation following the administration of naloxone, the level of consciousness and breathing of the affected person should be closely observed until full recovery has been achieved [49].

Naloxone's accessibility varies greatly among countries, but it should be available everywhere. It is a safe medication, devoid of psychoactive effects, and has already saved countless lives.

5.6.2.2 Pharmacological Interventions

People with OUDs require access to services for opioid detoxification, withdrawal management, and opiate-assisted treatment, usually with methadone or buprenorphine. Naltrexone is another approved medication useful for maintenance in completely detoxified patients (see Chaps. 14 and 15). Attention requires a gender perspective, particularly to support pregnant women or women with children who have difficulties attending medical services when they lack childcare options. Considering the comorbidity of OUDs and blood-borne infections, health care should also include HIV and hepatitis C testing. Pharmacological interventions are more effective when other comorbid psychiatric diseases are treated.

5.6.2.3 Psychosocial Interventions

Medication-assisted therapy is more effective when combined with evidence-based inpatient or outpatient psychosocial interventions. These interventions include cognitive behavioral therapy; social skills training; individual, group, or family therapy; and contingence management, among others [50].

5.6.2.4 Adequate Pain Management

Low- and middle-income countries need sufficient access to licit morphine for pain management. Also, there is a need for periodically revised clinical guidelines. Updated clinical guidelines require input from the best health-care professionals without the intervention of the pharmaceutical companies that manufacture pain medications [51].

In addition, monitoring prescription programs should be common in all countries where misuse of opioid analgesics occurs. From the first decade of the twenty-first century, the US government has implemented prescription drug monitoring programs in several States and strengthened those already in operation to prevent the diversion of opioid analgesics. These programs are databases in which pharmacists record every time they dispense controlled substances. This helps care providers to detect if a patient asking for an opioid prescription has recently received opioids from other providers [52].

It is imperative to provide the best available treatments to people in pain, including not only pharmacological but also non-pharmacological approaches (see Chaps. 10 and 11).

5.6.2.5 Structural Interventions to Address Inequalities

Access to treatment for persons with OUDs should be part of a comprehensive health-care system that pays attention to social determinants. Access to education, social services, work opportunities, and housing support are some of the factors needed to improve the lives of people affected with these disorders [53].

5.7 Final Considerations

Opioids are effective drugs and can be safely used under medical supervision. These drugs are often invaluable tools to minimize suffering and preserve human dignity during end-of-life care in patients with terminal illnesses. However, opioids are also potent psychoactive drugs, which can lead to OUDs and overdoses. Measures to avoid opioid misuse include training health providers in safe prescription analgesic use, warning the general public on the risks of using opioids at higher doses and more extended periods than recommended, and establishing monitoring prescription programs. In addition, it is necessary to utilize pharmacological and non-pharmacological approaches for pain management instead of imposing excessive restrictions on opioid availability. Finally, to contain the current opioid crises and avoid new ones, it is necessary to implement structural interventions for people with OUDs and address social factors contributing to drug use.

References

1. Kunnumpurath S, Julien N, Kodumudi G, Kunnumpurath A, Kodumudi V, Vadivelu N. Global supply and demand of opioids for pain management. Curr Pain Headache Rep. 2018;22:34. https://doi.org/10.1007/s11916-018-0689-1.
2. UNODC. World drug report, United Nations publication, 2022. https://www.unodc.org/unodc/en/data-and-analysis/world-drug-report-2022.html
3. Mark TL, Parish W. Opioid medication discontinuation and risk of adverse opioid-related health care events. J Subst Abus Treat. 2019;103:58–63. https://doi.org/10.1016/j.jsat.2019.05.001.
4. Patrick SW, Schumacher RE, Benneyworth BD, Krans EE, McAllister JM, Davis MM. Neonatal abstinence syndrome and associated health care expenditures: United States, 2000–2009. JAMA. 2012;307:1934–40. https://doi.org/10.1001/jama.2012.3951.
5. Kidd K, Weinberg T, Caboral-Stevens M. The 21st century opioid addiction: a concept analysis and implications for nursing. J Addict Nurs. 31:17–22. https://doi.org/10.1097/JAN.0000000000000321.

6. Kiang MV, Humphreys K, Cullen MR, Basu S. Opioid prescribing patterns among medical providers in the United States, 2003–17: retrospective, observational study. BMJ. 2020;368:l6968. https://doi.org/10.1136/bmj.l6968.
7. SAMSHA. Key substance use and mental health indicators in the United States: results from the 2019 National Survey on Drug Use and Health. HHS Publ No PEP19-5068, NSDUH Ser H-54. 2019;170:51–8.
8. NIDA. Opioid overdose crisis [Internet]. Natl Inst Drug Abus. 2021.
9. CDC. Understanding the epidemic [Internet]. Centers Dis Control Prev Natl Cent Inj Prev Control. 2022.
10. Volkow ND. Making addiction treatment more realistic and pragmatic: the perfect should not be the enemy of the good. Heal Aff Forefr. 2022. https://doi.org/10.1377/forefront.20211221.691862.
11. El-Bassel N, Shoptaw S, Goodman-Meza D, Ono H. Addressing long overdue social and structural determinants of the opioid epidemic. Drug Alcohol Depend. 2021;222:108679. https://doi.org/10.1016/j.drugalcdep.2021.108679.
12. Porter J, Jick H. Addiction rare in patients treated with narcotics. N Engl J Med. 1980;302:123. https://doi.org/10.1056/nejm198001103020221.
13. Leung PTM, Macdonald EM, Stanbrook MB, Dhalla IA, Juurlink DN. A 1980 letter on the risk of opioid addiction. N Engl J Med. 2017;376:2194–5. https://doi.org/10.1056/NEJMc1700150.
14. Baker DW. History of the Joint Commission's pain standards: lessons for today's prescription opioid epidemic. JAMA. 2017;317:1117–8. https://doi.org/10.1001/jama.2017.0935.
15. Jones CM. Heroin use and heroin use risk behaviors among nonmedical users of prescription opioid pain relievers – United States, 2002–2004 and 2008–2010. Drug Alcohol Depend. 2013;132:95–100. https://doi.org/10.1016/j.drugalcdep.2013.01.007.
16. Quinones S. Dreamland: the true tale of America's opiate epidemic. New York: Bloomsbury Publishing; 2015.
17. Stoicea N, Costa A, Periel L, Uribe A, Weaver T, Bergese SD. Current perspectives on the opioid crisis in the US healthcare system: a comprehensive literature review. Medicine (Baltimore). 2019;98:e15425. https://doi.org/10.1097/MD.0000000000015425.
18. Dowell D, Haegerich TM, Chou R. CDC guideline for prescribing opioids for chronic pain – United States, 2016. MMWR Recomm Rep. 2016;65:1–49. https://doi.org/10.15585/mmwr.rr6501e1.
19. Ciccarone D. The triple wave epidemic: supply and demand drivers of the US opioid overdose crisis. Int J Drug Policy. 2019;71:183–8. https://doi.org/10.1016/j.drugpo.2019.01.010.
20. Jones W, Kaoser R, Fischer B. Patterns, trends and determinants of medical opioid utilization in Canada 2005–2020: characterizing an era of intensive rise and fall. Subst Abuse Treat Prev Policy. 2021;16:65. https://doi.org/10.1186/s13011-021-00396-5.
21. CCSA. Prescription opioids. Encycl Personal Individ Differ. 2020:4015.
22. Usman MA. Opioid crisis: the African perspective. Disaster Med Public Health Prep. 2020:1–2. https://doi.org/10.1017/dmp.2020.259.
23. Belzak L, Halverson J. The opioid crisis in Canada: a national perspective. Health Promot Chronic Dis Prev Can. 2018;38:224–33. https://doi.org/10.24095/hpcdp.38.6.02.
24. Special Advisory Committee on the Epidemic of Opioid Overdoses. Opioid- and Stimulant-related Harms in Canada. Ottawa: Public Health Agency of Canada; June 2022. https://health-infobase.canada.ca/substance-related-harms/opioids-stimulants/.
25. Kurth AE, Cherutich P, Conover R, Chhun N, Bruce RD, Lambdin BH. The opioid epidemic in Africa and its impact. Curr Addict Rep. 2018;5:428–53. https://doi.org/10.1007/s40429-018-0232-9.
26. Klein A, Patwardhan S, Loglo MGA. Divergences and commonalities between the US opioid crisis and prescription medicine mis/use in West Africa. Int J Drug Policy. 2020;76:102640. https://doi.org/10.1016/j.drugpo.2019.102640.
27. International Narcotics Control Board. Report 2017 Press material. United Nations Inf Serv. 2018;1100.

28. Dean L, Kane M. Tramadol therapy and CYP2D6 genotype [Internet]. Med Genet Summ. 2012.
29. UNODC. 2020 global synthetic drugs assessment, regional overview Africa [Internet]. 2020.
30. UNODC. World drug report 2019 (United Nations publication, Sales No. E.19.XI.8). [Internet]. 2019.
31. Harker N, Lucas WC, Laubscher R, Dada S, Myers B, Parry CD. Is South Africa being spared the global opioid crisis? A review of trends in drug treatment demand for heroin, nyaope and codeine-related medicines in South Africa (2012–2017). Int J Drug Policy. 2020;83:102839. https://doi.org/10.1016/j.drugpo.2020.102839.
32. Salm-Reifferscheidt L. Tramadol: Africa's opioid crisis. Lancet (London, England). 2018;391:1982–3. https://doi.org/10.1016/S0140-6736(18)31073-0.
33. Peprah P, Agyemang-Duah W, Appiah-Brempong E, Akwasi AG, Morgan AK. "With tramadol, I ride like a Jaguar": a qualitative study of motivations for non-medical purpose tramadol use among commercial vehicle operators in Kumasi, Ghana. Subst Abuse Treat Prev Policy. 2020;15:49. https://doi.org/10.1186/s13011-020-00292-4.
34. Ambekar A, Agrawal A, Rao R, Mishra AK, Khandelwal SK, Chadda RK. Magnitude of substance use in India. New Delhi Minist Soc Justice Empower Gov India; 2019.
35. Schmall E, Galofaro C. How tramadol, touted as safer opioid, became 3rd world peril [Internet]. AP News. 2019.
36. Singh B, Rao R. Perspectives-is there an opioid epidemic in India? J Public Health (Oxf). 2021;43:ii43–50. https://doi.org/10.1093/pubmed/fdab322.
37. Rostam-Abadi Y, Gholami J, Amin-Esmaeili M, Safarcherati A, Mojtabai R, Ghadirzadeh MR, et al. Tramadol use and public health consequences in Iran: a systematic review and meta-analysis. Addiction. 2020;115:2213–42. https://doi.org/10.1111/add.15059.
38. Nasiri N, Abedi L, Hajebi A, Noroozi A, Khalili M, Chegeni M, et al. Population size estimation of tramadol misusers in urban population in Iran: synthesis of methods and results. Addict Health. 2019;11:173–82. https://doi.org/10.22122/ahj.v11i3.240.
39. Alblooshi H, Hulse GK, El Kashef A, Al Hashmi H, Shawky M, Al Ghaferi H, et al. The pattern of substance use disorder in the United Arab Emirates in 2015: results of a National Rehabilitation Centre cohort study. Subst Abuse Treat Prev Policy. 2016;11:19. https://doi.org/10.1186/s13011-016-0062-5.
40. Chen T-C, Chen L-C, Knaggs RD. A 15-year overview of increasing tramadol utilisation and associated mortality and the impact of tramadol classification in the United Kingdom. Pharmacoepidemiol Drug Saf. 2018;27:487–94. https://doi.org/10.1002/pds.4320.
41. Sørensen AMS, Rasmussen L, Ernst MT, Mogensen SH, Laursen MV, Jimenez-Solem E, et al. Use of tramadol and other analgesics following media attention and risk minimization actions from regulators: a Danish nationwide drug utilization study. Eur J Clin Pharmacol. 2021;77:617–24. https://doi.org/10.1007/s00228-020-03016-6.
42. World Health Organization. Critical review report: tramadol. Crit Rev Rep Tramadol. 2018:19–22.
43. European Monitoring Centre for Drugs and Drug Addiction. European drug report 2020: trends and developments. EMCDDA. Luxembourg: Publications Office of the European Union; 2020.
44. UNODC. Monitoreo de Plantíos de Amapola 2018–2019 (MEXK54, 2021) [Internet]. United Nations; 2021. https://doi.org/10.18356/9789214030508.
45. Fleiz C, Arredondo J, Chavez A, Pacheco L, Segovia LA, Villatoro JA, et al. Fentanyl is used in Mexico's northern border: current challenges for drug health policies. Addiction. 2020;115:778–81. https://doi.org/10.1111/add.14934.
46. Cruz SL, Carranza-Aguilar CJ, Pérez-García IP. Sodium chloride injection to treat opioid overdose; Does it work? A preclinical study. Neurotoxicology. 2021;87:24–9. https://doi.org/10.1016/j.neuro.2021.08.012.
47. UNODC. Understanding the global opioid crisis. Glob SMART Updat. 2019;21:12.

48. Blanco C, Wiley TRA, Lloyd JJ, Lopez MF, Volkow ND. America's opioid crisis: the need for an integrated public health approach. Transl Psychiatry. 2020;10:167. https://doi.org/10.1038/s41398-020-0847-1.
49. World Health Organization. (2014). Community management of opioid overdose. World Health Organization. https://apps.who.int/iris/handle/10665/137462
50. Dugosh K, Abraham A, Seymour B, McLoyd K, Chalk M, Festinger D. A systematic review on the use of psychosocial interventions in conjunction with medications for the treatment of opioid addiction. J Addict Med. 2016;10:93–103. https://doi.org/10.1097/ADM.0000000000000193.
51. Furlan AD, Harvey AM, Chadha R. Warning from Canada: Latin America, South Africa and India may face an opioid epidemic in the coming years. J Glob Health. 2020;10:010324. https://doi.org/10.7189/jogh.10.010324.
52. Davis C, Sargent W, Vick J, Kuoh H, Cote P-O, editors. Leveraging Prescription Drug Monitoring Program (PDMP) data in overdose prevention and response. 2021.
53. European Monitoring Centre for Drugs and Drug Addiction. Opioids: health and social responses. 2021:1–11. https://doi.org/10.2810/273937.
54. CDC. Centers for Disease Control and Prevention (CDC), National Center for Health Statistics, Drug overdose data [Internet]. 2020.
55. The Lancet. A time of crisis for the opioid epidemic in the USA. Lancet. Elsevier Ltd. 2021;398:277. https://doi.org/10.1016/S0140-6736(21)01653-6

Chapter 6
Persons Who Misuse Opioids

Clara Fleiz, David Fajardo-Chica, Jorge Villatoro, and María Elena Medina-Mora

Abstract This chapter analyzes the living conditions of persons who inject drugs (PWIDs) in Tijuana, a border city in the northern part of Mexico, as an example of the adversities associated with drug use. The social and emotional problems of PWIDs include poverty, restricted access to health-care services, malnourishment, low education, and few job opportunities. In addition, stigma, discrimination, police brutality, traumatic experiences, and sexual violence are also common among opioid users. This chapter seeks to give voice to opioid users through personal narratives, emphasizing the need for a person-centered perspective to address the current opioid crisis effectively. Although most of the data presented here were collected in Tijuana, this chapter also analyzes some differences and commonalities of PWIDs in other parts of the world, through testimonies obtained from the literature.

Keywords Opioid · Person · Life conditions · Tijuana · Mexico

---

C. Fleiz (✉) · J. Villatoro
National Institute of Psychiatry, Ramon de la Fuente Muñiz (INPRM), Mexico City, Mexico

Opioids Working Group, Global Studies Seminar, Faculty of Medicine, National Autonomous University of Mexico (UNAM), Mexico City, Mexico
e-mail: fleiz@imp.edu.mx

D. Fajardo-Chica
Opioids Working Group, Global Studies Seminar, Faculty of Medicine, National Autonomous University of Mexico (UNAM), Mexico City, Mexico

Mental Health and Psychiatry Department, Faculty of Medicine, UNAM, Mexico City, Mexico

M. E. Medina-Mora
National Institute of Psychiatry, Ramon de la Fuente Muñiz (INPRM), Mexico City, Mexico

School of Psychology, UNAM,, Mexico City, Mexico

Global Mental Health Center (INPRFM), Mexico City, Mexico

S. L. Cruz (ed.), *Opioids*, https://doi.org/10.1007/978-3-031-09936-6_6

6.1 A Person-Centered Perspective

A person who uses opioids is more than just a biological organism interacting with chemical substances. The concept of person is a multidimensional one. It involves different levels at which a person exists. According to Cassell, "A person is an embodied, purposeful, thinking, feeling, emotional, reflective, relational, human individual always in action, responsive to meaning, and whose life in all spheres points both outward and inward" [1]. Thus, a person-centered perspective is a good starting point for a multidisciplinary and contextualized analysis of drug use and abuse. Opioid misuse occurs in a complex personal context involving all those levels mentioned by Cassell. Those levels determine a person's opioid misuse, trajectories, and consequences. The context created by those levels plays a major role in people's narratives. That is why paying attention to those contexts is of central importance for any approach that tries to capture the personal subjective experience holistically.

This chapter focuses on "persons who inject drugs" (PWIDs henceforth) living in the northern Mexican city of Tijuana who agreed to participate in in-depth interviews as part of a study we conducted in 2018 on heroin users. The Tijuana case illustrates an opioid crisis in a marginalized context.

The methodological aspects and results of the study have been previously published [2]. Presenting a general picture of persons who inject drugs in Tijuana serves two main objectives. First, it offers a detailed illustration of life experiences as related by their protagonists. There is recognition in the literature that personal experiences undergone by this population are lacking [3]. Shifting the focus to a personal perspective in these narratives decreases the moral risk of not respecting their dignity [4]. Second, it establishes some general guidelines helpful in conceptualizing on generalities and particularities of being a person who misuses opioids. Personal narratives derive from some central and recurrent topics. In the final section, we try to establish some parallelisms between the narratives of PWIDs in marginalized conditions in Tijuana and their counterparts in other countries. Problematic conditions are traceable through geographical locations such as hardship, marginalization, and stigma. Nonetheless, opportunities for solidarity and public attendance exist as well.

6.2 Life Conditions of Persons Who Inject Drugs (PWIDs) in Tijuana

6.2.1 *The City*

Man, 47 years old, Tijuana, México:

… For me, living here in Tijuana is living like a beggar because of the life I lead, but at the same time I brought it on myself. I am sure that if I did not use drugs, it would be an excellent city to live in because there is work, money, and opportunities, but I don't have them because of my addiction.

Mexico's northern border has an area of 3234 km^2 comprising the border between the states of California, Arizona, New Mexico, and Texas on the US side and the Mexican states of Baja California, Sonora, Chihuahua, Coahuila, Nuevo León, and Tamaulipas. This border strip encompasses over 25 cities, of which Tijuana is the largest binational metropolitan area in Mexico.

Tijuana is in the state of Baja California in the northwest of Mexico on the southern US border (San Diego, California). It has an urbanized area of 498 km^2. It is home to just over 1,751,000 inhabitants, and it is estimated that by 2030, this figure will exceed 2,335,000 [5]. It is a territory with high population mobility and transportation for migratory, commercial, and tourist purposes. It is also the largest border city in Mexico and the most popular one because of its artistic and cultural offerings. Nightclubs and restaurants in the city are always in high demand from people from Mexico and foreigners, mainly from the United States. For this reason, services are paid for in dollars and Mexican pesos, and English and Spanish are spoken.

The permanent economic activity of the city generates sources of income for all its inhabitants and for the drug user population engaged in the informal economy, many of whom are national or international migrants or were deported from the United States.

Several people in the city inject drugs, mostly heroin alone or in combination with crystal. This population lives in a permanent state of tension because of the constant operations of the state, municipal police, and army.

Woman, 37 years old, Tijuana, México:

> Well, that time [the police] arrived, there was an operation, they also came with the soldiers and arrived at the shooting gallery. They wanted to know who were selling drugs. At that time, they were taking a lot of people to the rehabilitation center, and since my partner was the one who was selling [drugs] at that time, they [the police] spotted him.

6.2.2 Drug-Using Spaces in the City

In Tijuana and along Mexico's northern border, most places where people use drugs, known as shooting galleries, are abandoned houses, neighborhoods, or rundown vacant lots, with unhygienic conditions. Some of these places also serve as drug outlets and are known as *conectas*. Due to demand, they are open 365 days a year, 24 hours a day. In the *conectas*, black tar-type heroin (known as "*chiva*"), white powdered heroin (known as "China white"), and methamphetamine crystal (known as "crico or ice") are widely available to meet the drug demand of PWIDs. In these places, PWIDs also obtain paraphernalia such as syringes and bowls, small containers users employ to mix and dissolve drugs in water. Most of these instruments are shared several times. The population is often forced or pressured to buy at certain *conectas* and sometimes to replace their drug of choice with what is available on the market. This situation is one of the primary triggers of violence in the area and the *conectas* and shooting galleries. In the operations targeting these spaces, the police detain and frequently beat users, who often, after several operations, decide to go elsewhere.

Woman, 43 years old, Tijuana, México:

I live in a shooting gallery, I live where they sell drugs, that is, just imagine, in a house, in a shooting gallery where the cops [police] go every day. In other words, you are not calm, you have to be alert all the time, you cannot relax because they turn up, I mean, f**k !, the worst, it's terrible, it's nasty here [...].

These spaces have an organization for their operation. One of the prominent figures is the leader. He sets the rules for the place; for example, he controls who comes in and out and the schedules, and if there are fights, he sorts things out. He often injects other users, because of his expertise to inject into different veins, and charges for this service. Moreover, from his experience with drug use, he knows how much a person should be injected with and how to prepare a dose, and if someone has an overdose, he knows how to revive the person by injecting saltwater intramuscularly, hitting them, placing ice on their genitals, or administering naloxone. The leaders of the arena also allow access to non-governmental organizations (NGOs) to carry out harm reduction programs, such as exchanging syringes, distributing condoms, and using rapid tests to detect the human immunodeficiency virus (HIV) and hepatitis C virus (HCV).

It is worth mentioning that women run some shooting galleries, although not very many. Some users also live alone or with their partners. Sometimes children, who are the offspring of this population, are born and grow up in these spaces, have poor nutrition, do not attend school, and in many cases reproduce the drug use of their parents since their environment limits the opportunities for better development.

6.2.3 Demographic Characteristics

In Tijuana, most of the population that use heroin and other opioids are men, and over half of the users are single and have completed junior high school. Just over a third went to senior high school, and a low percentage has college degrees. Three out of ten are ages 40–49, with an average age of 34 for women and 41 for men [2].

6.2.4 Housing

In this city and throughout the northern border region of Mexico, the heroin user population lives in extreme poverty. Most users live on the streets, in alleys, or in sheet or cardboard houses built under bridges, lacking basic electricity and drinking water services. PWIDs also live in parks, abandoned cars, shelters, migrant shelters, shooting galleries, or the sluices of the Tijuana sewage canal.

Man, 44 years old, Tijuana, México:

This is where my friends came to die; they sleep outdoors and die, or are killed or run over by a patrol car. There was a sort of craze; they would sprint and get run over by police cars.

Others have lived in penitentiary centers, while those with more income rent rooms in hotels or old houses.

Man, 21 years old, Tijuana, México:

Well, right now, I don't have a place to live, I'm living in the *conecta* where the shooting gallery is. Right there, next to the shooting gallery, there are rooms for rent. They charge you very little a week, $7 dollars per room, for yourself, or you can share with one or more friends. It is very cheap, but you risk the police arriving as it is a *conecta*, a shooting gallery because they have no qualms about breaking in […].

6.2.5 Work and Income

Users' daily lives revolve around drug use, so their priority is getting financial resources to buy the dose they need approximately every 3 hours. Another part of their income is allocated for food and sometimes for lodging. PWIDs work long hours during the day and at night at various recycling facilities, in sex work (mainly women), in restaurants or nightclubs in the city, cleaning the entrances, or shining shoes. Other activities include bricklaying, selling candy, and cleaning windshields at international checkpoints or zebra crossings with heavy traffic.

Many of these activities depend on what people on the streets want to give them or on low salaries from the contractors or bosses, which is why most of the population has a low monthly income. For example, a third of the population earns from approximately $100 to $250, and just over a third earns from $250 to $500.

Woman, 37 years old, Tijuana, México:

My job is to recycle, not just anyone puts their hand there, in so much sh*t [garbage], next to cans of Coca Cola [..]. I recycle throughout this area here, I sell twice a day now, once at night and once in the morning, and then I buy one [dose] right now and another in the morning, for two dollars [each]. It really is a struggle to get a fix.

6.2.6 Nourishment

Most of the population is undernourished. PWIDs eat once a day, mainly the remains of food they collect on the street, what people give them, or at soup kitchens for homeless people. Consequently, PWIDs are underweight, and their body mass index is lower than the average for the general population.

Man, 28 years old, Tijuana, México:

Sometimes I go hungry; if I eat twice a day, that's a lot. If I eat once a day, that's good, often people give me food or invite me to eat on the street. I eat what they give me or what my colleagues bring me.

At the same time, PWIDs consume high amounts of candy because they say that sugar helps them counteract withdrawal symptoms.

These data found in Tijuana concur with those reported in international studies, which report that malnutrition is more severe among those with more severe addiction and affects women who sleep on the streets more, together with those who have severed their family and social ties [6].

6.2.7 Access to Health-Care Services

In addition to this severe dependence, the drug user population suffers from various diseases requiring specialized medical care. However, they have various barriers to health care because services are scarce, they are a long way from the places where they are located (especially those focusing on the treatment of HIV and HCV and methadone clinics), and many of them cannot afford the transport fares, while those with a disability find it extremely difficult to move around.

Another significant barrier is the strong stigma and discrimination from health professionals, which discourages the population to seek for medical care.

Man, 47 years old, Tijuana, México:

[…] Well, they scared me a lot (in the hospital), they even told me (the doctors), the medicine is very expensive, we don't have it, God bless you. Can't you even give me an aspirin to sleep? And they said no. I also imagine this was because they saw that (the skin abscess) was due to drug addiction.

The suspension of activities further hampered access to health services due to the SARS-CoV-2 pandemic. Some of these care centers restricted their operating hours, while others were closed down. This situation lead NGOs with very limited resources to treat some of the diseases presented by PWIDs, such as tuberculosis, urinary tract infections, respiratory symptoms, fractures, and skin abscesses.

6.2.8 Violence

In addictions as severe as those developed by PWIDs on Mexico's northern border, violence and trauma are part of their life stories and have often triggered their drug use. In Tijuana, over 50% have suffered some violence, particularly sexual abuse, domestic violence, and police violence. Many of these violent acts occurred in early childhood and had major consequences for the population's mental health, such as post-traumatic stress, depression, and anxiety.

6.2.8.1 Sexual Abuse

Sexual abuse is a traumatic event that is more common among women, occurs in early childhood, and often lasts for long periods until women run away or leave home as a form of resistance to sexual violence.

Woman, 40 years old, Tijuana, México:

From the time I was 8 years old, I suffered sexual abuse, violent abuse in my family (from her biological father), and that was not easy. I felt a lot of anger and impotence because I could not do anything because he is my father, so, to escape, I came to Tijuana when I was 13. I left the first time, and then they (her family) found me and took me back (to Sonora), and then I came when I was 15 years old (again to Tijuana) and that time they did not find me, I am not going back.

These events change the trajectories of their lives entirely because, in addition to severing the family bond, they stop studying and decide, driven by circumstances, to live alone at a young age and without supportive social networks. Such conditions place them in a state of extreme vulnerability because they grow up without parental protection, without the opportunity to attend school, in violent environments with readily available drugs, without a fixed place of residence, and in general with few opportunities to develop.

Woman, 40 years old, Tijuana, México:

I studied until the second year of junior high school, but since I was left on my own because my mother went to Sonora, I no longer wanted to study, and then I came to Tijuana. It is very hard to be alone [she cries], but life goes on. I have no one here except God.

Some PWIDs work in activities that they often do not like or put them at risk of violence, such as sex work.

Woman, 40 years old, Tijuana, México:

When I arrived in Tijuana, I started going to table dances, and as drugs took over, I became a prostitute. I do it [sex work] for a living and get fed up with the people I have to put up with; that is one of the main reasons I would like to stop using drugs [...]. Once he [the client] held a screwdriver to my neck and took the money he had given me and even what I had brought, and hit me twice, I managed to escape, but I cut my hands when I jumped over a wall.

Opioid use, therefore, becomes a way to relieve the emotional pain caused by these traumatic experiences and the losses they have suffered.

Woman, 40 years old, Tijuana, México:

Maybe I do it because I want to stop suffering, you want to blot out everything from the past, there are many things in my mind, the abuse, wanting to have my girl with me. I lost another child, and I don't know if he is alive or dead, they took him away from me at the hospital, I was going crazy.

Women who have been sexually assaulted often decide not to report their assailants, choose not to tell anyone, isolate themselves from their families, and fail to seek professional help. This isolation leads them to experience the trauma by themselves, with states of emotional distress exacerbated in those living on the street.

Woman, 40 years old, Tijuana, México:

The loneliness makes me get high. I have been living on the street a lot, and lately, I have been very depressed because this is not for me, because I am drowning, and I have not found a way out.

6.2.8.2 Domestic Violence

One of the recurring events in the stories of heroin users is having been witnesses to and victims of domestic violence for a long time. The narratives of a significant portion of the users consistently show how they have witnessed verbal and physical aggression mainly committed by their fathers against their mothers from a young age.

> *Man, 45 years old, Tijuana, México:*
> My father beat my mother, forced her to make him food, and he did not bring anything home; my brother and I, as we were 9 or 10 years old, used out to beg for money. Once my father hit me and I fell. He hit me with his belt, and I said, one day I'm going to grow up, and I'm going to do the same to you, and yes, once I laid hands on him.

Parents also directed this physical violence toward their children. Among the childhood memories of users are the beatings they received and the acts of negligence. They generally report that their parents did not take care of them, left them alone, did not provide them with food or shelter, and did not send them to school. All these forms of violence had a strong impact on the mental health of the user population since they frequently report that all of this led them to start using and transition to dependency.

> *Man, 45 years old, Tijuana, México:*
> My mother separated from my father before I was born, and she hitched up with another person to give us the last names. He [the stepfather] became a millionaire overnight. I was filthy when I was little, and I had to go to school, but they didn't send me [...]. They put me to work, and what I wanted was to leave there [home].

One of the most common characteristics in the family histories of heroin users is the degree of dependence on different drugs of their parents [7, 8]. Studies consistently show that between a third and half of heroin users had at least one substance-dependent parent, as well as high rates of psychopathology [9–12]. In PWIDs on Mexico's northern border, we frequently found these experiences; some even started using them because they saw their parents do so or because their parents injected them for the first time.

> *Woman, 44 years old, Tijuana, México:*
> My mom, my brother, and I are the only ones who use heroin. She was the one who got me started; she injected me for the first time when I was 10 years old. She did it with coke and heroin, and here in my arm [her mother injected her]. This scar is where my mom used to inject me. I have no idea why she did it. I just remember that my feet bent, and that was all, and suddenly I fell asleep, that was all I felt. She also asked me to go to the *conecta*.

6.2.8.3 Arrests (Police)

In Tijuana and the rest of the northern border, the police, using excessive force and violating human rights, has arbitrarily detained nearly all PWIDs. One widespread practice is the confiscation of injection paraphernalia, which the police destroy. They then extort them for carrying syringes or just because of their appearance. They also remove and dispose of the few belongings they have.

Man, 23 years old, Tijuana, México:

Once, I really needed to go to the toilet, and I was just going home to inject myself, and I still hadn't even bought the drugs, but I had two dollars on me. The police stopped me and took my money, they brought me up, and I said, 'I want to go to the toilet, let me go to a toilet!' They started laughing, they handcuffed me and put me in the sun, they walked me around for like four hours like that. I held on so they wouldn't make fun of me, and they beat me.

Sexual violence against women is one of the most serious acts police members have committed and remains unpunished. None of the victims denounce their aggressors for fear of reprisals and because no state apparatus protects their rights.

Woman, 38 years old, Tijuana, México:

The police really have a nerve; they pick you up, and then they say so what! How much do you charge? They are disgusting ... Why do they pick me up? I don't have anything, just because they liked you, but when I was younger [the police] always picked me up. I told them, take me, I'm not going to give you anything and they told me, you're going to kiss me, and you're going to give me a blow job.

Another of the recurring acts of the police is to take users involuntarily and violently to rehabilitation centers, which often do not have the necessary infrastructure or trained health professionals. In Tijuana, nearly 40% of the population has been involuntarily sent to these centers.

Woman, 37 years old, Tijuana, México:

Well, that time [the policemen] arrived, there was a huge operation because they came with the soldiers as well and they arrived at the *conecta* with all the state policemen, and they wanted to know who was selling drugs there. That was when a lot of people were taken to the rehabilitation center.

In 2012, because of police harassment, users from different border cities and the north of the country, accompanied by various NGOs, academics, and the National Human Rights Commission, produced the First Human Rights Booklet for People who Inject Drugs. The document includes 16 fundamental rights, such as access to education, health, justice, and social development services granted by the state and the right to carry injection instruments, without this constituting a reason for physical, sexual, or psychological abuse, deprivation of liberty, extortion, or confiscation of these instruments. It also stipulates the right to carry a personal dose in accordance with what is allowed in the General Health Law of Mexico. This has been one of the first coordinated efforts between users and other government institutions to manage public policies that promote the human rights of PWIDs and mitigate police abuse [13].

6.2.9 Risks Associated with Opioid Use

Throughout the northern border of Mexico, PWIDs experience various risks associated with their use that considerably compromise their lives, especially overdoses and infections such as HIV, HCV, and skin abscesses. These diseases exacerbate by the precarious conditions in which they live and the limited access to medicines and health services.

6.2.9.1 Overdose

Opioid overdoses in PWIDs are related to the combination of drugs and their administration after being in a treatment or detention center because once they leave, they consume the same amount or more than before, and as their tolerance is lower, they frequently overdose. In Tijuana, almost 70% of the population has experienced at least one overdose, but the average is four in their lifetime and three in the past 12 months.

> *Woman, 37 years old, Tijuana, México:*
> Well, I was with the person I started using heroin with, we injected ourselves, and I thought that many of them were going to have an overdose in that short time, but I didn't, we walked half a block, and when we were going to go down the street, my knees buckled and I no longer remember when I fell to the floor, and when I woke up, I was all wet.

In Tijuana and other Mexican border cities, the method of choice for users to save others from an overdose is injecting saltwater into the veins or muscle, putting ice on the genitals, and hitting the feet with a hard object such as wood [2]. However, with the penetration of fentanyl, these ways of managing risks are not effective due to the potency of this synthetic opioid [14]. According to users, overdoses have increased in the city due to the expansion of this opioid [15].

> *Man, 65 years old, Tijuana, México:*
> I have revived three or four because they have been about to die from an overdose, and I have brought them round, I have helped them recover, I have injected salt into their veins so that they return to their normal rhythm, I hit them on their chests, I hit them on their backs, I squeeze their stomachs so that they breathe in.

Users require naloxone to deal with overdoses. However, few have access to it because it enters the country through international donations, and with the closure of borders due to the SARS-CoV-2 epidemic, it has been impossible to transfer it to NGOs, which dispense this medicine to the community. In addition, the General Health Law of Mexico currently misclassifies naloxone as a psychotropic drug [16], which means that those who require it can only do so in pharmacies with a prescription, making it impossible for PWIDs to get it. Another factor that exacerbates the lack of access to the population is police operations.

> *Woman, 28 years old, Tijuana, México:*
> Right now, I am not carrying naloxone; I left it in my room. I always used to carry a box, but since the last time the police grabbed me with the entire damage reduction kit, I no longer want to bring syringes or anything like that because it is a risk. Yes, I should have brought naloxone, but because there is a risk of the police, I don't feel confident enough to bring the whole kit with me, where I used to keep my syringes, water, and naloxone.

6.2.9.2 HIV/AIDS

The most recent data from Tijuana show that HIV prevalence in PWIDs in treatment centers is 6.6% and 7.3% in drug-use sites, which is higher than the rates found in other border cities such as Ciudad Juárez (1.5%) [2] and the national average registered in the general Mexican population (0.26%) [17].

In Tijuana, one of the factors most closely related to the presence of HIV and other sexually transmitted infections (STIs) in PWIDs is sharing syringes. It has been documented that over 60% of users have given or loaned their syringes, and in turn, nearly 70% have used a syringe and paraphernalia that someone had already used [2]. Risky sexual behaviors, such as not using a condom, exchanging sex for drugs, and having sex with casual partners, are also associated with HIV [18].

Woman, 40 years old, Tijuana, México:

I know I got HIV from syringes or who knows, maybe from sex. I don't know how I got it.

On the other hand, various studies have found that gender plays a significant role in the risk of acquiring HIV [18]. For example, in Tijuana, women have a higher prevalence than men (10.2% vs. 3.5%) [19]. These data are consistent with international research, which shows that, although women PWIDs represent a smaller proportion of PWIDs globally (3.5 million out of 16 million) [20], they tend to have a higher HIV prevalence than men [21].

These differences are related to socio-structural factors that include more significant stigma among women injecting drug users, physical and sexual violence, dependence on male partners to obtain and inject drugs, and engaging in sex work, which increases their risk of contracting HIV [18].

Woman, 52 years old, Tijuana, México:

Sometimes I don't [use a condom in sex work] because there are people who don't like to use it and they offer me more money, two or four dollars more, and I usually accept, to get more money for drugs. My husband does not really have a proper job he washes cars. I take more money home for the doses for both of us.

In addition, a significant proportion of heroin users in the city engage in polydrug use with amphetamine-type stimulants and, to a lesser extent, cocaine. Several studies have shown that people with these combined consumption patterns, with different routes of administration such as injected and smoked, have a higher risk of acquiring these infections [22].

Health services are limited in Tijuana, which has contributed to the city having one of the highest rates of HIV in Mexico. The city is also located on a route of drug trafficking, migrants, and deportees [22]. In addition, syringe confiscation and drug arrests are common in Tijuana, increasing HIV prevalence among PWIDs [23].

In 2017, an intervention for the prevention of HIV in PWIDs was implemented in the city, by training police officers in harm reduction and legal aspects of drug policies in Mexico [24]. The study found that police forces had low levels of knowledge about legal and health issues. Accordingly, 37% mistakenly believed that carrying syringes was illegal, even though Mexican criminal law has never penalized the possession of syringes. In addition, fewer than 30% knew that heroin possession had been decriminalized 6 years before the study. The authors of this study pointed out that it is of utmost importance to link health and safety policies. Improving the public health knowledge and attitudes of law enforcement officers will help reduce arrests and confiscation of syringes and encourage referral to HIV care and other health and social programs [23, 24].

Man, 41 years old, Tijuana:

Well, the police have detained me just for not bringing identification; they have never caught me with drugs. They have never even found me with a dose or a syringe. I don't go out on the street with any of that; it's a crime here in Tijuana.

6.2.9.3 HCV

Globally, PWIDs are one of the main groups at risk of HCV transmission [25]. Specifically in Tijuana, this infection is extremely prevalent: nine out of ten users in drug-use sites and seven out of ten in treatment centers have HCV. Similar rates exist in other northern border cities, such as San Luis Río Colorado and Ciudad Juárez [2].

Man, 47 years old, Tijuana, México:

I had hepatitis C a few months ago, but it was not sexually transmitted. The thing is that I did not have syringes that day and I was very desperate.

The saturation of hepatitis C infection among PWIDs on the northern border of Mexico is a high-risk factor for the spread of HIV [26]. It has been shown that more than 90% of HIV-infected PWIDs are coinfected with HCV [27, 28]. Risky drug use practices encourage this blood-borne transmission. Tijuana has the highest number of shooting galleries in the country, which contributes to the high prevalence of these infections in this population [24].

Man, 47 years old, Tijuana, México:

It was a weekend, and here [the NGO] was closed on Sunday, so I went to a store and noticed the trash can and saw a cloth. I grabbed it and there were lots of semi-new syringes; they were not very used, they looked clean. It seemed so easy! I took one of them, chose the newest one, and used it, and then I felt the symptoms. My eyes turned really yellow; I was urinating horribly.

6.2.9.4 Comorbidity Between Opioid Use and Other Mental Disorders

The comorbidity of opioid use with different mental health disorders has been widely documented [29–31]. In particular, the user community analyzed here suffers from various emotional disorders associated with depression, anxiety, and suicidal thoughts. Moreover, many live with trauma due to childhood adversity, the various types of violence they have experienced, and social disadvantages. This situation complicates and often explains their heavy dependence, particularly on heroin, an opioid that has analgesic effects and produces sedation, feelings of peace, and euphoria through the parenteral route. Injecting opioids could, therefore, be a way of anesthetizing the emotional suffering they have accumulated for several years and even generations [32].

Woman, 53 years old, Tijuana, México:

With heroin, nothing hurts me. I forget what happened to me as a child, and only in dreams do I remember that the police showed me the bodies of my children, they were pieces of meat, like ground meat and bones.

It is worth noting that the structural violence in the area, associated with drug use and trafficking, has had a major impact on all the ailments mentioned by the population.

Woman, 53 years old, Tijuana, México:

At the time my children were killed, and because of the violent way they were killed, I did not know how to cope with the pain, and when you are a coward, the easiest thing is drugs, supposedly they calm the pain, and yes, they do it for a moment, but look at me now.

These are often turning points in the lives of PWIDs because it is just when they start to use drugs, and because of the deep sadness they feel, they easily transition to polydrug use and more severe routes of administration.

Woman, 53 years old, Tijuana, México:

It was when I started taking pills [when her children died], the family doctor prescribed them for me, and they immediately had an effect […] I started working, then I had to buy them illegally. Then I started taking crystal and there I got lost […], I started wandering in the streets […], I went crazy. I don't remember when I started taking heroin, […] when I remember, I was shooting up in a shooting gallery and then I did not want to inject myself again, but because of the withdrawal syndromes, I could not stop, and they explained (the users who injected it) that I had to do it and keep on doing it.

6.2.9.5 Other Health Problems (Skin Abscesses)

Injecting drug use also causes severe damage to the skin of PWIDs, such as skin abscesses. These wounds have various causes, such as sharing syringes, lack of cleanliness at the injection site, mixing drugs, unhealthy environmental conditions, and lack of lighting in the places where drugs are used. These abscesses are extremely painful. They begin by erupting, often become larger and deeper, and can be complicated by other diseases unless suitable treatment is provided [33, 34]. These abscesses are extremely common in PWIDs along the northern border, mainly those that combine heroin with crystal. Most of the population receives medical care at NGO Wound Healing Clinics.

Woman, 54 years old, Tijuana, México:

I have a nasty abscess that I didn't take care of at the time; afterwards they wanted to amputate my foot at the General Hospital.

6.3 Opioid Use in Tijuana

The heroin use pattern in PWIDs involves high frequency and amounts. Everyone uses an average of five times a day. The main route of using is through injection. The population mainly consumes black tar heroin, followed by brown powdered heroin and white powdered heroin.

Woman, 40 years old, Tijuana, México:

With my partner, I consume four or five times a day, and at night. If we get the chance, we inject ourselves and go on like that for a few hours […].

The most used drug in this city, besides heroin, is crystal. The combination of these two drugs is a common practice among PWIDs. Until 2018 in Tijuana, nearly 50% of the population preferred to consume only heroin, while 40% mixed heroin with crystal methamphetamine. Although the population realizes that they are exposed to various health risks, combining the two drugs enables them to feel stronger effects. In addition to enhancing their effects, the population mixes these two drugs to function in their daily activities and improve their sexual performance.

Woman, 53 years old, Tijuana, México:
I mix the crystal with heroin in a spoon [...]. In fact, if I just take heroin, I fall asleep, and if I just take crystal, it makes me hysterical; I have to mix them up.

Cocaine is used less by PWIDs because of its high cost. Only a few report that they combine it with heroin, a mixture known as "speedball." PWIDs have generally used a variety of drugs throughout their lives. They have all tried marijuana, which often serves as a gateway drug. Over 80% have used anxiolytics and benzodiazepines combined with heroin to relax more, although according to their perception, it is a risky practice that increases the frequency of overdose.

Man, 45 years old, Tijuana, México:
I have often [overdosed] [...]. Heroin brings you down, and so does clonazepam. Once I injected myself and bent over, there were forty lines [on the syringe], I still didn't even take half of it, I only had about fifteen lines inside, suddenly I started to feel as if cars were passing by, I took out the needle, put it to one side, and then told [my friend] I feel bad, my sight began to blur, I felt that half my body was paralyzed, I was scared, crying, thinking that I was going to go blind, and my head and brain buzzed, and suddenly I lost consciousness, I fell down to the bottom of the river, they grabbed me, and they were making me react, and it took them about two hours to bring me to.

PWIDs have also used other non-prescription opioids, such as tramadol, propoxyphene (Darvon®), methadone, oxycodone, buprenorphine, and hydrocodone. The effects of these substances include mood changes, relaxation, and relief of severe pain. Some users report that they used an opioid for the first time under strict medical prescription after having suffered an accident. However, they developed dependence and were unable to obtain the medications after treatment. Thus, they chose to seek heroin, which is cheaper and available and counteracts withdrawal syndromes.

The heroin consumption market has changed at the border in recent years due to the emergence of fentanyl as a cutting agent, mainly for heroin in white powder and crystal, which has triggered an increase in overdoses throughout the region. Before 2018, the population mainly consumed heroin, and a third combined it with crystal. Most were unfamiliar with fentanyl, and only a few who had lived in the United States had tried it there. However, in 2019, this synthetic opioid began to displace black tar [15]. Since then, PWIDs have been aware that they are using fentanyl, because of which they say that overdoses have increased. It is important to note that use of this opioid has spread to other population groups who are not homeless and use drugs such as cocaine.

Woman, 40 years old, Tijuana, México:

Fentanyl has a similar effect to heroin, but more powerful. You take heroin, so you don't get sick and fentanyl, so you feel pleasure. It allows me to be functional. It makes me feel like a human being.

6.4 Persons Who Misuse Opioids in Other World Regions

6.4.1 Societal Isolation and Health Risk Behavior

Although affectations of health and life occur in any consumption context, this chapter focuses on the misuse of opioids in a highly marginalized context. The opioid crises have impacted every socioeconomic context, and there are a considerable number of persons who misuse opioids in privileged settings or at least not in contexts of homelessness and poverty. It is worth noticing that the misuse of opioids has a trajectory and that the context of consumption may be affected by consumption practices themselves. The cumulative negative consequences may ameliorate the buffering effect related to economic advantages. As many middle-class Americans' testimonies have already shown us, isolation happens regardless of social origin [35].

Social isolation, proceeded by the stigma and lack of social support, is one of the major obstacles to access resources that would make these people's lives easier, such as health care, psychological support, medical assistance, public recognition of their rights, and political recognition. In other words, social isolation prevents the full recognition of these people's citizenship. In those circumstances, one may find that the social dimension of the person is crucially affected, particularly the social relationships between the PWIDs and people who do not consume. In addition, social repulsion gets expressed when PWIDs do not become a priority for the community:

Female, 57 years old, Rural West Virginia, USA:

But a lot of them don't care if you die. I heard one say, 'You know, I'm fooling around with you, dope users, when I could be out there, helping somebody that needs help, having a hard time.' And I thought that was terrible. We're all people, you know… And they look at you like you're trash. Like you have a motive in mind for everything. And you don't. I don't. I don't want to be out here, I'm just trying to better myself [36].

As discussed previously, Tijuana consumption tends to be a group activity of up to 30 people. Couples and small groups consume together to avoid overdose risks. However, this is not a global reality. Many PWIDs around the world consume on their own. Many harm reduction programs include the explicit recommendation to "don't consume alone" [37, 38]. Group consumption favors an adequate response in the case of overdose by the intervention of their peers (sometimes by hitting them, using ice, or injecting them a salt solution), but it generates a civil responsibility in case of fatal overdose. Many people find that scenario problematic enough to not engage in group consumption.

For some users, there are good reasons to stay on their own and have lonely consumption. Some of these reasons are economic:

Male, 25 years old, Baltimore, USA:

I'm not going to be able to do it – $25 is not stretching with two people. I'm not giving nobody nothing. Like my stomach is rumbling. My body hurts from sleeping outside. Today is not the day to be sharing with somebody. And I don't have to share with nobody, because I know where to get it myself. I don't just use with people. I'm not going to be in a shooting gallery. I'm not going to be in a crack house. That's not what I do. Normally I get high in a house, but lately I've been getting high outside, and there is no need for me to be getting high with anybody else [39].

6.4.2 Police Brutality, Economic Extortion, and Sexual Violence

For PWIDs all around the world, police brutality seems to be a common experience. The marginality of this population makes them vulnerable to many types of violence. Similarly, just as in Tijuana, police violence, harassment, and brutality have been documented in Nigeria [40], Ukraine [41, 42], Bangladesh [43], Russia [44], Baltimore [45], Puerto Rico [46], and the Kyrgyz Republic [47].

The similarities between all stories about the police stance toward PWIDs are striking. The stigma attached to PWIDs by the police force is a major cause of distress and the basis for a diversity of mistreatments.

Male, 33 years old, Uyo, Nigeria:

The police see people who take illegal drugs as very bad people. They see them as not deserving to work on the streets with others... If they find out that you inject drugs, then you are treated as the worst person on earth. You are seen as a very bad person because you inject [40].

The reported bad attitude is not only dispositional. The police often exercise certain forms of active psychological and emotional affectation toward PWIDs due to harassment and arbitrariness.

Male, Bishkek, Kyrgyz Republic:

I don't want to carry around used needles with me. Police will find them, then will start questioning you. Maybe they don't really need anything from you, but they will just 'spoil your blood' [make your life difficult]. So much stress! [47].

PWIDs report being insulted and mistreated by police officials, and expressions of violence accompany their encounters with them. A prominent form is a physical intimidation.

Female, 29 years old, Uyo, Nigeria:

You know how the cops (police) behave. They don't know how to be peaceful. Everything is force, force. They carry gun and harass anybody they think has committed crime. Nigerian police do not know how to do things in a calm way [40].

PWID, unknown age, North Caucasus, Russia:

I don't know of one single drug user who has not been beaten up by police [44].

Verbal mistreatment and physical abuse usually escalate to symbolic and material exploitation. Moreover, in some places, PWIDs are also victims of economic extortion:

PWID, unknown age, North Caucasus, Russia:

A drug user is a source of income for the police. I had to provide money to avoid being put in jail. If I hadn't paid, I would have been put in jail. Even though those weren't my drugs that were on me, they simply put them in my pocket, just because I was a drug user. They have a quota to fill [44].

Female, 28 years old, Uyo, Nigeria:

Police is part of the problem we face as drug users. Let me put it that way. Why I say so? Is because when police catch you and take all your money, you have nothing to take care of yourself. Plus the way you feel about how they treated you, so you will say let me get high so I can feel better. Or let me use much drugs to forget these problems. You will continue to use drugs, instead of trying to stop [40].

Sometimes, such violence involves sexual violence, which affects women who inject drugs disproportionally compared to men.

Male, unknown age, St. Petersburg, Russia:

There's no question about police sexual violence toward drug users, that's routine. If they catch a female drug user, she will 100% service them, and then they will return to business [44].

6.4.3 Opportunities, Community, and Solidarity

There is deep uncertainty in the life of PWIDs. Overdose is an everyday mortal risk, a direct menace to existence. If PWIDs choose abstinence, the pains that come from withdrawal syndrome are sometimes unimaginable. Additionally, abandoning self-identification as an "addict" may be a source of suffering in contexts that lack the elements to give meaning to life and self [48]. Diverse forms of suffering related to opioid misuse in the physical, emotional, social, and existential spheres may only be alleviated through social intervention in situations framing such consumptions.

Many international experiences of overdose education and naloxone programs involve a multi-aspect intervention frequently found in the harm reduction programs (see Chap. 7) [49]. These programs educate on safer consumption practices and health prevention; they provide some naloxone kits (the overdose antidote, an opioid antagonist that does not have any other property) and often involve peers of the same community instructing how to act in the case of an overdose. These programs also try to involve public service providers such as firemen, police, and health-care workers. The challenges are profound. For instance, people who work in the law forces may be reticent to collaborating in harm reduction strategies.

Policeman participant, unknown age, Ottawa, Canada:

We are seen as a voice for residents. They do look to us for leadership on this issue. That's why it's very important that we're part of the table because, you know, the discussion we're having now is the discussion we have every day when we're out on the road, talking to the businesses, talking to communities, talking to the addicts, talking to the shelters. We do, in a way, represent a large segment of the population, and a large segment that are completely opposed to this [50].

Nevertheless, we strongly emphasize that reorienting the public services attitude toward this population benefits all the parties: PWIDs, the community, and service providers. For example, many international initiatives that started by offering overdose education and organizing naloxone administration programs [51, 52] have now built supervised consumption sites [53–55].

Residents and the community who shares public spaces with PWIDs perceive some advances:

Resident, unknown age, Toronto, Canada:

I think anything that's making it safer for the people doing drugs, and safer for the community at large is good. And so I'd support it [supervised consumption facilities]. It strikes me that, like when they started the needle exchange programme, it didn't take very long to not see spent needles everywhere, in parks. And it was so glorious! [56].

PWIDs, users of these public services, recognize that these places open opportunities for improving the relationship between the local police and them.

Service users, unknown ages, Toronto, Canada:

Participant 1: "Cops are finally starting to wake up a little bit, I think. They understand harm reduction, some of them. I met a lot of cops now that actually understand fucking harm reduction. It's mind-blowing, but they're starting to get it. They realize, you know, their kids, whoever, everybody knows somebody that does dope."

Participant 2: "And people are dying all over the friggin' place" [57].

6.5 Final Considerations

Persons who misuse opioids in marginalized context are, among others, organisms interacting with some substance. By doing so, they come into contact with a plethora of elements: the satisfaction of physical craving, the avoidance of the abstinence syndrome, the feeling rush, a relief for the abstinence syndrome, but also an opportunity to have a group of peers, a social identity, and a sense of personal meaning. All these elements are sources of value for them [48]. Moreover, these human peers found value in their consumption regardless of being in the worst possible living conditions.

Persons are complex realities. People have lives. Throughout their lives, they go through diverse situations, behave in many different ways, and suffer a range of physical and mental disorders. It is important to notice that the concept of "addict" is stigmatizing for several reasons. First, it labels a person by her illness; second, by labeling her as such, it denies her humanity; and third, it essentializes the addictive disorder, inhibiting change and health improvement. Therefore, many scientific, medical, and professional associations do not recommend referring to people as "addicts."

Frequently, PWIDs are judged from a moral perspective that treats them as morally responsible for their suffering because of a weakness in their character. Confronting this situation, the brain disease model of addiction tells us that these people suffer from a neurobiological compulsive disorder [58]. However, we are now in a position that allows a more complex picture. A person in addiction is in a complex relation to the substance and may find a positive extrinsic value on it, i.e., something important for the user that does not depend on the substance itself but in an element related to them. We should be aware that these people face situations in which opioid misuse has a source of value according to them, even in the face of its negative, painful outcomes [48].

One of the challenges we face as a society for dignifying PWIDs' lives is building environments, social structures, and institutions that produce valuable elements for life construction even more valuable than the value found in the substance consumption by PWIDs. Some of the features linked with PWIDs in marginalized contexts worldwide include child trauma, social injustice, police mistreatment, stigma, poverty, sexual abuse, and incarceration. By recognizing the social dimension of the problem of this population, we are confronted with a moral demand to change the circumstances and the contexts of those for whom opioid misuse is a better life. Every human being deserves an environment of nourishment in which her or his potentialities can flourish with complete recognition of their rights.

Supporting civil initiatives that help PWIDs build healthier habits, avoid risks, and receive social recognition can have large-scale effects. First, the community hosts a space that has been previously denied to them. Second, PWIDs share spaces for their personal development and find benefits when participating in them. Finally, service providers may find an opportunity to play a crucial role in helping communities.

References

1. Cassell E. The person in medicine. Int J Integr Care. 2010;10(5):50–2. https://doi.org/10.5334/ijic.489.
2. Fleiz C, Domínguez M, Villatoro JA, Vázquez F, Zafra E, Sánchez R, et al. Cuqueando la Chiva: Contextos del consumo de heroína en la frontera norte de México. Ciudad de México, México: INPRFM; 2019. ISBN: 978-607-98474-0-1.
3. Buchman D, Orkin A, Strike C, Upshur R. Overdose education and naloxone distribution programs and the ethics of task shifting. Public Health Ethics. 2018;11(2):151–64. Available from: https://doi.org/10.1093/phe/phy001.
4. Charon R. The ethicality of narrative medicine. Narrative research in health and illness. In: Hurwitz B, Greenhalgh T, Skultans V, editors. London: Blackwell BMJ Books. p. 24–35.
5. Comité de Planeación para el Desarrollo del Estado, Gobierno de Baja California. Proyecciones de Población Municipios de Tijuana, Tecate y Playas de Rosarito 2020–2030, México. [Internet]. México; 2019. Available from: http://www.copladebc.gob.mx/publicaciones.html.

6. Neale J, Nettleton S, Pickering L, Fischer J. Eating patterns among heroin users: a qualitative study with implications for nutritional interventions. Addiction. 2011;107(3):635–41. https://doi.org/10.1111/j.1360-0443.2011.03660.x.
7. Degenhardt L, Dierkerb L, Chiu WT, Medina-Mora ME, Neumark Y, Sampson N, et al. Evaluating the drug use "gateway" theory using cross-national data: consistency and associations of the order of initiation of drug use among participants in the WHO World Mental Health Surveys. Drug Alcohol Depend. 2010;108:84–97. https://doi.org/10.1016/j.drugalcdep.2009.12.001.
8. Darke S, Torok M, Ross J. Developmental trajectories to heroin dependence: theoretical and clinical issues. J Appl Soc Psychol. 2017;47(3):165–71. https://doi.org/10.1111/jasp.12434.
9. Kessler RC, Wang PS. The descriptive epidemiology of commonly occurring mental disorders in the United States. Annu Rev Public Health. 2008;29:15–129. https://doi.org/10.1146/annurev.publhealth.29.020907.090847.
10. Slade T, Johnston A, Browne O, Andrews G, Whiteford H. 2007 National survey of mental health and wellbeing: methods and key findings. Aust N Z J Psychiatry. 2009;43(7):594–605. https://doi.org/10.1080/00048670902970882.
11. Darke S, Ross J. The relationship between suicide and heroin overdose among methadone maintenance patients in Sydney, Australia. Addiction. 2001;96(10):1443–53. https://doi.org/10.1046/j.1360-0443.2001.961014438.x.
12. Roy A. Risk factors for attempting suicide in heroin addicts. Suicide Life Threat Behav. 2010;40(4):416–20. https://guilfordjournals.com/doi/pdf/10.1521/suli.2010.40.4.416.
13. Comisión de los Derechos Humanos del Estado de Chihuahua, Programa Compañeros A.C., Centros de Integración Juvenil, Comisión de Salud Fronteriza. Cartilla de derechos humanos para personas usuarias de drogas [Internet]. México: FECHAC; 2020. Available from: https://yociudadano.com.mx/noticias/presentan-cartilla-de-derechos-humanos-de-las-personas-usuarias-de-drogas/.
14. Cruz SL, Carranza-Aguilar CJ, Pérez-García IP. Sodium chloride injection to treat opioid overdose; does it work? A preclinical study. Neurotoxicology. 2021;87:24–9. https://doi.org/10.1016/j.neuro.2021.08.012.
15. Fleiz C, Arredondo J, Chavez A, Pacheco L, Segovia LA, Villatoro JA, et al. Fentanyl is used in Mexico's northern border: current challenges for drug health policies. Addiction. 2019;115(4):778–81. https://doi.org/10.1111/add.14934.
16. Ley General de Salud. Diario Oficial de la Federación, Última reforma publicada DOF 01-06-2021. 2006 (7 Feb 1984).
17. Centro Nacional para la Prevención y Control del VIH y el Sida. Boletín Día Mundial del Sida 2020 [Internet]. Mexico; 2020 Dic. Available from: https://www.gob.mx/censida/documentos/boletin-dia-mundial-del-sida-2020.
18. Jain JP, Abramovitz D, Strathdee SA, Gonzalez-Zuniga P, Rangel G, West BS, et al. Sex work as a mediator between female gender and incident HIV infection among people who inject drugs in Tijuana, Mexico. AIDS Behav. 2020;24(9):2720–31. https://doi.org/10.1007/s10461-020-02828-y.
19. Strathdee SA, Lozada R, Ojeda VD, Pollini RA, Brouwer KC, Vera A, et al. Differential effects of migration and deportation on HIV infection among male and female injection drug users in Tijuana, Mexico. PLoS One. 2008;3(7):e2690. https://doi.org/10.1371/journal.pone.0002690.
20. Mathers BM, Degenhardt L, Phillips B, Wiessing L, Hickman M, Strathdee SA, et al. Global epidemiology of injecting drug use and HIV among people who inject drugs: a systematic review. Lancet. 2008;372(9651):1733–45. https://doi.org/10.1016/S0140-6736(08)61311-2.
21. Des Jarlais DC, Feelemyer JP, Modi SN, Arasteh K, Hagan H. Are females who inject drugs at higher risk for HIV infection tan males who inject drugs: an international systematic review of high seroprevalence areas. Drug Alcohol Depend. 2012;124(1–2):95–107. https://doi.org/10.1016/j.drugalcdep.2011.12.020.
22. Bucardo J, Brouwer KC, Magis-Rodríguez C, Ramos R, Fraga M, Perez SG, et al. Historical trends in the production and consumption of illicit drugs in Mexico: implications for the pre-

vention of blood borne infections. Drug Alcohol Depend. 2005;79(3):281–93. https://doi.org/10.1016/j.drugalcdep.2005.02.003.
23. Baker P, Arredondo J, Borquez A, Clairgue E, Mittal ML, Morales M, et al. Municipal police support for harm reduction services in officer-led referrals of people who inject drugs in Tijuana, Mexico. Harm Reduct J. 2021;18(1):1–10. https://doi.org/10.1186/s12954-021-00513-4.
24. Strathdee SA, Arredondo J, Rocha T, Abramovitz D, Rolon ML, Patiño E, et al. A police education programme to integrate occupational safety and HIV prevention: protocol for a modified stepped-wedge study design with parallel prospective cohorts to assess behavioural outcomes. BMJ Open. 2015;5(8):e008958. https://doi.org/10.1136/bmjopen-2015-008958.
25. Marquez LK, Fleiz C, Burgos JL, Cepeda JA, McIntosh C, Garfein RS, et al. Cost-effectiveness of hepatitis C virus (HCV) elimination strategies among people who inject drugs (PWID) in Tijuana, Mexico. Addiction. 2021;116(10):2734–45. https://doi.org/10.1111/add.15456.
26. White EF, Garfein RS, Brouwer KC, Lozada R, Ramos R, Firestone-Cruz M, et al. Prevalence of hepatitis C virus and HIV infection among injection drug users in two Mexican cities bordering the US. Salud Publica Mex. 2007;49(3):165–72. https://www.medigraphic.com/cgi-bin/new/resumenI.cgi?IDARTICULO=16689.
27. Maier I, Wu GY. Hepatitis C and HIV co-infection: a review. World J Gastroenterol. 2002;8(4):577–9. https://doi.org/10.3748/wjg.v8.i4.577.
28. Miller CL, Wood E, Spittal PM, Li K, Frankish JC, Braitstein P, et al. The future face of coinfection: prevalence and incidence of HIV and hepatitis C virus coinfection among young injection drug users. JAIDS. 2004;36(2):743–9. https://journals.lww.com/jaids/Fulltext/2004/06010/Factors_associated_with_persistent_high_risk.00012.aspx.
29. Wang L, Panagiotoglou D, Min JE, DeBeck K, Milloy MJ, Kerr T, et al. Inability to access health and social services associated with mental health among people who inject drugs in a Canadian setting. Drug Alcohol Depend. 2016;168:22–9. https://doi.org/10.1016/j.drugalcdep.2016.08.631.
30. Crawford V, Crome IB, Clancy C. Co-existing problems of mental health and substance misuse (dual diagnosis): a literature review. Drugs. 2003;10(1):1–74. https://doi.org/10.1080/0968763031000072990.
31. Adams M, Sionean C, Broz D, Lewis R, Wejnert C. Serious mental illness among young people who inject drugs: an assessment of injection risks and healthcare use. J Infect Dis. 2020;222(Suppl 5):S401–9. https://doi.org/10.1093/infdis/jiaa238.
32. Maté G. Addiction: childhood trauma, stress and the biology of addiction. J Restor Med. 2012;1(1):56–63. https://doi.org/10.14200/jrm.2012.1.1005.
33. Murphy EL, DeVita D, Liu H, Vittinghoff E, Leung P, Ciccarone DH, Edlin BR. Risk factors for skin and soft-tissue abscesses among injection drug users: a case-control study. Clin Infect Dis. 2001;33(1):35–40. https://doi.org/10.1086/320879.
34. Calderón L, Arroyo ME, Tirado A, Bonifaz A, Ponce RM. *Múltiples abscesos piógenos en usuario de cocaína* [Multiple pyogenic abscesses in a cocaine user]. Dermatol Rev Mex. 2014;58:178–83.
35. The Opioid Diaries. TIME. [Internet] [cited 2021 Sep 3]. Available from: https://time.com/james-nachtwey-opioid-addiction-america/.
36. Allen ST, Grieb SM, O'Rourke A, Yoder R, Planchet E, Hamilton White R, Sherman S. Understanding the public health consequences of suspending rural syringe service program: a qualitative study of the experiences of people who inject drugs. Harm Reduc J. 2019;16(33):1–10. https://doi.org/10.1186/s12954-019-0305-7.
37. Bardwell G, Kerr T, McNeil R. The opioid overdose epidemic and the urgent need for effective public health interventions that address men who use drugs alone. AJMH. 2019;13(3):1557988319859113. Available from: https://doi.org/10.1177/1557988319859113.
38. Clark AK, Wilder CM, Winstanley EL. A systematic review of community opioid overdose prevention and naloxone distribution programs. J Addict Med. 2014;8(3):153–63. https://doi.org/10.1097/ADM.0000000000000034.

39. Winikier A, Tobin K, Gicquelais R, Owczarzak J, Latkin C. "When you're getting high... you just don't want to be around anybody" A qualitative exploration of reasons for injecting alone: perspectives from young people who inject drugs. Subst Use Abuse. 2020;55(13):1–8. Available from: https://doi.org/10.1080/10826084.2020.1790008.
40. Nelson E, Brown AS. Extra-legal policy strategies and HIV risk environment: accounts of people who inject drugs in Nigeria. Drugs. 2020;27(4):312–9.
41. Mazhnaya A, Bojko M, Marcus R, Filippovych S, Islam Z, Dvoriak S, Alice F. In their own voices: breaking the vicious cycle of addiction, treatment and criminal justice among people who inject drugs in Ukraine. Drugs. 2016;23(2):163–75. Available from: https://doi.org/10.3109/09687637.2015.1127327.
42. Kutsa O, Marcus R, Bojko M, Zelenev A, Mazhnaya A, Dvoriak S, Fikippovych S, Altice F. Factors associated with physical and sexual violence by police among people who inject drugs in Ukraine: implication for retention on opioid agonist therapy. J Int AIDS Soc. 2016;19(Supl 3):20897. Available from: https://doi.org/10.7448/IAS.19.4.20897.
43. Kahn SI, Irfan SD, Khan M. "I held on the police's leg for mercy": experiences of adversity, risk and harm among people who inject drugs during an anti-drug drive in Dhaka, Bangladesh. Int J Drug Policy. 2021;92:103299. Available from: https://doi.org/10.1016/j.drugpo.2021.103299.
44. Lunze K, Raj A, Cheng D, Quinn E, Lunze F, Liebschutz JM, Bridden C, Walley A, Blokhina E, Krupitsy E, Samet J. Sexual violence from police and HIV risk behaviors among HIV-positive women who inject drugs in St. Petersburg, Russia – a mixed methods study. J Int AIDS Soc. 2016;19(Supl 3):20877. Available from: https://doi.org/10.7448/IAS.19.4.20877.
45. Park JN, Linton SL, Sherman SG, German D. Police violence among people who inject drugs in Baltimore, Maryland. Int J Drug Policy. 2019;64:54–61.
46. Abadie R, Gelpi-Acosta C, Davila C, Rivera A, Welch-Lazoritz M, Dombrowsky K. Int J Drug Policy. 2017;51:121–7. https://doi.org/10.1016/j.drugpo.2017.06.011.
47. Deryabina A, El-Sadr WM. Uptake of needle and syringe program service in Kyrgyz Republic: key barriers and facilitators. Drug Alcohol Depend. 2017;179:180–6. Available from: https://doi.org/10.1016/j.drugalcdep.2017.07.002.
48. Pickard H. What we're not talking about when we talk about addiction. Hastings Cent Rep. 2020;20(4):37–46. Available from: https://doi.org/10.1002/hast.1172.
49. Hawk M, Coulter RWS, Egan JE, Fisk S, Friedman MR, Tula M, Kisnky S. Harm reduction principles for healthcare settings. Harm Reduct J. 2017;14:70. https://doi.org/10.1186/s12954-017-0196-4.
50. Watson TM, Barnaby L, Bayoumi A, Challacombe L, Wright A, Strike C. 'This is health service. Leave it alone': service user and staff views on policing boundaries involving supervised consumption services. Addict Res Theory. 2021;29(1):1–9. Available from: https://doi.org/10.1080/16066359.2020.1730821.
51. Devries J, Raffie S, Polston G. Implementing an overdose education and naloxone program in a health system. J Am Pharm Assoc. 2017;57(2):S154–60.
52. Giglio RE, Li G, DiMaggio CJ. Effectiveness of bystander naloxone administration and overdose education programs: a meta-analysis. Inj Epidemiol. 2015;2(10):1–9. https://doi.org/10.1186/s40621-015-0041-8.
53. Ng J, Sutherland C, Kolber M. Can Fam Physician. 2017;63(11):866. Available from: https://doi.org/10.1016/j.japh.2017.01.002.
54. Kennedy MC, Kerr T. Overdose prevention in United States: a call for supervised injection sites. Am J Public Health. 2017;107(1):42–3. Available from: https://doi.org/10.2105/AJPH.2016.303523.
55. Gagnon ML. It's time to allow assisted injection in supervised injection sites. CMAJ. 2017;189(34):E1083–4. Available from: https://doi.org/10.1503/cmaj.170659.
56. Kolla G, Strike C, Watson TM, Jairam J, Fischer B, Bayoumi A. Risk creating and risk reducing. Community perceptions of supervised consumption facilities for illicit drug use. Health Risk Soc. 2017;19(1–2):91–111.
57. Watson TM, Bayoumi A, Kolla G, Penn R, Fischer B, Luce J, Strike C. Police perceptions of supervised consumption sites (SCSs): a qualitative study. Subst Use Misuse. 2012;47(4):364–74. https://doi.org/10.3109/10826084.2011.645104.

Chapter 7
The Need for Structural Interventions for Persons Who Misuse Opioids

Claudia Rafful and Carlos Magis-Rodríguez

Abstract This chapter analyzes structural approaches that emerged as a community and later as evidence-based and policy strategies to improve the health and well-being of persons who use drugs (PWUDs) in general and opioids (PWUOs) in particular.

The first section introduces social determinants of health, structural violence, and structural vulnerability concepts that have served as frameworks for social epidemiologists, medical anthropologists, sociologists, and behavioral researchers to understand and intervene in contexts that harm PWUDs. The second section reviews structural interventions that positively impact PWUDs, including involvement of peers to address substance use and infectious disease risk behaviors, housing and economic assistance programs, medication-assisted treatments, and syringe exchange programs. This chapter also includes unintended consequences, concerns, and considerations when implementing each intervention in different settings.

Keywords Structural violence · Social determinants of health · Persons who use opioids

C. Rafful (✉)
Faculty of Psychology, National Autonomous University of Mexico (UNAM), Mexico City, Mexico

Opioids Working Group, Global Studies Seminar, Faculty of Medicine, UNAM, Mexico City, Mexico

Center for Global Mental Health Research, National Institute of Psychiatry Ramón de la Fuente, Mexico City, Mexico
e-mail: crafful@comunidad.unam.mx

C. Magis-Rodríguez
Opioids Working Group, Global Studies Seminar, Faculty of Medicine, UNAM, Mexico City, Mexico

Faculty of Medicine, UNAM, Mexico City, Mexico

S. L. Cruz (ed.), *Opioids*, https://doi.org/10.1007/978-3-031-09936-6_7

7.1 Overview

Conceptualization of opioid and other substance use disorders (SUD) as chronic brain diseases [1] instead of a moral fault has helped reduce stigma and promote a less judgmental approach to people seeking treatment. While valuable and evidence-based, this biomedical paradigm can be reductionist. Behavioral interventions, mainly cognitive-behavioral therapy approaches, have helped motivate and sustain behavioral changes essential to decrease and cease drug-seeking behaviors.

Priority setting in science – through funding allocation – has allowed significant advances in psychopharmacology and neuroscience, but the importance of the social context can sometimes be left behind. For instance, clinicians and neuroscientists have developed opioid-assisted treatments (OAT) using opiate agonists (methadone or buprenorphine), opioid antagonists (naloxone or naltrexone), and long-delivery opioids that, in theory, would disincentivize persons to use opioids in different quantities or through routes of administration other than those prescribed initially. However, there are social determinants that hinder access to these treatments.

In the past 40 years, two epidemics have shaken the concept of health, health promotion, and interventions. First, the HIV epidemic rapidly evidenced health disparities across countries and socioeconomic status, adding layers of complexity and intersectionality of social risk factors that decrease the odds of health and well-being for those living in poverty, gender minorities, racialized communities, and persons who use drugs (PWUDs). Second, the current epidemic of opioid-related deaths started with the unethical prescription of opioids for chronic pain and later shifted to street opioids, including fentanyl. The inadequate and late response of health systems to the opioid epidemic also exposed health disparities that non-governmental organizations (NGOs) and affected communities have effectively addressed but that need to be supported by structural interventions to have long-lasting effects. The evidence of the structural determinants of the opioid crisis is now so overwhelming that Dr. Nora Volkow, the director of the National Institute on Drug Abuse (NIDA), has recently acknowledged the crucial role of social pain in the opioid crisis [2]. In her commentary, Dr. Volkow expressed the importance of isolation, despair, economic inequities, social exclusion, rejection, and stigma (more severely suffered by racialized populations) as risk factors for opioid use disorders.

Another institutional shift toward a more comprehensive approach to SUD is the recently approved funding for the most significant implementation science in the history of SUD in the United States. This approach includes structural factors such as changes in opioid prescription practices, increased availability of drug treatment programs, naloxone to counteract opioid overdoses, and improved linkage to care [3]. Although these structural components may help reduce opioid-related deaths, they are still insufficient to truly change the structures that account for the despair context in which the opioid overdose has taken place.

There was a significant shift in the public acknowledgment of the crucial role of structural interventions for opioid use in the past decade. In particular, health and public institutions that historically supported abstinence-only treatments changed to a more "real-life" paradigm, mainly harm reduction [4] and medication-assisted programs.

7.2 Conceptual Frameworks

Theoretical and conceptual frameworks help understand the conditions in which people have misused opioids. With such understanding, it is possible to improve the quality of life – and not only promote drug abstinence – of persons who use opioids.

From a structural perspective, health is a product of social structures and processes, which may be affected by political, legal, and cultural contexts. As such, health outcomes depend less on individual behaviors and more on social processes and structural vulnerabilities [5]. The benefit of using structural frameworks in health research is that adequate interventions and policy reforms can modify structural factors that aim to reduce vulnerability and create healthier environments [6].

The intersection of social disadvantage, isolation, and pain cannot be addressed only through primary care [7] or pure biomedical approaches. Poverty and substance use problems act in synergy, reinforcing mental health problems and unstable housing and employment [7]. Structural variables can predict area-level vulnerability to opioid misuse, overdose, and the syndemic (synergistic epidemic) of opioid use and infectious diseases such as human immunodeficiency virus (HIV) and hepatitis C virus (HCV) [6]. In the following paragraphs, we explain key theoretical concepts.

7.2.1 Social Determinants of Health

Social determinants of health are structural forces that affect health outcomes from personal to global levels, including socioeconomic status, inequality, institutional policies, global trade agreements, and political forces [2, 8]. Structural determinants of health are architectural, economic, and political frameworks that create barriers to improving health or perpetuating social inequities. The structural determinant framework helps explain the opioid crisis and may also guide the policy and structural changes needed to improve community health [7].

Social and structural determinants of opioid misuse include stigma, racism, discrimination, heterosexualism, family structure, socioeconomic status, community engagement, and social support systems [9]. In addition, some opioid users are aging population with pain and disabled, under economic distress, that lack social cohesion and can have comorbid psychiatric disorders [7]. For these reasons, to improve public health approaches to the opioid epidemic, it is necessary to address the social determinants of health with person-centered approaches, implementation science, and improving care systems [10]. Additionally, structural changes within the policy agenda must include universal healthcare, education, and social services.

7.2.2 *Structural Violence*

Structural violence is the social arrangement that may harm persons and populations and is embedded in the global socioeconomic organization [11]. Specifically, the current opioid crisis in the United States and Canada was detonated by overprescription within the context of economic and social distress, especially among persons suffering physical and psychological trauma, inequality, isolation, and hopelessness [7]. Opioid fatal overdoses have also been considered deaths of despair [12, 13]. This concept refers to poverty, income inequality, unemployment related to deteriorating labor markets, reduced social capital, and high social isolation as the root causes for some deaths from opioid overdose [9]. In deprived communities, manufacturing and service jobs dominated the job market. Most of these works entail physical hazards and potential injuries that may lead to chronic pain conditions resulting in disability, poverty, and a perfect set for the quest for prescription and self-medication [7, 14].

7.2.3 *Structural Vulnerability*

Structural vulnerability refers to suffering-constrained individuals based on their social position within the hierarchical social structures [11]. It encompasses society's multiple overlapping and mutually enforcing power hierarchies, including institutional and policy-level status that may limit a person's ability to access healthcare and engage in healthy lifestyles [8]. As social determinants of health, structural vulnerabilities also highlight how the individuals' agency is constrained within socioeconomic and political processes such as income, housing, discrimination, and experiences of colonization [8, 15]. Quantitative assessment of structural vulnerabilities includes financial security, residence, risk environment, food access, social network, legal status, education, and discrimination [8].

This concept has helped understand the intersectionality of overlapping and inter-related vulnerabilities that put specific PWUDs at a higher risk of police brutality in more comprehensive social and economic hierarchies [16]. Other findings are the interactions between poor housing, neighborhood conditions, and scarce treatment alternatives, which may disrupt behaviors leading to opioid use disorders [17, 18]. Structural vulnerabilities related to barriers to prevent overdosing include lack of overdose prevention sites, potential eviction, and criminalization of drug use [15]. A significant barrier is the fragmentation of care, lack of behavioral health services within primary care settings, and lack of wraparound services for people who use opioids [18].

7.3 Structural Interventions

Structural interventions involve policy and law reforms, changes in administrative procedures, advocacy, and community organization, among others [19]. Public health interventions should ensure full participation of PWUDs in overdose prevention programs, developing and implementing all the structural changes to respond to the overdose crisis worldwide [15]. Implementing structural interventions at local, municipal, regional, and national levels impacts individual behaviors [19].

7.3.1 Peer Involvement

Involving peers and persons with lived experience to address substance use and infectious disease risk behaviors is extensively recommended, especially in under-resourced (including human resources) settings [20]. Efforts that do not include persons with lived experience in the design and implementation usually encounter difficulties that "real-life" experts can prevent. These experts play a fundamental role in overdose prevention, which tends to be overshadowed by public health partners [21], harm reduction activists, and academic partners.

A systematic review of low- and middle-income countries found substance use interventions that included peers in Ukraine, Russia, Vietnam, Thailand, Senegal, China, Malaysia, Georgia, South Africa, Iran, Kenya, India, Puerto Rico, and Zimbabwe [20]. However, most of the data of peers in substance use and, specifically, in opioid use interventions are from high-income countries: mainly Canada, Europe, and the United States.

The Vancouver Area Network of Drug Users (VANDU) and Downtown Eastside SRO Collaborative in British Columbia [22] is an example that deserves proper consideration. VANDU began in the 1990s as a community response to the opioid and HIV syndemic. Since its establishment, VANDU has led interventions and worked with other local and international activists, academia, and public officers. In addition, VANDU has played a fundamental role in all the structural interventions in Vancouver East Side (syringe exchange programs, safe consumption sites – in all the available versions – and civic actions) [21]. VANDU has also been vocal and is one of the main stakeholders and presenters in improving the health and well-being of PWUDs.

While the participation of persons with lived experience is of great benefit, there are some criticisms regarding the burden that implies the task-shifting in healthcare. The balance between community participation and task-shifting is still a challenge. In the current opioid crisis, most of the response has been provided by peers that have intensely worked in naloxone distribution and overdose prevention training of other PWUDs. Some unintended consequences for PWUDs can be fear of arrest and physical and mental health comorbidity [21]. Persons with lived experience that work in harm reduction services still belong to one of the most marginalized groups in any setting. There is a risk of perpetuating oppression by considering its role as volunteering, low-paid jobs, and no acknowledgment when working with persons with a higher educational level and more overall social capital.

7.3.2 Housing Programs and Income Assistance

Providing access to stable and dignified housing and income assistance are critical structural interventions. Housing interventions reduce overdose deaths [23] and are especially important for PWUDs and those who have comorbid mental health conditions, marginalized youth, women, and people recently released from prison [24–26]. Housing is such a complex problem that it goes beyond ensuring stable housing and considering the needs of the tenants. There must be a balance between basic public regulations and realistic regulations that may benefit persons. That is, regulations usually prohibit drug use on premises. However, in addition to being unrealistic, compliance with these regulations may put persons at greater risk if they use drugs in public venues or other unfamiliar locations. As a basic need, housing should be stable before drug treatment or any other intervention. It is not possible to require persons' drug abstinence when they lack stable housing.

Housing programs often include PWUDs as tenants, who are usually threatened with eviction if found using drugs on the premises. Eviction [27] and even changes in unstable housing arrangements [2] correlate with increased risk of HIV, overdose, spatial patterns of drug use changes, shifts in substance use, financial hardship, and changes in drug supply.

Structural interventions that ensure income security and employment may also reduce overdose deaths [23], especially among people with mental health disorders and recently imprisoned [24]. Cash transfer programs have been implemented among vulnerable populations, including PWUDs. However, a randomized clinical trial has found that payment days are associated with increased substance use and related harms, including overdose [28]. Innovative interventions have tested desynchronized monthly and biweekly payments, and, although they showed decreased odds of increased drug use, they also found an increase in exposure to violence [29].

7.3.3 Access to Essential Services

Many healthcare systems have inadequately responded to the international opioid crisis, with a slow response to excessive opioid prescription and pharmaceutical marketing of opioids, and a lack of timely treatment response. As such, healthcare systems have acted as structural determinants of the opioid crisis [18].

In general, healthcare services have been unwilling to identify appropriate intervention points and care delivery strategies [18]. The response to the prescription opioid crisis was inadequate because of punitive and controlling measures for patients and providers. Patients can be suspected of pretending pain to get opioids, and providers are burdened with institutional and insurance paperwork and constant prescription monitoring. This situation has also affected racialized populations that have restricted access to pain medication and promotes mistrust between patients and providers [7].

7.3.4 Syringe Exchange Programs (SEP)

SEP, also known as needle/syringe exchange programs, provide users with sterile syringes and injection equipment to reduce transmission of blood-borne diseases. These programs were implemented in the 1980s as a community intervention to prevent HIV and HCV transmission [30–32].

To date, SEP is one of the structural interventions most used by people who inject drugs due to its efficacy to prevent PWUDs' morbidity and mortality. Implementation and coverage of this intervention are the results of decades of work. According to the 2020 Harm Reduction Report [30], by 2020, 86 countries had at least 1 SEP. However, the same report also stressed the concern for the lack of funding and political willingness dependence that most organizations that run SEP face.

One of the most common barriers to open a SEP has been the "Not in my backyard" community opposition [33]. Communities usually show resistance toward having a business (for or non-profit) that serves PWUDs for fear of increases in crime rates, adverse consequences for social cohesion, bad prestige for the zone, and the concern for what children may see in public venues. However, evidence shows that there has not been an increase in delinquency or any other negative consequence in the neighborhoods where SEP or any other harm reduction service has opened [34]. Moreover, harm reduction services that include SEP open in areas where PWUDs are already located. Therefore, neighbors' concern is baseless, considering that the stigmatized population is already in that location.

7.3.5 Safe Consumption Services (SCS)

SCS are centers where drug consumption is allowed. SCS aim to reduce the risk of infection transmission and prevent paraphernalia sharing and inadequate injection patterns (e.g., neck injections). They also intend to prevent overdose deaths and refer drug users to health and social services if needed or requested [35, 36].

SCS serve as a linkage to care for marginalized persons that would not be in contact with healthcare providers, including staff members [21, 37]. Contrary to the previous and current community attitude toward SCS, there is no evidence of such locations increasing drug injection incidence, drug trafficking, or crime in the surrounding neighborhoods [35].

The first SCS, referred to as safe consumption room, was established in 1986 in Bern, Switzerland, as part of the public health response to drug-related deaths and the HIV epidemic among PWUDs. This SCS opened in conjunction with other harm reduction services such as needle exchange programs and OAT [36, 38]. Other European countries have opened more SCS and are currently operating as integrated, specialized, and mobile services [38].

The integrated SCS are usually embedded in other drug services, including HIV testing, needle exchange programs, wound treatment, other medical care services, psychosocial care, and social services, including shelters. Staff usually controls access and allows a limited number of persons, usually adults. Some centers have kitchen services, showers, washing machines, a sitting area, OAT, inpatient services, detoxification, and several consumption rooms for injection, inhaling, and oral consumption. Professional healthcare providers are available in all rooms and are trained for overdose prevention and to provide referrals for other services.

Specialized SCS are usually part of NGOs that provide a range of services close to the location, but not in the same building. Clients are allowed a specific time slot, and health providers are also available in need and referrals.

Mobile SCS are provided in established small drug scenes in large cities across different locations. Clients are registered and learn the weekly schedule of the vans. A common restriction is also for people registered in OAT who may not be allowed to use the SCS.

One of the most studied SCS is Insite, the first safe injection facility opened in the Americas. Dozens of Insite-related research articles have been published since 2003, supporting the overall benefit SCS provides to the communities. For example, there was a 35% decrease in overdose mortality in Vancouver after implementing SCS and an increase in drug treatment uptake among SCS users compared to other PWUDs [39, 40].

With the recent opioid overdose epidemic in North America, Canada implemented overdose prevention sites (OPS) in 2016 [41, 42]. There are some differences between OPS and SCS. While SCS require an exemption to operate under Canadian federal law, OPS operate under provincial regulations, are more peer-driven, and do not necessarily provide clinical services [22].

Compared to SCS, OPS have a lower cost, are easier to implement, and can be located in tents, trailers, containers, NGOs, and housing facilities [42], among other easily accessed spaces located where PWUDs usually are. Both services aim to provide a safe environment; while SCS initially aimed to prevent HIV and other infectious diseases, they also provided overdose prevention. OPS specifically provide immediate overdose response and other harm reduction services as secondary.

Restrictions implemented in existing facilities include access only to adults, toward occasional or first-time clients, persons in OAT, residents, and intoxicated persons. Other restrictions include unique schedules and centers for women, and it is forbidden to undergo open transactions [38]. In addition, some of the concerns of implementing OPS include legal protection from being arrested on site, agreements with law enforcement for referrals, the confidentiality of users, and anonymity (e.g., not asking for identification or use of security tapes) [43].

As of July 2021, SCS officially operated in 13 countries: Switzerland, Germany, the Netherlands, Spain, Norway, Luxembourg, Denmark, Greece, France [36], Portugal [44], Belgium [45], Australia, and Canada [46]. Ireland has been working with a permission granted in 2017 [36], but in 2021, it was deemed invalid [47]. In the United States, Rhode Island is the first state to allow SCS within harm reduction premises [48]. California, New York, and Philadelphia have also presented law initiatives, and at least one unlicensed center is running in the United States [49]. Some of the anticipated barriers in the United States are fear of police interaction, privacy, data confidentiality, trust, and transportation [43].

Unofficially, according to the International Network of Drug Consumption Rooms, there are more SCS currently operating in Austria, Brazil, Bulgaria, Colombia, Czechia, Finland, Hungary, Iceland, Iran, Italy, Mexico, Poland, Romania, Serbia, Slovakia, Sweden, the United Kingdom, and the United States [50].

7.3.6 Opioid-Assisted Treatment (OAT)

OAT, also known as opioid maintenance treatment, medication-assisted treatment, medication for opioid use disorders (MOUD), and substitution therapy, refers to prescribing specific opioid drugs to persons who use opioids. OAT is preferred over MOUD because persons may not necessarily fulfill the psychiatric diagnoses for opioid use disorder.

OAT have mainly been used to reduce harms associated with opioid injection, such as HIV/HCV transmission through injection risk behaviors [19], to improve HIV treatment adherence [51], to reduce the risks related to street opioid use, and to reduce the odds of opioid overdose deaths [52].

OAT include prescription of opioid agonist (e.g., heroin, hydromorphone, morphine, methadone, and buprenorphine), opioid antagonists (naloxone, naltrexone), or combinations (e.g., suboxone). To date, 84 countries have at least 1 OAT service [30], most of them run by privately funded NGOs. The most prescribed agonist is methadone, which has been used for over 50 years, followed by buprenorphine, which has been used for approximately 20 years (Chap. 14).

Although OAT is effective, access and relapse are major concerns [10, 53]. Types of structural implementations related to OAT include expanding treatment options and services, improving funding and regulation, and intervening in public perception and attitudes toward persons who use opioids [19].

Geographic treatment availability is an essential structural determinant. OAT is mainly provided in high-income countries and urban settings. That leaves most of the persons that need treatment far from them. In countries with affordable or universal healthcare that includes opioid treatment, rural settings should invest more financially and socially to access treatment than their urban counterparts. The United States and Canada have OAT, and they are the countries where the opioid epidemic has hit the hardest. Unfortunately, there is almost no access to OAT in the rest of the Americas. Even more, the few resources spent on harm reduction and OAT have been provided by international NGOs. This funding is not sustainable and reliable because priorities and interests may shift while population needs remain unaddressed.

Regulatory systems dispensing OAT limit flexibility and responsiveness of the programs; bureaucracy imposes excessive administrative paperwork and costs that do not guarantee the quality of care. Most of the OAT interventions include a coordinated care model, in which at least two healthcare professionals shared care responsibilities [53]. That is, there are minimum staff requirements that restrain treatment availability. In the United States, the COVID-19 epidemic served as a circumstantial background to a more flexible and take-home OAT prescription [54].

OAT have poor retention rates [10], are underutilized, and suffer from prescription limitations [55]. In particular, retention in methadone and buprenorphine treatments is low; recent data suggest that injectable diacetylmorphine and hydromorphone may be more successful for those with low adherence to previous treatment efforts [23]. Injectable OAT has been used mainly in Europe (e.g., the United Kingdom, Switzerland, Germany, Denmark, and the Netherlands) and more recently in Canada [42, 56]. Injectable OAT is effective for persons who inject drugs, especially for persons with treatment-refractory opioid use disorders; successful outcomes include less opioid use, less criminal behavior, and more well-being [42, 57]. Low-threshold programs, such as oral, snorted, or injected hydromorphone provision, nested within existing drop-in services, and dispensing machines, may provide a viable alternative for OAT provision [55].

Researchers and providers cyclically face political resistance to incorporate injectable OAT; until the more recent overdose epidemic in Canada, provinces have been working toward the incorporation, through constitutional challenges, to extend prescription of injectable opioids to study participants [42]. Injectable OAT is effective but unsuitable for some settings and communities since it requires human and financial resources, infrastructure, and specialized staff training [55].

Regardless of intrinsic difficulties, there is a clear need for comprehensive strategies to reduce illicit opioid supply; expand OAT [58]; scale up low-barrier opioid distribution programs, including hydromorphone prescription; disrupt illegal drug supply; and avoid fatal overdose [55].

7.3.7 Naloxone Availability

Naloxone is an effective opioid receptor antagonist that can be delivered intranasally, as a spray, *or* as an injection (Chaps. 5, 8, and 14). The FDA approved naloxone in 1971 to prevent constipation among persons with prescription opioid use [59]. Naloxone has been available for almost half a century but mostly in care settings [60, 61]. In the 1990s and early 2000s, take-home naloxone programs were implemented in the European Union due to the heroin epidemic [59].

Several studies have been performed to understand and design the best practices for naloxone delivery to reduce the odds of overdose. However, new challenges have been found, such as the increased potency of opioids mainly due to heroin adulteration with fentanyl and other synthetic opioids. For instance, in 2016, the proportion of fatal opioid overdose was higher for synthetic opioids than for prescription opioids in the United States [62]. In addition, the use of synthetic opioids, either by choice or accidentally, implies that the usual dose of naloxone may not be as effective as with less potent opioids.

Overdose reversal using naloxone as antidote is a structural intervention for several reasons. First, naloxone is still not approved for over-the-counter purchases in most countries. Second, it is not available in most low- and middle-income countries. Finally, even in countries where naloxone is not a controlled substance, it is hardly available when and where it is most needed.

Some of the barriers to using naloxone include low availability and fear of police encounters. Public-sponsored naloxone distribution programs need to be scaled up in countries in which they are already available (e.g., Canada) and implemented in countries where they do not exist (e.g., Mexico).

Overdose prevention education and naloxone kits have been made available at the community level through health centers, first responders (e.g., firefighters, law enforcement, paramedics), persons who use drugs, peers, and relatives of persons who use drugs, among others.

Law enforcement officers face opioid overdoses in their daily activities, making them an ideal group to receive overdose prevention training and naloxone kits. However, officers' attitudes toward users may need to be changed [63]. In 2010, the National Drug Control Strategy in the United States included working with law enforcement officers to reduce overdose deaths [63, 64]. Also, all states and the District of Columbia have access to naloxone [65], including immunity for prescribers, laypeople who may administer it, and dispenser organizations, among others. Moreover, federal entities encourage first responders to carry and use naloxone in case of opioid overdose [66]. In Vancouver, public-funded naloxone programs in private low-income housing buildings hired peer tenants to provide naloxone training and distribute it to other residents [22].

When a fatal overdose occurs, witnesses may respond inadequately and end up harassed or arrested at the scene charged for possession [67]. Fear of legal consequences is a barrier to call for help in case of an overdose [63, 68]. Therefore, there must be an agreement not to charge callers for drug possession or use or even murder in case of an overdose. Good Samaritan laws that legally protect potential bystanders of an overdose and overdose prevention training programs are essential, together with access to emergency departments and SCS, among others [67]. Naloxone distribution also needs training and constant reminders to PWUDs to carry it with them all the time [23].

Naloxone's availability is not synonymous with its administration. Therefore, take-home naloxone adoption within a community needs to be understood and explored more deeply than providing administration training and quantifying the number of naloxone kits distributed and used [60]. Naloxone administration involves close and trusting relationships among different actors. It also needs a policy context that allows PWUDs to acquire and use naloxone freely and first responders who know what to do in case of an opioid overdose. Take-home naloxone programs also hold responsible other PWUDs for saving the lives of their peers [60, 69].

In sites and countries where opioid overdoses are increasing but have not yet reached the epidemic levels of the United States and Canada, some persons who use opioids may be reluctant to use naloxone because of the unpleasant withdrawal effects [60]. These effects can be minimized by carefully monitoring and titrating the naloxone dose through injection [60].

Other considerations include that newly abstinent persons are at higher risk of opioid overdose due to their tolerance loss and would benefit from carrying naloxone with them [70, 71].

7.3.8 Drug Checking

Europe introduced drug checking in the 1990s, originally thought of as a harm reduction service for nightlife and partying settings [42, 72]. Drugs can be checked with low-cost portable devices or more expensive stationary technologies.

Drug checking immediately informs persons before their drug consumption whether the substance they intend to use is what they thought they purchased and provides information about drug quality, purity, and potential harms. A second effect is information gathering for service users and the general population through public warnings [42].

In the case of the opioid epidemic, drug checking services have been mostly used to prevent overdosing due to drug adulteration with fentanyl and other synthetic opioids. Thanks to drug checking in the community setting, fentanyl has been found not only in heroin but also in combination with stimulants (i.e., cocaine, methamphetamine) [73, 74].

As it happens with other interventions, drug checking has nuances; in particular, this service has a limited effect in decreasing overdose rates. Although it may dissuade persons from using substances containing fentanyl, at this point, it is unclear whether they would have such effect [23]. Some factors related to vulnerabilities of the populations may limit the success of drug testing services. Some examples are having to give up a sample, time constrains, discrepancies, inaccuracy, ambivalence toward overdose risk, and availability of drug checking technologies [75]. Also, users may prefer using more potent substances (i.e., fentanyl).

Legal exemptions and implementation barriers are also structural challenges for drug checking [15]. In addition, safety and potential consequences need to be considered when introducing drug checking or any other intervention to which PWUDs may be unfamiliar.

Finally, it may be that that substance is the only one available in a particular location, and individuals may prefer to accept the risk than suffer withdrawal. It is not only knowledge and preference that relate to the use of contaminated/altered samples. Drug checking services may, by themselves, alter drug-using behaviors and lead to drug disposal in some settings [75] but not others, based on the vulnerability and poverty levels of the persons who use drugs [75]. For this, engagement of drug dealers in drug checking may be an option in settings in which users may not be willing to through contaminated drugs and where criminalization and struggle to obtain the substance result in a significant sacrifice for users [75].

7.3.9 Safe Drug Supply

As previously stated, some persons who use opioids have expressed a preference for fentanyl use [43]. This information should not be disregarded in overdose prevention and other opioid-related interventions. If research, policy, and the overall community efforts genuinely intend to improve public health and well-being, it must be taken into account that drug criminalization impedes and constrains the public health response to drug use [15]. Safe supply and drug reform to reduce opioid overdose is supported by activists and researchers [42]. However, it is still a bold step that will be deemed controversial for a long time but is already being discussed across the world. In the next years, implementation of drug law reforms will need to be tailored to the special needs and contextual characteristics of specific regions, including access to healthcare systems, infrastructure and human resources, current epidemic status, socioeconomic factors, and others (Chap. 4).

7.3.10 Opioid Prescription Regulations

Opioid prescription regulations can be a double-edged sword. For example, in the United States, they removed long-acting formulations of high-strength opioids to contain the epidemic of prescription opioid use unleashed by long-acting and highly concentrated oxycodone presentation. These regulations have also established monitoring systems and a shared database to oversee opioid prescription [76]. These measures immediately reduced opioid prescription but were followed by an unintended and inadequate increase in opioid discontinuation and tapering [77]. In addition, regulations had unintended spillover effects, including an increase in non-prescription opioid use [78], exposure to street heroin and fentanyl [76], and injection risk behaviors that led to HIV outbreaks among populations in which pain medication was misused. A clear example was the HIV outbreak in Scott County, Indiana [79]. Other effects of opioid prescription regulations paradoxically included increased overdose rates [76, 80] and more admissions to emergency care services [81]. Finally, policies intended to prevent opioid overdose by improving opioid prescription have been seen by PWUDs as propagating stigma, loss of autonomy, and reproducing and producing structural vulnerabilities [76].

Tamper (i.e., crushing or dissolving pills to snort or inject)-resistant formulations can reduce diversion of prescription opioids and fatal overdoses. However, modeling simulations [82] and retrospective studies [83] have found a modest effect on overdose prevention due to unintended consequences such as the increased use of heroin and increased stigma, marginalization, and feelings of "orphaned by the system." Stringent opioid prescription policies reduce the identity of persons that use opioid to "addict" and become powerless in pain management and opioid use [76]. Therefore, researchers and the pharmaceutical industry need to find a balance between marketing opioid formulations that become attractive for misuse and a human rights approach to pain management.

7.4 Conclusions

Person-centered approaches to drug use should take into account the persons' needs, values, and preferences [10]. As such, it is necessary to acknowledge that PWUDs are not necessarily interested in engaging in treatment [55] and require non-treatment options to ensure safer opioid use.

All the interventions described in this chapter can and should be expanded wherever they are needed. However, their implementation must be context-grounded, considering sociocultural context, safety, and legal challenges. A chain of associations must be developed in each set to identify where and which interventions may be more effective, when and how advocacy should focus, and the potential implementation and participation barriers [19]. Community engagement is essential to ensure the sustainable adoption of evidence-based programs to address opioid overdose and the root health inequities [84].

There is sound evidence that reducing drug availability [85] and adequate prescription guidelines are necessary but insufficient to prevent overdose. Additionally, much effort has been invested in overdose response programs, naloxone training and distribution, SCS, and other harm reduction services to address the public health crisis entailed by the opioid epidemic, but with modest progress [21, 23, 75]. Consequently, efforts are needed to address determinants of the opioid crisis. Community-driven interventions are key in implementing and sustaining culturally relevant treatment programs that may be more suitable in cases in which the context has an added importance [84].

To effectively address the opioid overdose crisis, stakeholders need responsive political environments incorporating harm reduction and drug policy experimentation (Strike & Watson, 2019). It is necessary to identify which structural elements may have a more significant impact on the health outcome, including the proximal behavioral risks and distal structural sources of these risks [19]. No single response or approach can have a long-lasting effect, for which a broad approach that targets social dynamics is needed. Where and how OAT is provided matter [53]. The diversity of the communities, resources available, values, and competing priorities may influence how engaged a community is in addressing a health problem [84].

Addressing the opioid epidemic requires addressing the social and structural determinants of mental health, HIV, HCV, and other comorbidities [9]. This includes a combination of several – if not all – interventions: overdose prevention sites, safe supply initiatives, drug decriminalization, housing first (i.e., stable housing without abstinence prerequisites) to prevent overdoses, and reducing stigma and shame associated with drug use and relapse [71].

Instead of temporary exemptions for organizations that provide services to PWUDs, and the constant threat of closing, sustained community and public efforts should be accompanied by law reforms that guarantee access to services and ensure providers will not face legal problems while saving lives.

Overdose prevention training and naloxone distribution aim to empower PWUDs, provide agency, and reduce health inequalities; however, unintended consequences include a deeper healthcare marginalization by segregating PWUDs [21].

Low-threshold care services have proven to successfully prevent and reduce HIV incidence among PWUDs [86]. Low-threshold treatments, including underage populations [87], need to be expanded to act as alternatives that replace illicit drug supply. Even these treatment programs have low retention rates and high relapse rates: for this, safe supply and harm reduction are essential interventions [23]. According to the healthcare system and accessibility, persons in need of OAT may not have the opportunity to access it. Under these circumstances, harm reduction, cultural competency, and low-threshold OAT at emergency settings are successful for economically disadvantaged populations, unstably housed, and with polysubstance use [88].

Changing the addiction paradigm from a will-failure disorder to a brain disease helped reduce stigma and develop effective medical treatments. However, it is time to recognize that many sociological determinants influence the initiation and progression of this disease. Only then, a multifactorial problem will be addressed with holistic approaches, including structural interventions other than OAT.

References

1. Leshner AI. Addiction is a brain disease, and it matters. Science. 1997;278:45–7.
2. Volkow ND. To end the opioid crisis, we must address painful social disparities. Drug Alcohol Depend. 2021;222:108678.
3. Chandler RK, Villani J, Clarke T, McCance-Katz EF, Volkow ND. Addressing opioid overdose deaths: the vision for the HEALing communities study. Drug Alcohol Depend. 2020;217:108329.
4. Strang J, Volkow ND, Degenhardt L, Hickman M, Johnson K, Koob GF, et al. Opioid use disorder. Nat Rev Dis Primers. 2020;6:1–28.
5. Heimer R, Clair S, Teng W, Grau LE, Khoshnood K, Singer M. Effects of increasing syringe availability on syringe-exchange use and HIV risk: Connecticut, 1990–2001. J Urban Health: Bull N Y Acad Med. 2002;79(4):556–70.
6. Perlman DC, Jordan AE. The syndemic of opioid misuse, overdose, HCV, and HIV: structural-level causes and interventions. Curr HIV/AIDS Rep. 2018;15:96–112.
7. Dasgupta N, Beletsky L, Ciccarone D. Opioid crisis: no easy fix to its social and economic determinants. Am J Public Health. 2018;108(2):182–6.

8. Bourgois P, Holmes SM, Sue K, Quesada J. Structural vulnerability: operationalizing the concept to address health disparities in clinical care. Acad Med: J Assoc Am Med Coll. 2017;92:299–307.
9. El-Bassel N, Shoptaw S, Goodman-Meza D, Ono H. Addressing long overdue social and structural determinants of the opioid epidemic. Drug Alcohol Depend. 2021;222:108679.
10. Blanco C, Wiley TRA, Lloyd JJ, Lopez MF, Volkow ND. America's opioid crisis: the need for an integrated public health approach. Transl Psychiatry. 2020;10:1–13.
11. Quesada J, Hart LK, Bourgois P. Structural vulnerability and health: Latino migrant laborers in the United States. Med Anthropol: Cross Cult Stud Health Illn. 2011;30:339–62.
12. Case A, Deaton A. Opioids. In: Deaths of despair and the future of capitalism. Princeton: Princeton University Press; 2020. p. 109–30.
13. Case A, Deaton A. Rising morbidity and mortality in midlife among white non-Hispanic Americans in the 21st century. Proc Natl Acad Sci U S A. 2015;112:15078–83.
14. Rigg KK, Monnat SM. Urban vs rural differences in prescription opioid misuse among adults in the United States: informing region specific drug policies and interventions. Int J Drug Policy. 2015;26:484–91.
15. Kolla G, Strike C. 'It's too much, I'm getting really tired of it': overdose response and structural vulnerabilities among harm reduction workers in community settings. Int J Drug Policy. 2019;74:127–35.
16. Friedman J, Syvertsen JL, Bourgois P, Bui A, Beletsky L, Pollini R. Intersectional structural vulnerability to abusive policing among people who inject drugs: a mixed methods assessment in California's central valley. Int J Drug Policy. 2021;87:102981.
17. Matthew DB, editor. Un-burying the lead: public health tools are key to beating opioid epidemic. 2018.
18. Terry NR. Structural determinism amplifying the opioid crisis : it's the healthcare, stupid !*. Northeast Univ Law Rev. 2018;11:315–71.
19. Heimer R, Bray S, Burris S, Khoshnood K, Blankenship KM. Structural interventions to improve opiate maintenance. Int J Drug Policy. 2002;13:103–11.
20. Satinsky EN, Kleinman MB, Tralka HM, Jack HE, Myers B, Magidson JF. Peer-delivered services for substance use in low- and middle-income countries: a systematic review. Int J Drug Policy. 2021;95:103252.
21. Jozaghi E. The opioid epidemic: task-shifting in health care and the case for access to harm reduction for people who use drugs. Int J Health Serv. 2020:1–8.
22. Bardwell G, Collins AB, McNeil R, Boyd J. Housing and overdose: an opportunity for the scale-up of overdose prevention interventions? Harm Reduct J. 2017;14:77.
23. Kerr T. Public health responses to the opioid crisis in North America. J Epidemiol Community Health. 2019;73:377–8.
24. Luchenski S, Maguire N, Aldridge RW, Hayward A, Story A, Perri P, et al. What works in inclusion health: overview of effective interventions for marginalised and excluded populations. Lancet. 2018;391:266–80.
25. Somers JM, Patterson ML, Moniruzzaman A, Currie L, Rezansoff SN, Palepu A, et al. Vancouver at home: pragmatic randomized trials investigating housing first for homeless and mentally ill adults. Trials. 2013;14:365.
26. Zabkiewicz DM, Patterson M, Frankish J, Somers JM. The Vancouver at home study: overview and methods of a housing first trial among individuals who are homeless and living with mental illness. J Clin Trials. 2012;2:1–7.
27. McNeil R, Fleming T, Collins AB, Czechaczek S, Mayer S, Boyd J. Navigating post-eviction drug use amidst a changing drug supply: a spatially-oriented qualitative study of overlapping housing and overdose crises in Vancouver, Canada. Drug Alcohol Depend. 2021;222:108666.
28. Richardson L, Dong H, Kerr T, Milloy MJ, Hayashi K. Drug-related harm coinciding with income assistance payments: results from a community-based cohort of people who use drugs. Addiction. 2021;116:536–45.

29. Richardson L, Laing A, Choi JC, Nosova E, Milloy MJ, Marshall BD, et al. Effect of alternative income assistance schedules on drug use and drug-related harm: a randomised controlled trial. Lancet Public Health. 2021;6:e324–e34.
30. Harm Reduction International. The global state of harm reduction. 7th ed. London: Harm Reduction International; 2020.
31. Des Jarlais DC, McKnight C, Goldblatt C, Purchase D. Doing harm reduction better: syringe exchange in the United States. Addiction. 2009;104(9):1441–6.
32. Gibson DR, Flynn NM, Perales D. Effectiveness of syringe exchange programs in reducing HIV risk behavior and HIV seroconversion among injecting drug users. AIDS. 2001;15(11):1329–41.
33. Davidson PJ, Howe M. Beyond NIMBYism: understanding community antipathy toward needle distribution services. Int J Drug Policy. 2014;25:624–32.
34. Chami G, Werb D, Feng C, DeBeck K, Kerr T, Wood E. Neighborhood of residence and risk of initiation into injection drug use among street-involved youth in a Canadian setting. Drug Alcohol Depend. 2013;132(3):486–90.
35. Potier C, Laprévote V, Dubois-Arber F, Cottencin O, Rolland B. Supervised injection services: what has been demonstrated? a systematic literature review. Drug Alcohol Depend. 2014;145:48–68.
36. European Monitoring Centre for Drugs and Drug Addiction, editor. Perspectives on drugs. Drug consumption rooms: an overview of provision and evidence. 2018.
37. Marshall BDL, Milloy MJ, Montaner JSG, Kerr T. Reduction in overdose mortality after the opening of North America's first medically supervised safer injecting facility: a retrospective population-based study. Lancet. 2011;377:1429–37.
38. Schäffer D, Stöver H, Weichert L, editors. Drug consumption rooms in Europe: models, best practice and challenges. 2014.
39. DeBeck K, Kerr T, Bird L, Zhang R, Marsh D, Tyndall M, et al. Injection drug use cessation and use of North America's first medically supervised safer injecting facility. Drug Alcohol Depend. 2011;113:172–6.
40. Marshall BDL, Milloy MJ, Wood E, Montaner JSG, Kerr T. Reduction in overdose mortality after the opening of North America's first medically supervised safer injecting facility: a retrospective population-based study. Lancet. 2011;377:1429–37.
41. Arkell C. Harm reduction in action: supervised consumption services and overdose prevention sites. CATIE-Canada's source for HIV and hepatitis C information. 2018:6.
42. Strike C, Watson TM. Losing the uphill battle? Emergent harm reduction interventions and barriers during the opioid overdose crisis in Canada. Int J Drug Policy. 2019;71:178–82.
43. Park JN, Sherman SG, Rouhani S, Morales KB, McKenzie M, Allen ST, et al. Willingness to use safe consumption spaces among opioid users at high risk of fentanyl overdose in Baltimore, Providence, and Boston. J Urban Health. 2019;96:353–66.
44. Pinto de Oliveiraa A, Gautier D, Nunes P, Correia V, Leite A, Taylor H, et al. First year of implementation of a drug consumption room in Lisbon: the client's profile. Eur J Public Health. 2020;30. https://doi.org/10.1093/eurpub/ckaa166.403.
45. Castillo T. A safe place to use drugs: lessons from Europe's supervised consumption sites the fix. 2019.
46. Drug Policy Alliance. Supervised consumption services. 2021.
47. MacNamee G. 'Devastating news': planning permission given to supervised injection centre is deemed invalid. TheJournal.ie. 2021.
48. Sutton M. Drug policy Alliance statement on Rhode Island becoming first in the nation to authorize harm reduction centers to prevent overdose deaths. New York: Drug Policy Alliance; 2021.
49. Kral AH, Davidson PJ. Addressing the nation's opioid epidemic: lessons from an unsanctioned supervised injection site in the U.S. Am J Prev Med. 2017;53:919–22.
50. International Network of Drug Consumption Rooms. Drug Consumption Rooms (DCRs) in the world. 2021.

51. Williams AR, Nunes EV, Bisaga A, Levin FR, Olfson M. Development of a cascade of care for responding to the opioid epidemic. Am J Drug Alcohol Abuse. 2019;45:1–10.
52. Fatseas M, Auriacombe M. Why buprenorphine is so successful in treating opiate addiction in France. Curr Psychiatry Rep. 2007;9:358–64.
53. Lagisetty P, Klasa K, Bush C, Heisler M, Chopra V, Bohnert A. Primary care models for treating opioid use disorders: what actually works? A systematic review. PLoS One. 2017;12:1–40.
54. Joudrey PJ, Adams ZM, Bach P, Van Buren S, Chaiton JA, Ehrenfeld L, et al. Methadone access for opioid use disorder during the COVID-19 pandemic within the United States and Canada. JAMA Netw Open. 2021;4(7):e2118223-e.
55. Ivsins A, Boyd J, Beletsky L, McNeil R. Tackling the overdose crisis: the role of safe supply. Int J Drug Policy. 2020;80:102769.
56. Fischer B, Rehm J, Kirst M, Casas M, Hall W, Krausz M, et al. Heroin-assisted treatment as a response to the public health problem of opiate dependence. Eur J Public Health. 2002;12:228–34.
57. Ferri M, Davoli M, Perucci CA. Heroin maintenance for chronic heroin-dependent individuals. Cochrane Database Syst Rev. 2011;12:1–55.
58. Jones MR, Viswanath O, Peck J, Kaye AD, Gill JS, Simopoulos TT. A brief history of the opioid epidemic and strategies for pain medicine. Pain Ther. 2018;7(1):13–21.
59. Campbell ND. Naloxone as a technology of solidarity: history of opioid overdose prevention. CMAJ. 2019;191:E945–E6.
60. Farrugia A, Fraser S, Dwyer R, Fomiatti R, Neale J, Dietze P, et al. Take-home naloxone and the politics of care. Sociol Health Illn. 2019;41:427–43.
61. McDonald R, Campbell ND, Strang J. Twenty years of take-home naloxone for the prevention of overdose deaths from heroin and other opioids—conception and maturation. Drug Alcohol Depend. 2017;178:176–87.
62. Jones CM, Einstein EB, Compton WM. Changes in synthetic opioid involvement in drug overdose deaths in the United States, 2010–2016. JAMA – J Am Med Assoc. 2018;319:1819–21.
63. Wagner KD, Bovet LJ, Haynes B, Joshua A, Davidson P. Training law enforcement to respond to opioid overdose with naloxone: impact on knowledge, attitudes, and interactions with community members. Drug Alcohol Depend. 2016;165:22–8.
64. Office of National Drug Control Policy. National Drug Control Strategy. The White House. 2010.
65. Lieberman A, Davis C, editors. Legal interventions to reduce overdose mortality: naloxone access laws. The Network for Public Health Law. 2021.
66. Guidance for law enforcement and first responders administering naloxone. 2020.
67. Davidson PJ, McLean RL, Kral AH, Gleghorn AA, Edlin BR, Moss AR. Fatal heroin-related overdose in San Francisco, 1997–2000: a case for targeted intervention. J Urban Health. 2003;80:261–73.
68. Wagner KD, Liu L, Davidson PJ, Cuevas-Mota J, Armenta RF, Garfein RS. Association between non-fatal opioid overdose and encounters with healthcare and criminal justice systems: identifying opportunities for intervention. Drug Alcohol Depend. 2015;153:215–20.
69. Buchman DZ, Orkin AM, Strike C, Upshur REG. Overdose education and naloxone distribution programmes and the ethics of task shifting. Public Health Ethics. 2018;11:151–64.
70. Bowles JM, Lankenau SE. "I gotta go with modern technology, so I'm gonna give 'em the Narcan": the diffusion of innovations and an opioid overdose prevention program. Qual Health Res. 2019;29:345–56.
71. Bowles JM, Smith LR, Mittal ML, Harding RW, Copulsky E, Hennessy G, et al. "I wanted to close the chapter completely … and I feel like that [carrying naloxone] would keep it open a little bit": refusal to carry naloxone among newly-abstinent opioid users and 12-step identity. Int J Drug Policy. 2021;94:103200.
72. Benschop A, Rabes M, Dirk JK. Pill testing, ecstasy and prevention: a scientific evaluation in three European cities. Amsterdam: Rozenberg Publishers; 2002. p. 138.

73. Fleiz C, Arredondo J, Chavez A, Pacheco L, Segovia LA, Villatoro JA, et al. Fentanyl is used in Mexico's northern border: current challenges for drug health policies. Addiction. 2020;115:778–81.
74. Centre on Drug Policy Evaluation. What's in Toronto's drug supply? Toronto's Drug Checking Service. 2021.
75. Bardwell G, Boyd J, Tupper KW, Kerr T. "We don't got that kind of time, man. We're trying to get high!": exploring potential use of drug checking technologies among structurally vulnerable people who use drugs. Int J Drug Policy. 2019;71:125–32.
76. Antoniou T, Ala-Leppilampi K, Shearer D, Parsons JA, Tadrous M, Gomes T. "Like being put on an ice floe and shoved away": a qualitative study of the impacts of opioid-related policy changes on people who take opioids. Int J Drug Policy. 2019;66:15–22.
77. Ranapurwala SI, Ringwalt CL, Pence BW, Schiro S, Fulcher N, McCort A, et al. State medical board policy and opioid prescribing: a controlled interrupted time series. Am J Prev Med. 2021;60:343–51.
78. Coffin PO, Rowe C, Oman N, Sinchek K, Santos GM, Faul M, et al. Illicit opioid use following changes in opioids prescribed for chronic non-cancer pain. PLoS One. 2020;15:1–14.
79. Strathdee SA, Beyrer C. Threading the needle. NEJM. 2015;373:397–9.
80. Oliva EM, Bowe T, Manhapra A, Kertesz S, Hah JM, Henderson P, et al. Associations between stopping prescriptions for opioids, length of opioid treatment, and overdose or suicide deaths in US veterans: observational evaluation. BMJ. 2020;368:1–10.
81. Mark TL, Parish W. Opioid medication discontinuation and risk of adverse opioid-related health care events. J Subst Abus Treat. 2019;103:58–63.
82. Wakeland W, Nielsen A, Geissert P. Dynamic model of nonmedical opioid use trajectories and potential policy interventions. Am J Drug Alcohol Abuse. 2015;41:508–18.
83. Benintendi A, Kosakowski S, Lagisetty P, Larochelle M, Bohnert ASB, Bazzi AR. "I felt like I had a scarlet letter": recurring experiences of structural stigma surrounding opioid tapers among patients with chronic, non-cancer pain. Drug Alcohol Depend. 2021;222:108664.
84. El-Bassel N, Gilbert L, Hunt T, Wu E, Oga EA, Mukherjee TI, et al. Using community engagement to implement evidence-based practices for opioid use disorder: a data-driven paradigm & systems science approach. Drug Alcohol Depend. 2021;222:108675.
85. Martins SS, Sampson L, Cerdá M, Galea S. Worldwide prevalence and trends in unintentional drug overdose: a systematic review of the literature. Am J Public Health. 2015;105:e29–49.
86. Wodak A, McLeod L. The role of harm reduction in controlling HIV among injecting drug users. AIDS. 2008;22:S81–92.
87. Marshall BDL, Green TC, Yedinak JL, Hadland SE. Harm reduction for young people who use prescription opioids extra-medically: obstacles and opportunities. Int J Drug Policy. 2016;31:25–31.
88. Kalmin MM, Goodman-Meza D, Anderson E, Abid A, Speener M, Snyder H, et al. Voting with their feet: social factors linked with treatment for opioid use disorder using same-day buprenorphine delivered in California hospitals. Drug Alcohol Depend. 2021;222:108673.

Part II
Pharmacological and Medical Aspects of Opioid Use

Chapter 8
Opioid Effects and Classification

Silvia L. Cruz, Miguel I. Paz-Ramos, Araceli Hernández-Mendoza, and César J. Carranza-Aguilar

Abstract This chapter describes the main groups of opioid receptor ligands based on their origin, chemical structure, and pharmacological actions. The first section describes natural, semisynthetic, and synthetic compounds, as well as endogenous opioid peptides and their precursors. The effects of morphine, the opioid prototype, are covered in the second section. The third part reviews the structure-activity relationships of main opioids and provides a short description of the structural characteristics of (a) morphine-like drugs (e.g., codeine and heroin), (b) morphinans (e.g., thebaine), (c) benzomorphans (e.g., pentazocine), (d) phenylpiperidines (e.g., fentanyl), and (e) diphenylheptanes (e.g., methadone). The last section explains the differences in opioid effects based upon differential affinities for constitutively active or inactive receptor subtypes.

Keywords Opioid effects · Opioid classifications · SAR (structure-activity relationships) · Endogenous opioids · Morphinans · Benzomorphans · Phenylpiperidines · Diphenylheptanes

S. L. Cruz (✉)
Department of Pharmacobiology, Center for Research and Advanced Studies (Cinvestav), Mexico City, Mexico

Opioids Working Group, Global Studies Seminar, Faculty of Medicine, National Autonomous University of Mexico (UNAM), Mexico City, Mexico
e-mail: slcruzm@cinvestav.mx

M. I. Paz-Ramos · A. Hernández-Mendoza · C. J. Carranza-Aguilar
Department of Pharmacobiology, Center for Research and Advanced Studies (Cinvestav), Mexico City, Mexico

S. L. Cruz (ed.), *Opioids*, https://doi.org/10.1007/978-3-031-09936-6_8

8.1 Opioid Receptor Ligands

8.1.1 Natural, Semisynthetic, and Synthetic Opioids

Morphine, codeine, and thebaine are naturally occurring opioids in the poppy plant *Papaver somniferum* (Chap. 1). Opium, morphine, and codeine produce similar effects but with different potencies. Thebaine is a weak pharmacological compound found in several *Papaver* species (e.g., *P. bracteatum* and *P. orientale*) and is used as a precursor to synthesize other opioids. Chemical changes in morphine structure produce semisynthetic opioids, which are molecules with minor modifications, such as the presence of an acetyl group or the exchange of an –OH radical for an oxygen in specific carbons. Examples of these drugs are diacetylmorphine (heroin), dihydromorphine, hydrocodone, oxycodone, dihydrocodeine, buprenorphine, nalorphine, nalbuphine, and naloxone.

Opioid receptor agonists bind to specific receptors and produce an effect. Antagonists also bind to opioid receptors but lack an effect. Agonist-antagonists are agonists of specific receptor subtypes and antagonists of others.

Synthetic opioids are chemically diverse compounds entirely produced in the laboratory. This group includes pethidine (also known as meperidine), methadone, fentanyl, fentanyl analogs, and many other compounds, some of which are sold in the darknet as new psychoactive substances (Chap. 16). These drugs exert their actions by binding to opioid receptors that mediate pain, and gastrointestinal motility, and activate rewarding brain pathways, among other effects.

8.1.2 Endogenous Opioid Peptides

The natural ligands to opioid receptors are enkephalins, endorphins, dynorphins, and other endogenous opioid peptides (Table 8.1). All of these peptides derive from three polypeptide precursors: proenkephalin (PENK), pro-opiomelanocortin (POMC), and prodynorphin (PDYN). These precursors are cleaved by proteases giving rise to different smaller compounds in a highly dynamic process that depends on physiological needs and the tissues where synthesis occurs.

Proenkephalin contains six copies of [Met]-enkephalin (Tyr-Gly-Gly-Phe-Met) and one of [Leu]-enkephalin (Tyr-Gly-Gly-Phe-Leu). Enkephalins are widely distributed in the brain and spinal cord regions related to pain perception (periaqueductal gray, trigeminal nucleus, laminae I and II of the spinal cord) and mood (striatum, amygdala, locus coeruleus, nucleus accumbens) and in the gastrointestinal tract, the adrenal medulla, and the olfactory areas (Table 8.1). POMC is the precursor of β-endorphins and other non-opioid bioactive peptides, including the adrenocorticotropic hormone (ACTH), β-lipotropin (β-LPH), and α-melanocyte-stimulating hormone (α-MSH). Prohormone convertases (PCs) in the anterior lobe of the pituitary gland cleave POMC into ACTH and β-lipotropin. In the hypothalamus and skin,

Table 8.1 Precursors, amino acid sequence, and receptor preference of selected endogenous opioid peptides

Precursor	Peptide	Amino acid sequence	Receptor affinity
Pro-enkephalin (PENK)	[Leu]-enkephalin [Met]-enkephalin Heptapeptide Octapeptide Adrenorphin (metorphamide)	**Tyr-Gly-Gly-Phe-Leu** **Tyr-Gly-Gly-Phe-Met** **Tyr-Gly-Gly-Phe-Met**-Arg-Phe **Tyr-Gly-Gly-Phe-Met**-Arg-Gly-Leu **Tyr-Gly-Gly-Phe-Met**-Arg-Arg-Val	$\delta \gg \mu$
Proopiomelanocortin (POMC)	β-Endorphin	**Tyr-Gly-Gly-Phe-Met**-Thr-Ser-Glu-Lys-Ser-Gln-Thr-Pro-Leu-Val-Thr-Leu-Phe-Lys-Asn-Ala-Ile-Ile-Lys-Asn-Ala-Tyr-Lys-Lys-Gly-Glu	μ
Prodynorphin (PDYN)	Dynorphin A (1–17) Dynorphin B α-Neoendorphin	**Tyr-Gly-Gly-Phe-Leu**-Arg-Arg-Ile-Arg-Pro-Lys-Leu-Lys-Trp-Asp-Asn-Gln **Tyr-Gly-Gly-Phe-Leu**-Arg-Arg-Gln-Phe-Lys-Val-Val-Thr **Tyr-Gly-Gly-Phe-Leu**-Arg-Lys-Tyr-Pro-Lys	κ
Unknown	Endomorphin-1 Endomorphin-2	**Tyr-Pro-Trp-Phe** **Tyr-Pro-Phe-Phe**	μ

further processing of β-lipotropin produces β-endorphins. POMC, PCs, and β-endorphins are also located in cells of the immune system, where they exert some inhibitory effects. There are several forms of endorphins, but β-endorphin(1–31) has strong analgesic effects and is 20–30 times more potent than morphine. Shorter forms, such as β-endorphin(1–27), are less active.

Stressors induce corticotropin-releasing factor (CRH) from the hypothalamus inducing ACTH and β-endorphin synthesis in the anterior pituitary. This effect has been associated with stress-induced analgesia [1, 2].

Prodynorphin, the precursor of dynorphins, can produce dynorphin A (Dyn A 1–17), dynorphin B (Dyn B 1–13), α-neoendorphin, big dynorphin (Dyn A 1–32), leumorphin (Dyn B 1–29), leucine-enkephalin-arginine (Leu-enkephalin-Arg), and, potentially, [Leu]-enkephalin (because it has three leu-enkephalin core opioid sequences) by differential post-translational processing. Dynorphins are abundant in the brain and spinal cord; other sites expressing dynorphins are the adrenal glands, testis, and anterior pituitary [3]. In addition to causing allodynia (pain produced by a stimulus that would not usually be painful), dynorphin promotes anxiety, stress, and dysphoria.

Endomorphins are analgesic tetrapeptides with high affinity and selectivity for the μ-opioid receptor. The precursor for endomorphins remains to be identified [4].

Endogenous peptide precursors are abundant in the central nervous system (CNS). They are synthesized in the nucleus and transported to the nerve terminals. Specific processing enzymes recognize double basic amino acid sequences positioned before and after the opioid peptide and cleave them. Each of the opioid precursors contains multiple active peptides, which can be modified differentially depending on the brain areas. They also often coexist with other neurotransmitters or neuropeptides. Physiological demands alter the processing of these peptides, and the final product produced by and stored within a given neuron depends not only on the precursor but also on the enzymes available to process it.

Endogenous opioid peptides play a regulatory role in the organisms' response to physiological and environmental demands. They are involved in pain processing, regulation of the hypothalamic-pituitary-adrenal axis, reward, and other neuroendocrine functions critical for survival [5].

8.2 Effects of Morphine

Morphine is the opioid prototype drug, from both the pharmacological and structural points of view. It has significant effects on the CNS, gastrointestinal tract (GIT), and cardiovascular, neuroendocrine, and immune systems (Fig. 8.1).

8.2.1 Central Nervous System

8.2.1.1 Analgesia

Morphine is the gold standard for severe pain management because it inhibits pain transmission. Pain is a complex phenomenon with emotional and physiological components, while nociception refers only to the neuronal processes that encode and process noxious stimuli. Therefore, it is preferable to use nociception when referring to "pain" without the emotional aspect. Generally speaking, three neurons participate in nociception transmission. The first one senses environmental stimuli and responds by activating a second neuron located in the dorsal horn of the spinal cord. This "second-order" neuron transmits sensory information to the thalamus. From there, a third neuron projects its axons to the cortex, making synapses with other brain cells that activate modulatory descending pain pathways. Thus, morphine has a dual action; it inhibits the neurons involved in the ascending pain pathways and promotes the release of various neurotransmitters in the descending pain pathways counteracting painful stimuli (see Chap. 10).

8.2.1.2 Euphoria

Morphine produces euphoria through indirect activation of the mesocorticolimbic pathway. The ventral tegmental area (VTA) is one of the major dopaminergic brain areas adjacent to the *substantia nigra* in the midbrain. Dopaminergic VTA neurons project to the *nucleus accumbens*, a small region in the ventral striatum, increasing

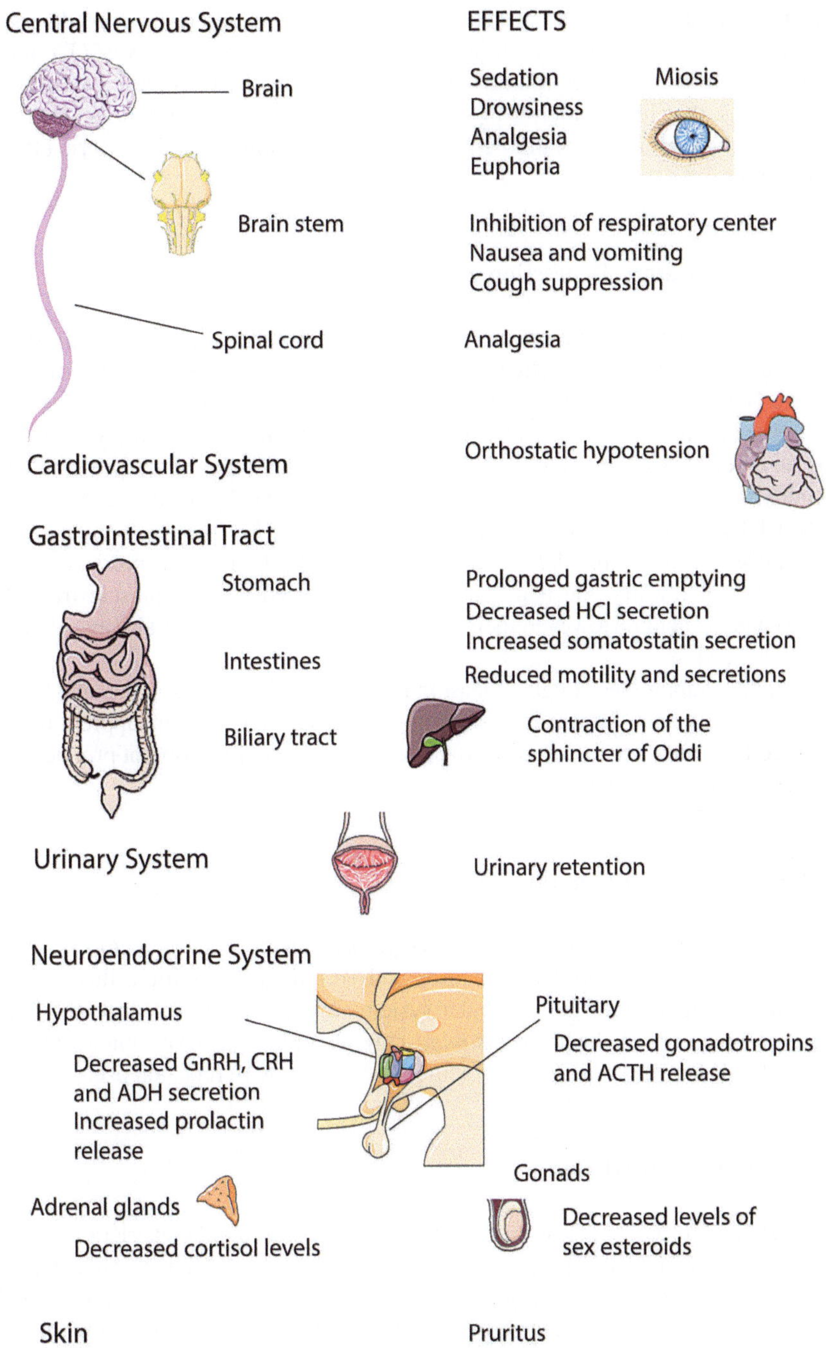

Fig. 8.1 Effects of morphine. (Figures courtesy of Smart Medical Art)

dopamine release. VTA also projects to other areas, including the amygdala and several cortex regions. GABA interneurons exert a tonic inhibition over VTA dopaminergic neurons. When morphine binds to μ-opioid receptors in these GABAergic interneurons, they reduce their firing rate, causing dopamine release. Repeated morphine administration results in tolerance to both the analgesic and euphoric effects of morphine.

8.2.1.3 Respiratory Depression and Cough Suppression

Morphine acts directly on respiratory centers by inhibiting neurons that respond to carbon dioxide tension. In particular, morphine inhibits a few clusters of neurons in the pre-Bötzinger complex and the parabrachial/Kölliker-Fuse nuclei. The pre-Bötzinger complex has pacemaker neurons that regulate the respiratory cycle and play an essential role in inspiration. When opioids inhibit neurons expressing neurokinin-1 receptors within this area, they cause respiratory rhythm arrest, apnea, and potential death. On the other hand, opioids acting on the Kölliker-Fuse and parabrachial nuclei produce irregular respiratory patterns [6]. Opioid receptors are abundant in alveolar walls, mechanosensory receptors, and smooth muscle from the trachea and bronchia. However, their exact role in impairing upper airway caliber is not clear [7, 8].

Morphine suppresses cough, at least in part through inhibition of the cough center in the brain stem. However, respiratory depression and cough suppression are not directly related because highly effective antitussive opioids do not produce significant respiratory depression.

8.2.1.4 Nausea and Vomiting

Morphine stimulates the chemoreceptor trigger zone in the medulla oblongata in the brain's lowest portion, producing nausea and vomiting. In addition, there is evidence that stimulation of the vestibular system and activation of neurokinin-1 and serotonin receptors in the *area postrema* may also play a role in opioid-induced emesis [9]. Tolerance to nausea and vomiting develops with repeated use.

8.2.1.5 Pupil Constriction

A persistent and easily recognizable effect of morphine is pupil constriction even in poor lighting conditions ("pinpoint pupils," or miosis). Pupil diameter depends on both the sympathetic and parasympathetic nervous systems. Opioids cause pupil constriction by indirectly activating the parasympathetic neurons located within the oculomotor nucleus in the brain stem, causing the iris sphincter muscle to contract [10]. Pupil contraction is pathognomonic (a clinical characteristic used to make a diagnosis) of opioid intoxication.

8.2.1.6 Seizures

Morphine produces convulsions, but only at doses much higher than those required to produce analgesia.

8.2.2 *Gastrointestinal System*

The enteric nervous system is rich in opioid receptors that inhibit cholinergic neurons, decrease GIT motility, and cause constipation. Tolerance to this effect is mild, and constipation is an adverse effect even in chronic opioid users. The "narcotic bowel syndrome" is a dysfunction caused by regular usage or escalating opioid doses, accompanied by chronic, frequent abdominal pain, intermittent vomiting, and weight loss [11]. In addition, morphine produces constriction of the sphincter of Oddi, increasing the pressure of the bile duct. For this reason, opioids should not be prescribed for biliary colic.

8.2.3 *Cardiovascular System*

Effects of morphine on the cardiovascular system vary significantly. Acute morphine administration produces orthostatic hypotension. This mild drop in blood pressure is not significant if the patient remains seated or reclined. Cardiovascular alterations, however, are important when people go through withdrawal. Discontinuation of opioids or counteracting their effects with specific antagonists produces a significant increase in blood pressure and heart rate, along with other sympathetic effects that contribute to the distress and discomfort experienced by patients that stop taking opioids.

Chronic opioid use often correlates with an increased risk of myocardial infarction and a lesser risk of coronary artery disease. Valvular endocarditis and cardioembolic stroke can also occur among injection drug users. Opioid overdose is associated with ischemic events, stroke, arrhythmias, and heart failure [12].

8.2.4 *Neuroendocrine System*

Hypogonadism and hypocortisolism are known side effects of chronic opioid use. Morphine inhibits the hypothalamus-pituitary-adrenal axis by activating μ-, δ-, and κ-opioid receptors. Even a single morphine administration decreases the hypothalamic secretion of corticotropin-releasing hormone (CRH) and antidiuretic hormone (ADH), the pituitary secretion of adrenocorticotropic hormone (ACTH), and the release of cortisol (in humans) or corticosterone (in animals) from adrenal glands [13].

Morphine also inhibits gonadotropin-releasing hormone (GnRH) secretion by activating opioid receptors in the hypothalamus, causing a decrease in gonadotropin release from the pituitary gland and sex steroid release from the gonads. The resulting hypogonadism causes gynecomastia, erectile dysfunction, decreased muscle mass in men, menstrual irregularities (dysmenorrhea) or amenorrhea in women, and reduced bone density. Morphine and other opioids increase prolactin release after a single administration, but long-term effects may vary. Prolactin increase can contribute to infertility, milk production (galactorrhea), and menstrual irregularities [14]. On the other hand, low cortisol levels produce various symptoms, such as anorexia, fatigue, and abdominal and general discomfort. Chronic opioid users also have an altered circadian cortisol rhythm [15].

8.2.5 Immune System

A complex interaction exists between the immune system (IS) and the nervous system (NS). Morphine has immunosuppressive effects due to the interaction with classical and non-classical opioid receptors. Opioid receptors have been identified in all types of immune cells. Activation of opioid receptors leads to desensitization of innate and adaptive immune signaling. The complexity of the opioid-mediated NS-IS interaction is also evident in the indirect effects of opioids on immune responses by hypothalamic-pituitary-adrenal axis (HPA) modulation, which leads to the release of cortisol and adrenaline (mediators that alter immune cell function).

The consequences of opioid actions on the IS are related to the essential physiological process of inflammation resolution and return to homeostasis and immunosuppressive conditions that predispose to infectious diseases, decrease tolerance to microbiota, or suppress recognition of malignant cell tumors. On the other hand, considering that inflammation is a process involved in numerous acute and chronic diseases, some specific opioid actions on the IS are now considered therapeutic tools to control intense and deleterious inflammatory disorders, such as the COVID-19-related cytokine storm. Those treatments aim to avoid a prolonged immunodeficient state that compromises a patient's recovery. Future research on biased ligands of non-classical opioid receptors and discoveries on the influence of age and history of previous health conditions on the effects of opioids on immune responses will help to understand the intricate opioid-mediated network of control of neuroimmune communication and its consequences on health and disease (Chap. 12).

8.2.6 *Miscellaneous*

8.2.6.1 Pruritus

Morphine produces intense generalized pruritus, mainly when used intrathecally or epidurally (neuroaxial analgesia). The incidence of pruritus ranges from 60% to 85% and varies among opioids. The onset of itching with morphine occurs within hours and remains for several days. With highly lipophilic opioids, such as fentanyl, pruritus appears within minutes and lasts several hours. Opioid-induced pruritus is different depending on the route of administration. Itching occurs less with oral than with injected morphine, although its incidence increases with prolonged use. Morphine acts on mast cells by activating a non-canonical opioid receptor, the mas-related G protein-coupled receptor X2 (MRGPRX2). Activation of this receptor results in histamine release from mast cells, causing itching at the site of injection [16]. Itching associated with neuroaxial opioid administration is mediated by several mechanisms that involve μ-opioid receptors.

Several studies indicate that itching occurs by activating a heterodimer formed by a specific isoform of the μ-opioid receptor (MORD1) and GRPR or gastrin-releasing peptide receptor, recently renamed BB2 [17]. Acute itching is primarily due to the release of histamine, prostaglandins, leukotrienes, tryptase, calcitonin gene-related peptide, and substance P, but chronic itching seems to have a different pathological basis.

Other opioid effects described on the skin are keratocyte proliferation, migration, and differentiation, wound healing, and inflammatory responses in the human epidermis [18].

8.2.6.2 Urinary Retention

Morphine increases the tone of the bladder sphincter but inhibits the urinary voiding reflex, causing urinary retention.

8.3 Chemical Classification of Opioid Compounds and Main Group Effects

To learn about the structure-activity relationship of opioids, it is helpful to remember common chemical moieties and functional groups (Box 8.1). Chemical moieties are parts of a molecule, which cannot stand on their own. Functional groups are specific groups of atoms that always react similarly regardless of the compound to which they are attached.

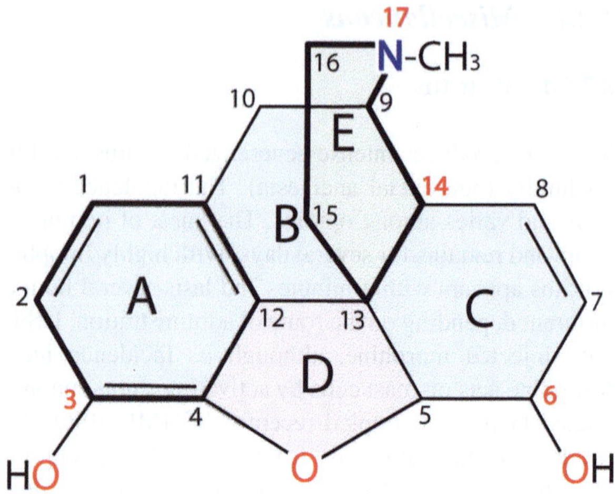

Fig. 8.2 Morphine chemical structure with the phenanthrene nucleus (A, B, and C rings), a furan ring (D), and the piperidine ring (E). Numbers correspond to the main atoms of the molecule. Changes in carbons 3, 6, and 14 and nitrogen 17 are characteristic of morphine-like drugs

Morphine has five rings (Fig. 8.2):

A. An aromatic phenyl ring with six carbon (C) and hydrogen atoms (H), alternating single and double bonds.
B. A cyclohexane ring with 6 C and 12 H atoms.
C. A ring of 6 C and 10 H atoms (cyclohexene) with a double bond between 2 C atoms.
D. A cyclic ether made of 4 C, 8 H, and one oxygen (O) atoms known as tetrahydrofuran or oxolane.
E. A piperidine ring formed by 5 C, 11 H, and 1 nitrogen (N) atoms. Another way to refer to this group is as azocycloalkane, i.e., cyclohexane with a nitrogen atom replacing a carbon atom.

A, B, and C rings together are usually referred to as a phenanthrene nucleus (blue rings, Fig. 8.2).

Numbering of the morphine molecule begins with the carbon indicated in ring A; follows down, right, and up along the carbon backbone of A, B, C, and D rings; and finishes with the N atom. Morphine has a hydroxyl (–OH) group in carbons 3 and 6 and a methyl (–CH_3) group attached to the nitrogen atom. The –OH group in position 3 is a phenolic hydroxyl group because it is attached to the phenyl group. The –OH group in position 6 is the alcoholic hydroxyl group. Another feature of morphine structure is an unsaturated bond between carbons 7 and 8.

Box 8.1 Chemical Groups Found in Opioids

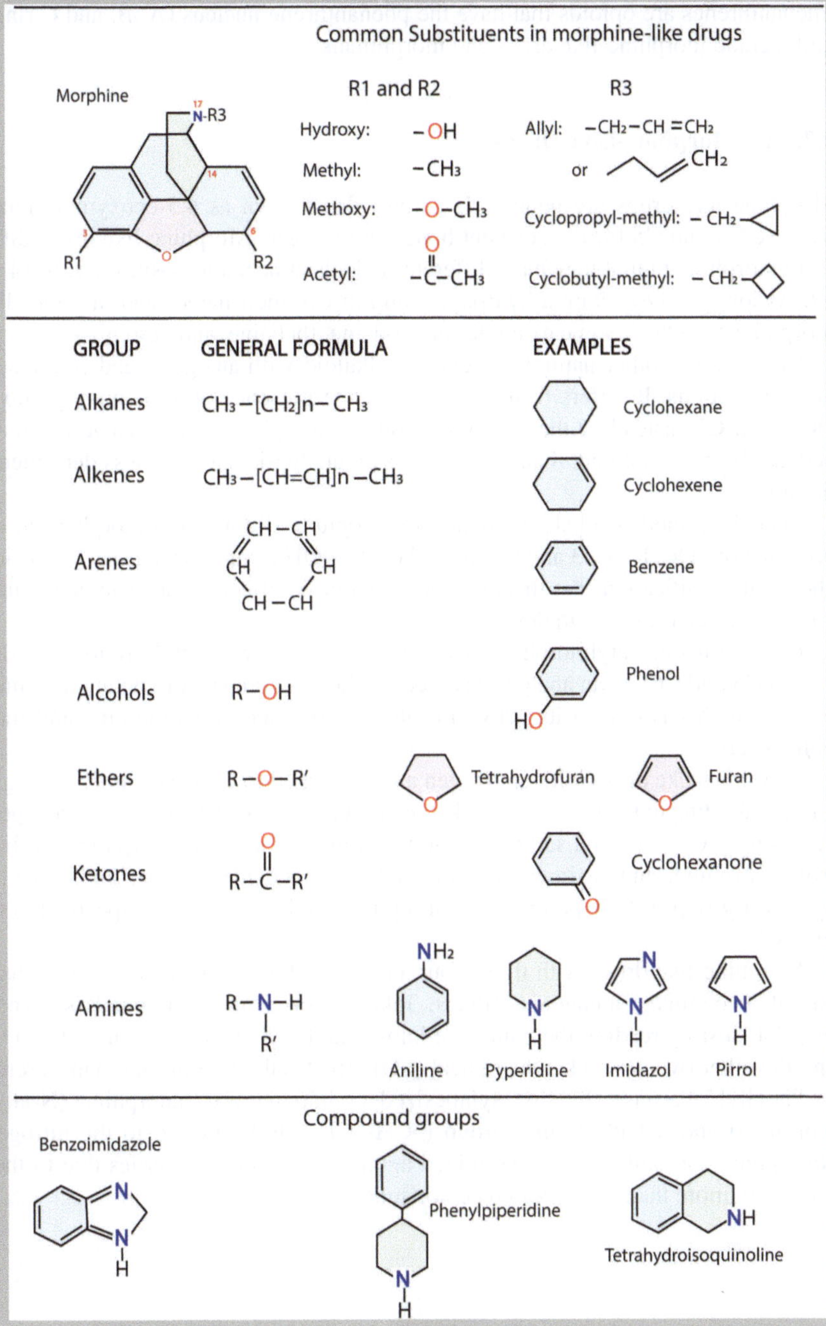

8.3.1 *Phenanthrenes*

Phenanthrenes are opioids that have the phenanthrene nucleus (A, B, and C rings) and include morphine-like drugs and morphinans.

8.3.1.1 Morphine-Like Drugs

Morphine-like drugs are pentacyclic drugs also known as 4,5-epoxymorphinans because C4 and C5 form a covalent bond with oxygen. Morphine-like drugs differ from morphine in that they have different radicals attached to positions 3, 6, 14, or 17. As long as they retain a hydroxyl group in C6, their names end in "*-ine*" like morphine or in "in." Some examples are codeine, thebaine, and heroin.

Codeine is another naturally occurring alkaloid with analgesic and cough suppressant actions. It differs from morphine in the presence of a methoxy group (–OCH_3) in C3 instead of the hydroxyl group in morphine. This change results in codeine being less potent than morphine as an analgesic and with less dependence liability.

Thebaine, another alkaloid isolated from opium, differs from morphine in the presence of –OCH_3 in C3 and C6 instead of the –OH groups. This is such a critical chemical modification that thebaine lacks analgesic effects, is a toxic compound, and causes seizures at high doses.

Heroin (aka diacetylmorphine or diamorphine) has two acetyl groups (see Box 8.1) in C3 and C6. As an analgesic and euphoriant, heroin is more potent than morphine, but this is due to its active metabolites 6-monoacetylmorphine and morphine itself.

Morphine-like drugs with an oxygen atom attached to C6 have names ending in "*-one*," alluding to the presence of a ket*one* group (–C=O). This structural change is paired with C7–C8 bond saturation. Some examples are hydromorphone, derived from morphine, and hydrocodone, derived from codeine. Former addition of a hydroxyl group in C14 produces oxymorphone and oxycodone, respectively (see Fig. 8.3).

Morphine-like drugs with three, four, or five carbons attached to the N17 have partial or complete antagonist effects. Take a look at the compounds shown in Fig. 8.4. Just by reading the name, we know that the first two have an –OH in C6 and the other two have a ketone radical (=O). The "Nal-" part of their names refers to "N-allyl" because the first synthesized compound was nalorphine (N-allylmorphine) and it had an allyl group (–CH_2–CH=CH_2) attached to the nitrogen. Nalorphine and nalbuphine have mixed agonist-antagonist properties due to their affinity to more than one opioid receptor subtype.

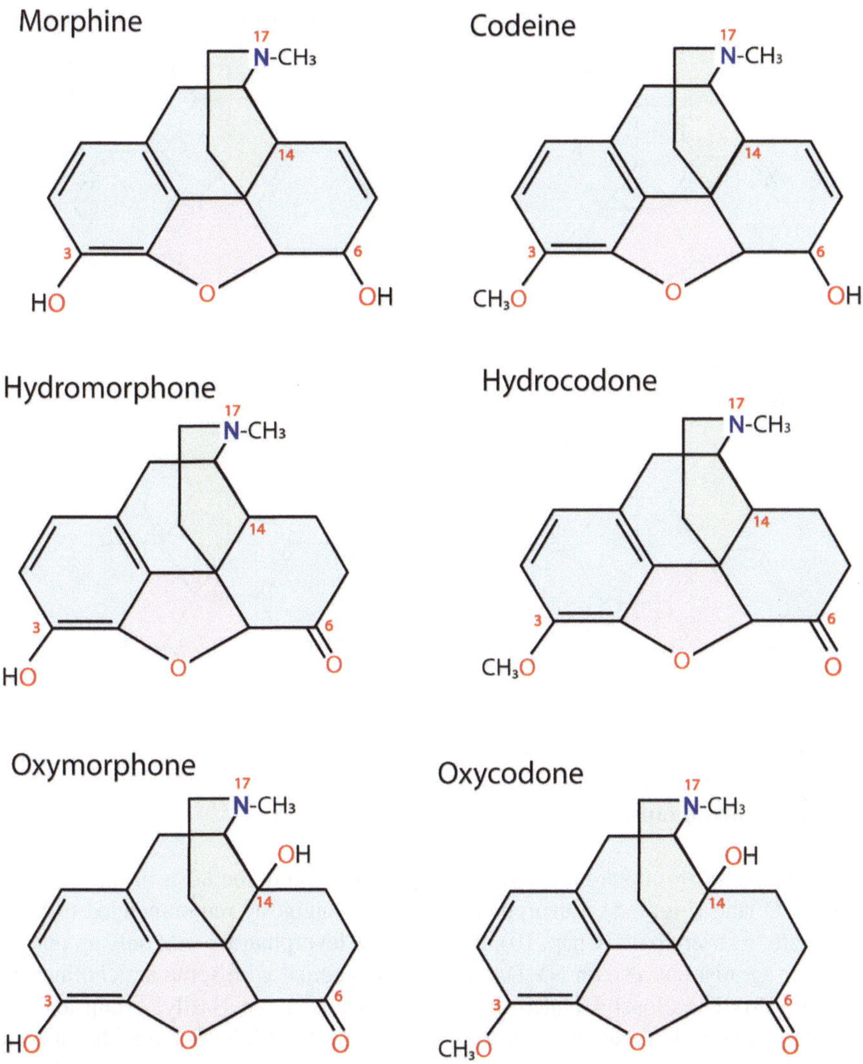

Fig. 8.3 Representative 4,5-epoxymorphinans (morphine-like drugs) with agonist effects

Naloxone and naltrexone were the first "pure" antagonists synthesized by substituting the –OH group in C6 with an oxygen, producing a ketone (Fig. 8.4). Naloxone and naltrexone are not pharmacologically active but effectively displace opioid agonists from their opioid receptors. The main difference between these two drugs is the duration of effects (see Table 8.2). Naloxone is the drug of choice to reverse opioid overdoses.

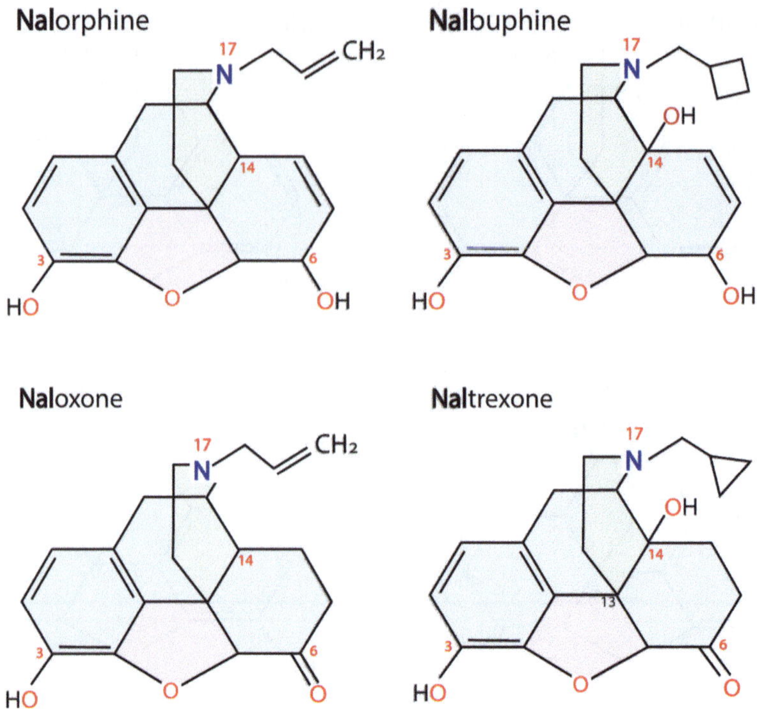

Fig. 8.4 Representative 4,5-epoxymorphinans with antagonist effects

8.3.1.2 Morphinans

Members of this group are tetracyclic compounds lacking the ether bridge of morphine's D ring (Fig. 8.5). Levorphanol is a good analgesic recommended to treat moderate to severe pain (Chap. 10). Interestingly, levorphanol is not only an opioid receptor agonist but also an NMDA receptor antagonist with some anticholinergic effects [19]. The closely related drug levallorphan has an N-allyl group and is, therefore, an opioid receptor antagonist. On the other hand, butorphanol has mixed agonist-antagonist effects and is used to treat moderate to severe migraine and pain (Fig. 8.5).

8.3.2 Benzomorphans

Further simplification of the morphine molecule produces tricyclic opioids that lack the phenanthrene ring. These compounds, known as benzomorphans, provided the first pharmacological tools to study kappa-opioid receptors. Activation of kappa-opioid receptors produces analgesia, dysphoria, and diuresis. Pentazocine and

Table 8.2 Clinically relevant opioids

	Affinity				
Agonists	μ	δ	κ	Use	Other targets
Morphine	+++	+	+	Analgesic Antitussive	
Heroin	+++	++	++	Analgesic Misused drug	
Codeine	+	+		Analgesic Prodrug	
Hydrocodone	+++		+	Analgesic Antitussive	
Oxycodone	+		++	Analgesic	
Oxymorphone	+++			Analgesic	
Hydromorphone	+++			Analgesic	
Levorphanol	+++	+	+	Analgesic	NMDA antagonist SERT and NET inhibitor
Tramadol	+			Analgesic	SERT and NET inhibitor
Tapentadol	++			Analgesic	NET inhibitor
Meperidine	+		+	Analgesic	ADRA2B (+)
Diphenoxylate	++			Antimotility agent[p]	
Loperamide	++			Antidiarrheal agent[p]	Ca^{+2} channel blocker
Fentanyl	+++			Analgesic Preanesthetic medication	
Sufentanil	+++	+	+	Anesthetic	
Alfentanil	+++			Anesthesia, maintenance Tx	
Remifentanil	+++			Anesthetic	
Methadone	+++			Analgesic Medication-assisted Tx	NMDA antagonist
Antagonists					
Naloxone	−−−	−−	−−	Opiate overdose	
Naltrexone	−−−	−−	−−	Opiate overdose Medication-assisted Tx	
Methylnaltrexone	−−−		−−	Opioid-induced constipation Tx[p]	
Naloxegol	−−−			Opioid-induced constipation Tx[p]	
Naldemedine	−−−	−	−	Opioid-induced constipation Tx[p]	
Agonist-antagonists					
Nalorphine	−−		+++	Analgesic Opiate overdose	

(continued)

Table 8.2 (continued)

	Affinity				
Agonists	μ	δ	κ	Use	Other targets
Nalmefene	−−	−−	P (+)	Alcohol dependence Tx Opiate overdose	
Pentazocine	P (−)	+	++	Analgesic Preanesthetic medication	
Butorphanol	P (−)		+++	Analgesic	
Nalbuphine	P (−)		++	Analgesic Pre- and postoperative medication	
Buprenorphine	P (+)	−	−−	Analgesic Medication-assisted Tx	

+++, ++, +, agonist strength; −−−, −−, −, antagonist strength; P (+), partial agonist; P (−), partial antagonist; [p] peripheral actions, *SERT* serotonin transporter, *NET* norepinephrine transporter

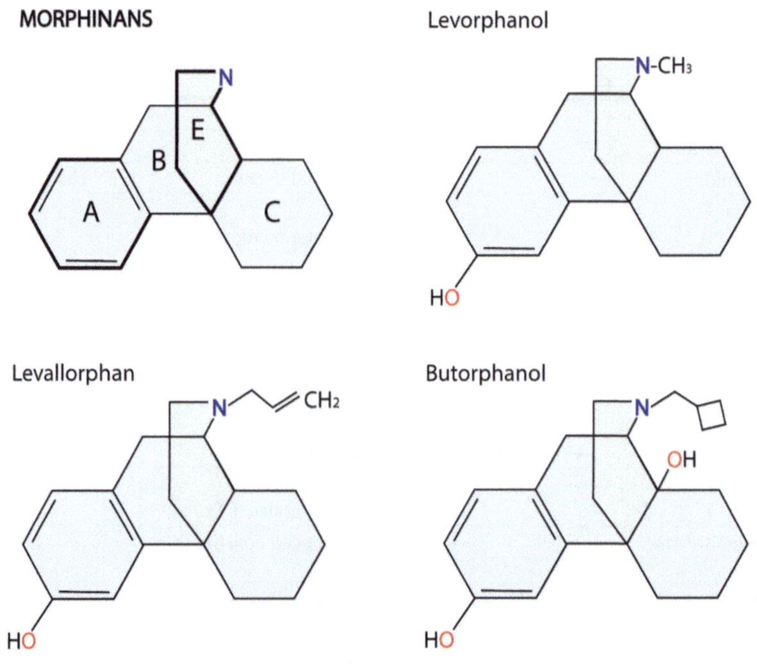

Fig. 8.5 Examples of morphinans (tetracyclic opioids): levorphanol (an opioid receptor agonist) and levallorphan and butorphanol (opioids with agonist and antagonist effects)

phenazocine are examples of these drugs. Pentazocine has mixed effects; if given alone, it produces analgesia but can counteract morphine's effects because of its weak antagonist actions at mu-opioid receptors. Phenazocine is also a potent analgesic but produces dysphoria, nightmares, and hallucinations (Fig. 8.6).

8.3.3 Phenylpiperidine Opioids

Phenylpiperidine opioids are bicyclic compounds with an aromatic phenyl ring (A) and a nitrogen-containing piperidine ring (E). Pethidine (also known as meperidine), diphenoxylate, and loperamide are some members of this group. On the other hand, fentanyl and fentanyl analogs, sometimes called anilino-piperidines, have the same A and E rings, but with an NH group joining them (Fig. 8.6). Among phenylpiperidine compounds, diphenoxylate and loperamide are very effective antidiarrheal opioids, while fentanyl and its analogs are among the most potent opioids with high analgesic efficacy, dependence liability, and respiratory depressant effects.

8.3.4 Diphenylheptane Opioids

This group has two phenyl rings, a tertiary N and a seven-carbon group. Methadone and propoxyphene are some of the best known examples of this group. In addition to their agonist activity at opioid receptors, both drugs act as NMDA receptor antagonists with varying potencies [20]. Methadone is a very effective analgesic with long-lasting effects (Chap. 10) commonly used as a medication for people with opioid dependence (Chaps. 14 and 15). Dextropropoxyphene is a weak analgesic drug with restricted clinical use.

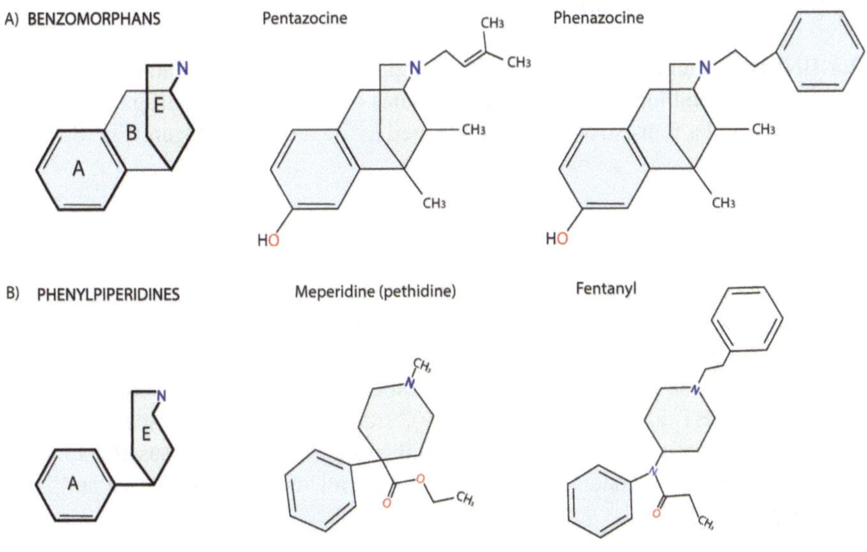

Fig. 8.6 (**a**) Examples of benzomorphans, the tricyclic opioids with high affinity for kappa-opioid receptors. (**b**) Examples of fentanyl and fentanyl analogs, highly potent phenylpiperidine opioids

8.3.5 *Miscellaneous*

8.3.5.1 Oripavine Derivatives

Oripavine is a naturally occurring opioid and a thebaine metabolite. Both compounds are abundant in another poppy plant species, *P. orientale*. They differ from other morphine-like opioids in their lack of analgesic effects and the presence of a diene system (two double carbon-carbon bonds) in ring C. Buprenorphine, another effective opioid used to treat persons with substance use disorders, is also an oripavine derivative [21].

8.3.5.2 Etonitazine and Other Benzimidazole Derivatives

In the late 1950s, the Swiss pharmaceutical company CIBA developed a series of compounds based not on the phenanthrene nucleus but on benzimidazole, formed by a benzene ring and an imidazole (an acyclic organic compound with two nitrogen and three carbon atoms; Fig. 8.7). Etonitazine is 60 times more potent than morphine in humans (1500× in rats). However, neither this drug nor other benzimidazole opioids became commercial analgesics because they did not offer advantages over opioids already available and had dependence liability. Unfortunately, some of these compounds have been recently incorporated to the new psychoactive substances marketed through the darknet (see Chap. 16).

8.3.5.3 Nor-binaltorphimine (Nor-BNI)

Nor-BNI is a potent and highly selective kappa-opioid receptor antagonist with a structure that combines two mirrored templates of morphine-like drugs (hence the prefix "bi") and a four-carbon radical attached to the tertiary nitrogen (Fig. 8.7).

8.3.5.4 Tramadol

Tramadol is a prodrug; therefore, it must be converted into its pharmacologically active metabolite O-desmethyl-tramadol by enzymatic action to produce analgesia. Tramadol, on its own, has little affinity for opioid receptors but acts as a serotonin and norepinephrine reuptake inhibitor. By blocking serotonin and norepinephrine transporters (SET and NET, respectively), tramadol activates descending pain inhibitory pathways and can act as a stimulant opioid at high doses (Chap. 10). These properties have made tramadol a preferred misused drug in Africa and Asia (see Chap. 5).

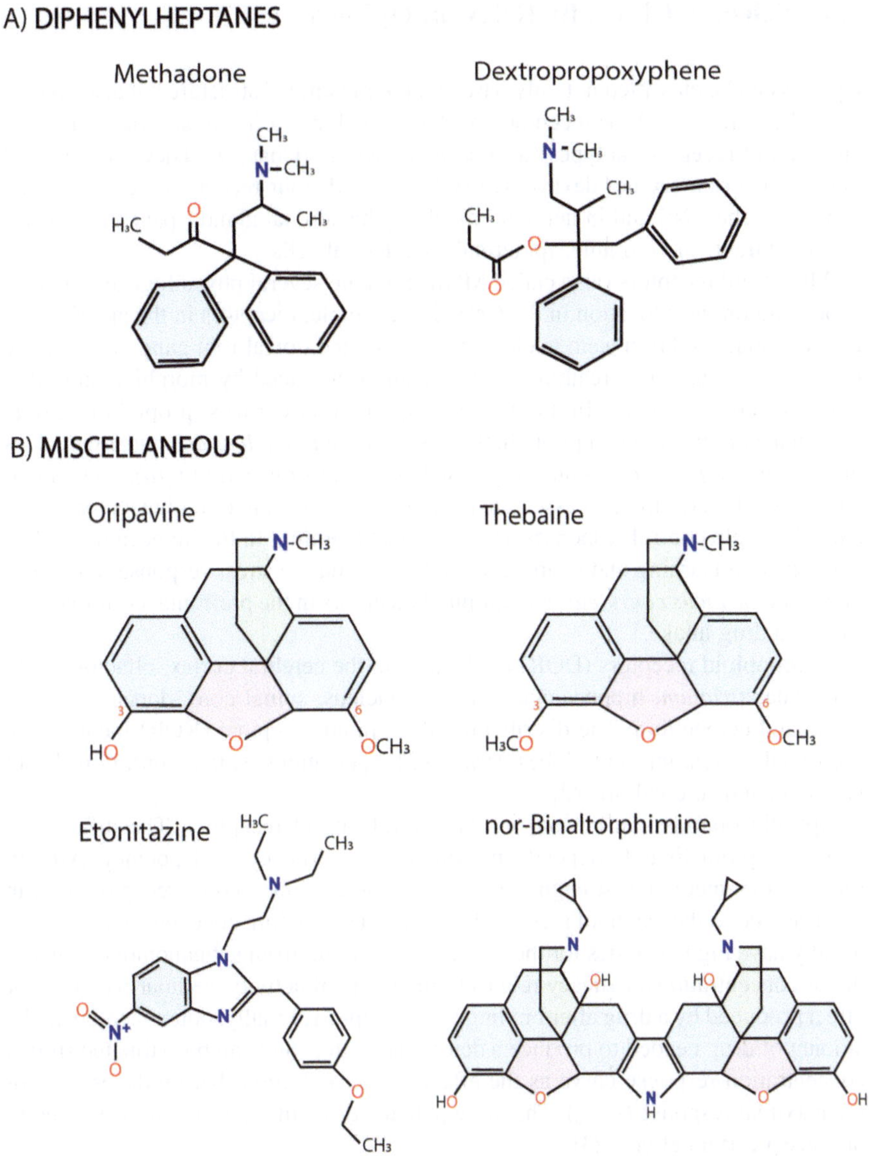

Fig. 8.7 (**a**) Chemical structures of methadone and dextropropoxyphene, two diphenylheptanes. Members of this group have affinity for opioid receptors and other non-opioid receptors. (**b**) Miscellaneous opioids

8.4 Selected Clinically Relevant Opioids

Opioids can be classified not only based on their chemical structure but also considering their effects and the receptors to which they bind to. There are four stereospecific opioid receptor subtypes (μ, δ, k, and ORL-1) (Chap. 9). They are activated only by levorotatory not dextrorotatory ligands. All four receptors are widely distributed in the CNS and in nerve terminals of the adrenal glands, pancreas, gastrointestinal tract, and immune, epidermal, and dermal cells.

Mu-opioid receptors (also called MOR) regulate several physiological functions depending on their location in the CNS. For example, receptors in the anterior cingulate, thalamus, brain stem nuclei, spinal cord, and dorsal root ganglia contribute to the analgesia, slow breathing, and relaxation produced by morphine and other opioid receptor agonists. In the parietal and temporal cortices, μ-opioid receptors participate in sensory perception. In the reward system (*nucleus accumbens*, nucleus of the *caudate-putamen*, ventral tegmental area, and *substantia nigra*), they modulate motivation, desire, and associative learning. In addition, opioid receptors in the amygdala help control responses and emotional learning. In the hippocampus, they participate in learning and neurogenesis. They modulate stress responses and withdrawal in the *locus coeruleus*, and μ-opioid receptors in the prefrontal cortex control food and drug intake.

Delta-opioid receptors (DOR) are located in the cerebral cortex, olfactory bulb, amygdala, *striatum*, hippocampus, pontine nucleus, spinal cord, dorsal root ganglion, and cerebellum. The distribution of κ-opioid receptors (KOR) includes the prefrontal cortex, temporal lobes, striatum, hippocampus, spinal cord, dorsal root ganglion, and cerebellum [22].

Specific opioids can act on one or several opioid receptors. The differences between opioid ligands depend on their affinity, efficacy, and potency. Affinity refers to the interaction strength between an opioid ligand and its receptor and can be measured as the extent of receptor binding at a given drug concentration. Opioids usually have high affinities for their receptors, ranging from subnanomolar to nanomolar concentrations. Efficacy relates to the grade of activity, cellular response, or effect produced by a drug after binding to its receptors. Finally, potency refers to the amount of drug needed to produce a determined effect and can be estimated from a concentration-response curve as the effective concentration that produces 50% of the maximal response (EC_{50}). The more potent a drug, the less amount is needed to produce a certain effect [23].

8.4.1 Opioid Receptor Agonists

Opioid agonists have both affinity and efficacy. They mimic the effects of endogenous opioid peptides when they bind to their receptors. Low-efficacy agonists need to occupy a large fraction of the available receptors to produce their effects, whereas high-efficacy agonists do so with less receptor occupancy. The efficacy of agonists depends on the maximum response that can be produced in a specific tissue (intrinsic activity) and on the receptor site where the drug binds.

A drug that binds to the same site as the endogenous opioids is called an orthosteric agonist, whereas it is an allosteric agonist if it binds to a different region (from the ancient Greek words "ortho" = right, straight; "allos" = other, different; and "steros" = form) [23, 24]. Other differences in activity and efficacy of opioid agonists are related to the stimulation of various opioid receptor subtypes and structural and genetic differences that change opioid receptor sensitivity (Chap. 9).

Morphine is the archetypal opioid receptor agonist. It binds to μ-, δ-, and κ-opioid receptors. All morphine-like agonists are analgesics and have some common adverse effects associated with μ-opioid receptor activation, such as tolerance and addiction development. The specific pharmacological profile of opioids depends on their relative potency, maximum effect, and duration of action. Table 8.2 shows clinically relevant opioids according to their pharmacology and function.

A complete characterization of clinically used opioids is beyond the scope of this chapter, but some data could be helpful to keep in mind. For example, heroin is more potent and lipophilic than morphine. Codeine and hydrocodone are prescribed primarily as antitussive agents but are also effective as pain relievers alone or combined with aspirin or acetaminophen. For moderate to severe pain management, oxycodone is available in continuous-release preparations (OxyContin®), alone or with acetaminophen or aspirin. Oxymorphone can be administered orally or as a suppository, and levorphanol is effective orally and injected [19]. Hydromorphone, a short-acting potent opioid drug, is used when other first-line treatments fail.

Tramadol is usually prescribed for postoperative pain due to its actions as a μ-opioid receptor agonist and as a norepinephrine and serotonin reuptake inhibitor [25]. A closely related compound, tapentadol is also a μ-opioid receptor agonist and norepinephrine reuptake inhibitor used as an analgesic [26]. The prolonged-release formulation of tapentadol is used to treat specific types of chronic pain with less tolerance development than other opioids [27].

Fentanyl, sufentanil, alfentanil, and remifentanil are short-acting agonists with high lipophilicity used as anesthesia adjuvants. Meperidine is an analgesic that produces less constipation than morphine but has similar dependence liability. In addition, meperidine's metabolite normeperidine can produce delirium, hyperreflexia, and seizures.

Loperamide or the combination of diphenoxylate with atropine lacks euphoric or analgesic effects when given orally. Both opioids are effective antidiarrheal drugs due to their inhibitory actions at the presynaptic μ-opioid receptors of the enteric nervous system [28].

Methadone is a μ-opioid receptor agonist and an N-methyl-D-aspartate (NMDA) receptor antagonist with a prolonged duration of action. Compared to morphine, methadone has a longer plasma half-life, improved analgesic efficacy, and reduced opioid tolerance and can be used for chronic pain treatment or as medication for opioid-assisted treatment in heroin-dependent persons (see Chaps. 14 and 15) [23, 24, 29, 30].

Overall, the use of illicit or prescription μ-opioid receptor agonists can produce euphoria, sedation, miosis, and analgesia. High doses can cause fatal respiratory depression, and frequent use leads to opioid dependence and withdrawal (Chap. 13).

8.4.2 *Opioid Receptor Antagonists*

Opioid antagonists are substances with affinity for opioid receptors but without pharmacological efficacy. These compounds can bind to the receptor at the primary site and exert direct competition with agonists (competitive antagonists) or bind to another location to block, reduce, or counteract the action of the agonists [23, 31].

The most common opioid receptor competitive antagonists are naloxone and naltrexone, two compounds with affinity for μ-, δ-, and κ-opioid receptors. Naloxone is the preferred antidote to counteract opioid-induced overdoses due to its rapid distribution and short half-life (Chaps. 5 and 7). Naltrexone has a much longer duration of action and is prescribed to decrease craving and blocks the rewarding effects of alcohol and heroin (Chaps. 14 and 15).

When naloxone is injected or administered as a nasal spray, it rapidly counteracts the respiratory depression caused by an opioid overdose because it displaces heroin or any other opioid agonist from opioid receptors. Antagonism of the central opioid receptors arouses respiratory drive, increases alertness, finishes analgesia and euphoria, produces mydriasis, and can precipitate a withdrawal syndrome [31, 32]. On the other hand, naloxone is poorly absorbed when given orally. This makes it an ideal drug to combine with potent opioid agonists because if they are crushed and injected, naloxone will block their psychoactive effects, acting as an abuse-deterrent.

Some antagonists do not readily cross the blood-brain barrier but exert their effects by blocking μ-opioid receptors at peripheral nerve terminals in the bronchial smooth muscle and the digestive tract. Antagonists such as methylnaltrexone, naloxegol, and naldemedine effectively reverse opioid-induced constipation without compromising opioid analgesic effects or causing withdrawal [33].

8.4.3 *Agonist-Antagonists and Partial Agonists/Antagonists*

Several drugs act as mixed agonist-antagonists due to the variety of opioid receptors they can target. These agents can be agonists of an opioid receptor type and antagonists of another. Nalorphine, for example, is a κ-opioid receptor agonist but also a μ-opioid receptor antagonist. This drug produces analgesia but counteracts the respiratory depression caused by opioid agonists and can precipitate withdrawal in opioid-dependent subjects when given at high doses [34].

Other opioids have high affinity for one or various opioid receptor subtypes but may have only partial efficacy. These drugs can occupy all available receptors in a given tissue without producing a maximum effect. For example, nalmefene, a naltrexone analog, is a μ- and δ-opioid receptor antagonist with a partial agonistic effect on κ-opioid receptors. This particular pharmacological profile makes it useful for alcohol dependence management and to treat acute opioid overdoses [35].

Pentazocine, butorphanol, and nalbuphine are agonist-antagonist opioids with high affinity for κ-opioid receptors but also act as partial antagonists of μ-opioid receptors. In addition, pentazocine has a weak agonistic effect at δ-opioid receptors. Pentazocine is a good analgesic with a rapid onset and an action duration shorter than morphine. It produces analgesia by activating κ-opioid receptors. In addition, many drugs can cause drug-drug interactions when they are co-administered with pentazocine. This possibility must be taken into account when prescribing it. Pentazocine is available in tablets, either alone or combined with naloxone.

As a nasal spray, butorphanol is effective for migraine treatment. It is also a good analgesic but may produce unpleasant effects such as dysphoria, nightmares, and anxiety.

Nalbuphine is another mixed agonist-antagonist opioid as potent as morphine, but it produces its analgesic effects via κ- not μ-opioid receptors' activation. High doses of pentazocine and nalbuphine can produce hallucinations and dysphoria. They can also partially antagonize the pruritus, respiratory depression, and analgesic effects of morphine and precipitate withdrawal in opioid users [34, 36].

Finally, buprenorphine, unlike the agonists-antagonists described above, is a partial agonist of the μ-opioid receptor with antagonistic actions at the δ- and k-opioid receptors. This opioid is prescribed for postoperative pain treatment, has a long-lasting analgesic effect in patients with chronic pain, and is effective in medication-assisted treatment for opioid use disorders (see Chaps. 14 and 15). Abrupt discontinuation of buprenorphine may cause opioid withdrawal syndrome. Unlike morphine, buprenorphine produces fewer side effects such as respiratory depression, euphoria, addiction, or dependence [30, 34, 37].

8.4.4 Inverse Agonists

The classical receptor theory postulates that receptors are inactive in the absence of ligands. Agonists are considered to bind to receptors and produce an effect. On the other hand, antagonists bind to the receptors but lack pharmacological efficacy. Accumulating evidence has shown that G protein-coupled receptors (GPCRs) are not inert but have varying degrees of basal, or "constitutive," activity in the absence of ligands [38, 39]. Moreover, according to the two-state model of GPCR activation, receptors can undergo a ligand-independent isomerization from an inactive (R) to an active (R*) state under specific circumstances [38, 40].

The two-state receptor theory redefines agonist drugs as ligands that bind to R and stabilize them as R*. Conversely, it proposes that "inverse agonists" bind with high affinity to R*, suppress their basal constitutive activity, and favor the R state. These compounds are also known as negative antagonists because, unlike neutral antagonists (which do not change the R/R* balance), they alter receptor signaling and produce effects opposite to those produced by agonists. Interestingly, inverse agonists behave as neutral antagonists in the absence of constitutive activity [24, 38].

Opioid receptors can be constitutively active. In particular, chronic exposure to morphine increases the ratio of R* μ-opioid receptors with respect to R, and this effect has been correlated with the development of tolerance and dependence [41, 42]. Moreover, the inverse agonist properties of some antagonists can only be revealed after opioid agonist exposure. For example, there is evidence that naloxone and naltrexone act as inverse agonists in opioid-dependent preparations by reducing opioid receptor constitutive signaling, whereas other compounds such as 6-β-naltrexol and CTOP behave as neutral antagonists [43, 44].

Although some preclinical studies have explored the therapeutic utility of inverse agonists, the results are still inconclusive [38]. Nevertheless, inverse agonism must be considered when using drugs classically considered neutral antagonists to interpret experimental observations adequately.

8.5 Final Considerations

Available natural, semisynthetic, and synthetic opioids are valuable tools to treat several ailments effectively. However, separating the desired from the undesired effects of opioids remains a significant challenge despite intensive research and new opioid development. On the other hand, the extensive work of opioid synthesis has provided invaluable medications as well as psychoactive drugs with high dependence liability that are now illegally marketed as new psychoactive substances in the darknet (Chap. 16). A better understanding of the pharmacological profiles of individual drugs, tolerance, and dependence development (Chap. 13) is mandatory to take advantage of the many benefits provided by opioids without harming people and risking lives.

References

1. Veening JG, Barendregt HP. The effects of beta-endorphin: state change modification. Fluids Barriers CNS. 2015;12:3. https://doi.org/10.1186/2045-8118-12-3.
2. Pilozzi A, Carro C, Huang X. Roles of β-endorphin in stress, behavior, neuroinflammation, and brain energy metabolism. Int J Mol Sci. 2020;22:338. https://doi.org/10.3390/ijms22010338.
3. Cahill C, Tejeda HA, Spetea M, Chen C, Liu-Chen L-Y. Fundamentals of the dynorphins/kappa opioid receptor system: from distribution to signaling and function. Handb Exp Pharmacol. 2021;271:1–19. https://doi.org/10.1007/164_2021_433.
4. Gu Z-H, Wang B, Kou Z-Z, Bai Y, Chen T, Dong Y-L, et al. Endomorphins: promising endogenous opioid peptides for the development of novel analgesics. Neurosignals. 2017;25:98–116. https://doi.org/10.1159/000484909.

5. Fricker LD, Margolis EB, Gomes I, Devi LA. Five decades of research on opioid peptides: current knowledge and unanswered questions. Mol Pharmacol. 2020;98:96–108. https://doi.org/10.1124/mol.120.119388.
6. Koo CY. Respiratory effects of opioids in perioperative medicine. Open Anesthesiol J. 2011;5:23–34. https://doi.org/10.2174/1874321801105010023.
7. Ehsan Z, Mahmoud M, Shott SR, Amin RS, Ishman SL. The effects of anesthesia and opioids on the upper airway: a systematic review. Laryngoscope. 2016;126:270–84. https://doi.org/10.1002/lary.25399.
8. Zebraski SE, Kochenash SM, Raffa RB. Lung opioid receptors: pharmacology and possible target for nebulized morphine in dyspnea. Life Sci. 2000;66:2221–31. https://doi.org/10.1016/S0024-3205(00)00434-3.
9. Saito R, Takano Y, Kamiya H-O. Roles of substance P and NK(1) receptor in the brainstem in the development of emesis. J Pharmacol Sci. 2003;91:87–94. https://doi.org/10.1254/jphs.91.87.
10. Larson MD, Behrends M. Portable infrared pupillometry: a review. Anesth Analg. 2015;120:1242–53. https://doi.org/10.1213/ANE.0000000000000314.
11. Khansari M, Sohrabi M, Zamani F. The usage of opioids and their adverse effects in gastrointestinal practice: a review. Middle East J Dig Dis. 2013;5:5–16.
12. Krantz MJ, Palmer RB, Haigney MCP. Cardiovascular complications of opioid use: JACC state-of-the-art review. J Am Coll Cardiol. 2021;77:205–23.
13. Fountas A, Van Uum S, Karavitaki N. Opioid-induced endocrinopathies. Lancet Diabetes Endocrinol. Elsevier Ltd. 2020;8:68–80. https://doi.org/10.1016/S2213-8587(19)30254-2.
14. Antony T, Alzaharani SY, El-Ghaiesh SH. Opioid-induced hypogonadism: pathophysiology, clinical and therapeutics review. Clin Exp Pharmacol Physiol. 2020;47:741–50. https://doi.org/10.1111/1440-1681.13246.
15. de Vries F, Bruin M, Lobatto DJ, Dekkers OM, Schoones JW, van Furth WR, et al. Opioids and their endocrine effects: a systematic review and meta-analysis. J Clin Endocrinol Metab. 2020;105:1020–9. https://doi.org/10.1210/clinem/dgz022.
16. Nguyen E, Lim G, Ross SE. Evaluation of therapies for peripheral and neuraxial opioid-induced pruritus based on molecular and cellular discoveries. Anesthesiology. 2021;135:350–65. https://doi.org/10.1097/ALN.0000000000003844.
17. Liu XY, Liu ZC, Sun YG, Ross M, Kim S, Tsai FF, et al. Unidirectional cross-activation of GRPR by MOR1D uncouples itch and analgesia induced by opioids. Cell. Elsevier Inc. 2011;147:447–58. https://doi.org/10.1016/j.cell.2011.08.043.
18. Bigliardi PL, Tobin DJ, Gaveriaux-Ruff C, Bigliardi-Qi M. Opioids and the skin – where do we stand? Exp Dermatol. 2009;18:424–30. https://doi.org/10.1111/j.1600-0625.2009.00844.x.
19. Prommer E. Levorphanol: revisiting an underutilized analgesic. Palliat Care Res Treat. 2014;8:7–10. https://doi.org/10.4137/PCRT.S13489.
20. Ebert B, Thorkildsen C, Andersen S, Christrup LL, Hjeds H. Opioid analgesics as noncompetitive N-methyl-d-aspartate (NMDA) antagonists. Biochem Pharmacol. 1998;56:553–9. https://doi.org/10.1016/S0006-2952(98)00088-4.
21. Hosztafi S. Recent advances in the chemistry of oripavine and its derivatives. Adv Biosci Biotechnol. 2014;05:704–17. https://doi.org/10.4236/abb.2014.58084.
22. Kibaly C, Xu C, Cahill CM, Evans CJ, Law P-YY. Non-nociceptive roles of opioids in the CNS: opioids' effects on neurogenesis, learning, memory and affect. Nat Rev Neurosci. Springer US. 2019;20:5–18. https://doi.org/10.1038/s41583-018-0092-2.
23. Trescot AM, Datta S, Lee M, Hansen H. Opioid pharmacology. Pain Physician. 2008;11:S133–53.
24. Cruz SL, Granados-Soto V. Opioids and opiates: pharmacology, abuse, and addiction. In: Pfaff DW, Volkow ND, editors. Neuroscience in the 21st century. New York: Springer New York; 2016. p. 3625–57. https://doi.org/10.1007/978-1-4939-3474-4_156.
25. Subedi M, Bajaj S, Kumar MS, YC M. An overview of tramadol and its usage in pain management and future perspective. Biomed Pharmacother. Elsevier. 2019;111:443–51. https://doi.org/10.1016/j.biopha.2018.12.085.

26. Raffa RB, Buschmann H, Christoph T, Eichenbaum G, Englberger W, Flores CM, et al. Mechanistic and functional differentiation of tapentadol and tramadol. Expert Opin Pharmacother. 2012;13:1437–49. https://doi.org/10.1517/14656566.2012.696097.
27. Deeks ED. Tapentadol prolonged release: a review in pain management. Drugs. Springer International Publishing. 2018;78:1805–16. https://doi.org/10.1007/s40265-018-1007-2.
28. Jain M, Wylie WP. Diphenoxylate and atropine [Internet]. In: StatPearls. StatPearls Publishing; 2021.
29. Friedman A, Nabong L. Opioids. Phys Med Rehabil Clin N Am. 2020;31:289–303. https://doi.org/10.1016/j.pmr.2020.01.007.
30. Inturrisi CE. Clinical pharmacology of opioids for pain. Clin J Pain. 2002;18:S3–13. https://doi.org/10.1097/00002508-200207001-00002.
31. Aronson JK. Opioid receptor antagonists. In: Meyler's side effects of drugs. 16th ed. Oxford: Elsevier; 2016. p. 381. https://doi.org/10.1016/B978-0-444-53717-1.01184-7.
32. Theriot J, Sabir S, Azadfard M. Opioid antagonists [Internet]. In: StatPearls. StatPearls Publishing; 2021.
33. Pergolizzi JV Jr, Christo PJ, LeQuang JA, Magnusson P. The use of peripheral μ-opioid receptor antagonists (PAMORA) in the management of opioid-induced constipation: an update on their efficacy and safety. Drug Des Devel Ther. 2020;14:1009–25. https://doi.org/10.2147/DDDT.S221278.
34. Vardanyan RS, Hruby VJ. Analgesics. In: Synthesis of essential drugs. Elsevier; 2006. p. 19–55. https://doi.org/10.1016/B978-044452166-8/50003-0.
35. Swift RM. Naltrexone and nalmefene: any meaningful difference? Biol Psychiatry. Elsevier. 2013;73:700–1. https://doi.org/10.1016/j.biopsych.2013.03.002.
36. Gress K, Charipova K, Jung JW, Kaye AD, Paladini A, Varrassi G, et al. A comprehensive review of partial opioid agonists for the treatment of chronic pain. Best Pract Res Clin Anaesthesiol. Elsevier Ltd. 2020;34:449–61. https://doi.org/10.1016/j.bpa.2020.06.003.
37. Urits I, Pham C, Swanson D, Berardino K, Bandi P, Amgalan A, et al. The utilization of buprenorphine in chronic pain. Best Pract Res Clin Anaesthesiol. Elsevier Ltd. 2020;34:355–68. https://doi.org/10.1016/j.bpa.2020.06.005.
38. Berg KA, Clarke WP. Making sense of pharmacology: inverse agonism and functional selectivity. Int J Neuropsychopharmacol. 2018;21:962–77. https://doi.org/10.1093/ijnp/pyy071.
39. Burford NT, Traynor JR, Alt A. Positive allosteric modulators of the μ-opioid receptor: a novel approach for future pain medications. Br J Pharmacol. John Wiley & Sons, Ltd. 2015;172:277–86. https://doi.org/10.1111/bph.12599.
40. Seifert R, Wenzel-Seifert K. Constitutive activity of G-protein-coupled receptors: cause of disease and common property of wild-type receptors. Naunyn Schmiedeberg's Arch Pharmacol. 2002;366:381–416. https://doi.org/10.1007/s00210-002-0588-0.
41. Meye FJ, van Zessen R, Smidt MP, Adan RAH, Ramakers GMJ. Morphine withdrawal enhances constitutive -opioid receptor activity in the ventral tegmental area. J Neurosci. 2012;32:16120–8. https://doi.org/10.1523/JNEUROSCI.1572-12.2012.
42. Wang Z, Bilsky EJ, Porreca F, Sadée W. Accelerated communication: constitutive μ opioid receptor activation as a regulatory mechanism underlying narcotic tolerance and dependence. Life Sci. 1994;54:PL339–50. https://doi.org/10.1016/0024-3205(94)90022-1.
43. Sirohi S, Dighe SV, Madia PA, Yoburn BC. The relative potency of inverse opioid agonists and a neutral opioid antagonist in precipitated withdrawal and antagonism of analgesia and toxicity. J Pharmacol Exp Ther. 2009;330:513–9. https://doi.org/10.1124/jpet.109.152678.
44. Cruz SL, Villarreal JE, Volkow ND. Further evidence that naloxone acts as an inverse opiate agonist: implications for drug dependence and withdrawal. Life Sci. 1996;58:PL381–9. https://doi.org/10.1016/0024-3205(96)00250-0.

Chapter 9
Opioid Receptors and Neuronal Signal Transduction

César J. Carranza-Aguilar, Zyanya P. Espinosa-Riquer, Frida L. Martínez-Cuevas, and Silvia L. Cruz

Abstract This chapter provides information on the classification, structure, and properties of the main opioid receptor subtypes, their canonical and non-canonical signaling pathways, and the variety of mechanisms influencing opioid signal transduction in neurons. Opioid receptors are members of the seven transmembrane G protein-coupled receptors. These receptors signal via inhibitory G proteins. The α subunit of G protein inhibits the adenylyl cyclase enzyme and reduces cyclic adenosine monophosphate and protein kinase A activity, altering gene expression. The βγ subunit increases K^+ conductance and reduces Ca^{2+} entrance, causing hyperpolarization. Furthermore, a subset of endosomal signals arises after receptor phosphorylation and β-arrestin activation. Biased agonists and allosteric modulators regulate opioid receptor effects. Alternative splicing, polymorphisms, dimerization, and epigenetic factors also determine the heterogeneity, expression, and function of opioid receptors.

Keywords Opioid receptor · Opioid signaling · G protein · β-Arrestin · Internalization

C. J. Carranza-Aguilar (✉)
Department of Pharmacobiology, Center for Research and Advanced Studies (Cinvestav), IPN, Mexico City, Mexico

Institute of Neurobiology, National Autonomous University of Mexico, Juriquilla, Querétaro, Mexico

Z. P. Espinosa-Riquer · F. L. Martínez-Cuevas
Department of Pharmacobiology, Center for Research and Advanced Studies (Cinvestav), IPN, Mexico City, Mexico

S. L. Cruz
Department of Pharmacobiology, Center for Research and Advanced Studies (Cinvestav), IPN, Mexico City, Mexico

Opioids Working Group, Global Studies Seminar, Faculty of Medicine, National Autonomous University of Mexico, Mexico City, Mexico

S. L. Cruz (ed.), *Opioids*, https://doi.org/10.1007/978-3-031-09936-6_9

9.1 Opioid Receptor Overview and Classification

Since 1967, pharmacological evidence suggested the existence of specific binding sites for opioids, but the identification of specific opioid receptors as biochemical entities with different subtypes occurred in the early 1990s (Chap. 1).

Opioid receptors are G protein-coupled receptors (GPCRs) belonging to the rhodopsin-like family or class A GPCRs [1]. Classical opioid receptor subtypes are μ (or MOR), δ (or DOR), and κ (or KOR) [2]. The fourth member of this group is the nociceptin/orphanin FQ peptide (NOP) receptor. NOP was originally considered an "orphan" opioid receptor because its endogenous ligand had not been identified at the time of its characterization. NOP has high homology with the classic opioid receptors, but distinct pharmacology [3–5].

All opioid receptors exhibit 49–58% primary sequence identity and are structurally similar [6]. They are distributed across the brain, spinal cord, skin, and gastrointestinal tract [7], mediating distinct effects.

Like other GPCRs, opioid receptors are dynamic structures that can assume diverse states between inactive and active conformations [8] and be regulated by binding ligands and small molecules, such as sodium ions and G proteins. In addition, opioid receptors display considerable functional complexity – oligomerization, constitutive activity, complex trafficking and recycling, biased agonism, and allosteric modulation – as discussed in the following sections.

9.1.1 Structure

Opioid receptors consist of seven transmembrane (TM) segments linked by three intracellular and three extracellular loops, an amino-terminus extracellular domain (N-terminal), and an intracytoplasmic carboxyl-terminal (C-terminal) domain. The N-terminal possesses glycosylation sites, and the C-terminal contains sites for other post-translational modifications such as phosphorylation, palmitoylation, or ubiquitination [6]. In addition, at the extracellular surface of the third transmembrane segment (TM3) and in the second extracellular loop, opioid receptors have a highly conserved pair of cysteine (Cys) amino-acid residues, which form a receptor-stabilizing disulfide bond (Fig. 9.1) [4, 9].

Opioid receptors have orthosteric and allosteric binding sites. The orthosteric site is where endogenous opioid peptides and receptor agonists bind and trigger downstream intracellular signaling pathways [10]. This site involves amino-acid residues in TM2, TM3, TM6, and TM7 segments, although some variants exist depending on the receptor subtype and ligand [11]. In general, the orthosteric binding site has a conserved anionic aspartic acid residue that forms a salt bridge with the positively charged amino group of opioid ligands, a cavity to accommodate substitutions of opioid molecules, and a flat surface to enable recognition of the phenolic moiety common to many opioids (Chap. 8) [12].

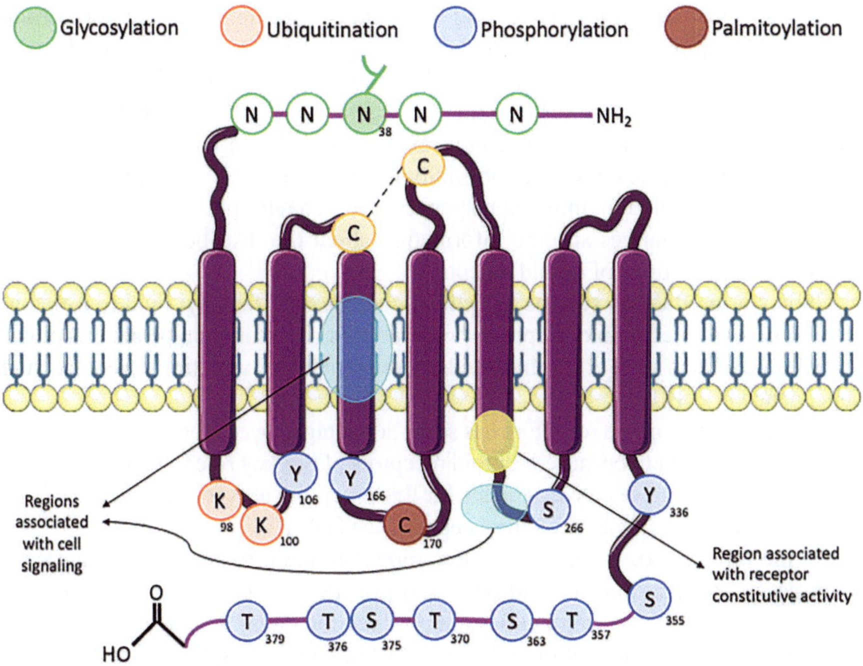

Fig. 9.1 Model of μ-opioid receptor structure showing the seven transmembrane segments with the N-terminal and C-terminal domains. Different colors indicate sites of post-translational modifications. Turquoise and yellow regions are associated with cell signaling and constitutive activity, respectively

On the other hand, allosteric sites are distinct pockets on the receptors, separated from the orthosteric binding site, located either in extracellular loops, TM segments and lipid interfaces, or intracellular regions [13]. Class A GPCRs, to which opioid receptors belong, have a conserved allosteric binding site that modulates receptors' activation by binding sodium ions (Na^+). This sodium binding site is formed by conserved amino acids located on TM2 and TM3 [14, 15].

9.1.2 Specific Features of Opioid Receptor Types and Subtypes

The three classic opioid receptor subtypes, μ (MOR), δ (DOR), and κ (KOR), as well as NOP are encoded by single multi-exonic genes designated as *Oprm1*, *Oprd1*, *Oprk1*, and *Oprl1*, also called MOR1, DOR1, KOR1, and ORL1, respectively [4, 16]. These four genes code for proteins of distinct lengths, depending on the number of amino acids: 398 for MOR, 372 for DOR, 380 for KOR, and 367 for NOP (IUPHAR/BPS) [001].

Central and peripheral neurons and neuroendocrine, immune, and ectodermal cells express opioid receptors. In the brain, μ-, κ-, and NOP receptors are located throughout the cortex, midbrain, and hindbrain; in contrast, δ-opioid receptors have a more focal distribution throughout the limbic and prelimbic brain regions. Different or overlapping cell populations express opioid receptor subtypes because they play essential roles in various physiological functions such as analgesia, mood, cardiovascular regulation, immune responses, and behavior [4].

Table 9.1 summarizes specific information about the distribution, ligands, and physiological functions of opioid receptors.

Clinically used drugs such as morphine, codeine, and fentanyl bind to μ-opioid receptors to produce antinociception; however, this receptor subtype also mediates undesirable effects such as respiratory depression, nausea, vomiting, tolerance, and dependence [17, 18].

Activation of δ-opioid receptors has some advantages over μ-opioid receptors for chronic pain treatment because δ-opioid receptors play a key role in regulating emotional responses, and selective agonists for this receptor subtype produce anxiolysis with fewer adverse effects, including dependence [18, 19].

On the other hand, κ-opioid receptors modulate pain, motor activity, and consciousness, produce diuresis, and have anti-pruritic effects [20, 21]. In addition, activation of κ-opioid receptors does not produce the common side effects associated with the other classical opioid receptors but causes dysphoria, sedation, and anxiety [21].

9.2 Signaling Pathways of Opioid Receptors

9.2.1 G Protein Pathway

Opioid receptors are coupled to the pertussis toxin-sensitive guanosine triphosphate (GTP)-binding proteins G_i/G_0, which are formed by the α_i, β, and γ subunits. When the receptors are in an inactive conformation, the α_i subunit binds to guanosine diphosphate (GDP), holding the three subunits together [22].

The binding of an opioid ligand to its receptor produces a conformational change that causes the exchange of GDP for GTP in the G protein's α_i subunit and disengages the βγ subunit complex [23]. Subsequently, opioid receptors uncouple from the G protein and remain momentarily unbound to any signaling system. Once freed, the α_i subunit inhibits the adenylyl cyclase (AC) enzyme, reducing cAMP production and protein kinase A (PKA) activity [23]. Such a reduction decreases transcription factor CREB (cAMP response element-binding protein) activation and protein synthesis. Also, the α_i subunit activation can lead to the blockade of TRPV1 (transient receptor potential cation channel subfamily V member 1), HCN (hyperpolarization-activated cyclic nucleotide-gated), ASIC (acid-sensing ion channels), and voltage-gated Na^+ (Na_v) channels [24]. These ion channels are implicated

Table 9.1 Distribution, ligands, and opioid receptor functions

Receptor	Distribution	Agonists	Antagonists	Physiological functions
μ MOR MOP (OP_3)	Thalamus Caudate Putamen Neocortex Nucleus accumbens Amygdala Dorsal horn of the spinal cord Periaqueductal gray Brain stem	Morphine Fentanyl Sufentanil DAMGO Buprenorphine Nalbuphine Codeine Levorphanol Methadone Meperidine *Endogenous:* *Endomorphine-1* *Endomophine-2* *β-Endorphin* *Enkephalins*	Naloxone Naltrexone Nalmefene β-FNA Nalorphine CTAP CTOP	Analgesia, mood, respiratory and cardiovascular functions, gastrointestinal motility, feeding, locomotor activity, thermoregulation, hormone secretion, immune functions
δ DOR DOP (OP_1)	Distribution similar to μ with high density in olfactory areas. Also present in the thalamus and hypothalamus	D-Ala-deltorphin I and II DPDPE SNC80 *Endogenous:* *Enkephalins* *β-Endorphin*	Naltrindole Naltriben Naltrexone Naloxone	Analgesia, gastrointestinal motility, mood and behavior, olfaction, cardiovascular regulation
κ KOR KOP (OP_2)	Cerebral cortex Amygdala Hypothalamus Pituitary Periaqueductal gray	Ketocyclazocine Bremazocine U-50488 U-69593 Butorphanol Salvinorin A *Endogenous:* *Dynorphin A* *Dynorphin B*	Nor-binaltorphimine GNTI Nalmefene Naloxone Naltrexone Buprenorphine	Regulation of nociception, diuresis, feeding, neuroendocrine and immune system functions
Nociceptin NOR NOP (OP_4)	Cortex, olfactory nucleus Ventral forebrain Hippocampus Hypothalamus Amygdala Ventral tegmental area Rostral ventromedial medulla Locus coeruleus Dorsal horn of the spinal cord	Ro64-6198 N/OFQ-(1–13) *Endogenous:* *Nociceptin/ orphanin FQ*	SSB612111 J-113397 UFP-101	Regulation of nociception, autonomic control of physiological processes

DAMGO (D-Ala2, N-MePhe4, Gly-ol)-enkephalin, *β-FNA* β-funaltrexamine, *CTAP* octapeptide (D-Phe-Cys-Tyr-D-Trp-Arg-Thr-Pen-Thr-NH_2), *CTOP* octapeptide (D-Phe-Cys-Tyr-D-Trp-Orn-Thr-Pen-Thr-NH2), *DPDPE* (2-D-penicillamine, 5-D-penicillamine)-enkephalin, *GNTI* guanidinonaltrindole, *N/OFQ-(1–13)* nociceptin/orphanin FQ(1–13)NH2, *UFP-101* [Nphe1, Arg14,Lys15]Nociceptin-NH_2

in transmission and pain modulation, neuronal firing and excitability, perception of a wide range of pH changes, and action potential initiation and propagation. Together, the inhibition of these channels causes neuronal hyperpolarization and analgesia.

On the other hand, the G protein's βγ complex increases K^+ conductance through G protein-gated inward rectifying potassium channels (GIRKs) and adenosine triphosphate-sensitive K^+ (K_{ATP}) channels. Other effects of the βγ subunit complex are a decrease in voltage-gated Ca^{2+} channel (Ca_v) conductance and the activity of heat-sensing transient receptor potential cation channel subfamily M member 3 (TRPM3). These effects also cause hyperpolarization, inhibit neurotransmitter release, and decrease cell responsiveness.

In addition, G protein's βγ complex mediates activation of mitogen-activated kinase (MAPK) cascades, including extracellular signal-regulated kinase (ERK), Jun N-terminal kinase (JNK), and p38. Finally, βγ can also activate phospholipase C (PLC) and increase the production of inositol triphosphate (IP_3) [24, 25].

9.2.2 Phosphorylation Patterns of Opioid Receptors

When opioid receptors uncouple from G proteins, the intracellular and C-terminal domains of the receptor remain exposed. The intracellular domains contain more than 20 serine, threonine, and tyrosine amino-acid residues that are potential phosphorylation sites for G protein receptor kinases (GRKs), calcium-activated protein kinase (PKC), Ca^{2+}/calmodulin-dependent protein kinase II (CAMKII), and cyclin-dependent kinase 5 (Cdk5). The phosphorylation pattern of receptors depends on the agonist bound to them and the kinases present in the cells' environment (Table 9.2) [26].

Phosphorylation facilitates β-arrestin (β-arr) recruitment, causing receptor desensitization, internalization, trafficking, and recycling. Phosphorylation is also a

Table 9.2 Kinases and amino acids involved in opioid receptor phosphorylation

Receptors	Kinases	Amino-acid phosphorylation sites	β-Arrestin recruited
μ	GRK (2, 3, 5, or 6), CaMKII, PKC (α, ε, or ζ), tyrosine kinase	Ser (363, 370, 375, 383, 394) Thr (180, 370, 376, 379, 394) Tyr (106, 166)	β-arr1, β-arr2
δ	GRK2, PKC, Cdk5	Ser (258, 344, 363) Thr (161, 358, 361)	β-arr2
κ	GRK (2, 3, or 5), PKC	Ser (369, 356, 358) Thr 357	β-arr1, β-arr2
NOP	GRK (2, 3, or 5)	Ser (356, 358, 369) Thr 357	β-arr3

critical step for signaling and biased agonism (see Sect. 9.3). These processes can occur from the first second to minutes after G protein uncoupling [6, 27, 28].

Specific phosphorylation "barcodes" are required to drive opioid receptor desensitization and internalization. In particular, phosphorylation of *Thr-Ser-Ser-Thr* and *Ser-Thr-Ala-Asn-Thr* motifs leads to agonist-induced desensitization, but phosphorylation of *Thr-Ser-Ser-Thr* or *Ser-Thr-Ala-Asn-Thr* motifs modulates β-arrestin interactions [26, 29]. In addition, specific opioid receptor agonists may induce various degrees of phosphorylation. For instance, DAMGO, methadone, fentanyl, and etorphine induce β-arrestin recruitment and μ-opioid receptor phosphorylation and internalization, but morphine does not.

There is evidence that several μ-opioid receptor agonists strongly promote β-arrestin binding and subsequent phosphorylation in the vicinity of Thr370 to Thr379 amino-acid residues. On the other hand, the δ-opioid receptor subtype has an *Asn-Pro-X-X-Tyr* motif that is phosphorylated in response to enkephalin [27, 28, 30].

9.2.3 *β-Arrestin Pathway*

As previously mentioned, receptor phosphorylation promotes β-arrestin binding in the first 3–5 min after G protein uncoupling. There are four types of β-arrestins, identified with numbers from 1 to 4. While μ-, δ-, and κ-opioid receptors bind β-arr1 and β-arr2, NOP binds β-arr3. β-Arrestins block further G protein activation by sterically hindering access to it. Once assembled, the opioid receptor-β-arrestin complex can remain in the cell membrane or start the endocytosis process [31, 32].

The opioid receptor-β-arrestin complex recruits the adaptor complex (AP2) and clathrin proteins. β-Arrestins act as scaffolds to assemble clathrin-coated pits (CCPs). CCPs invaginate into the cytosol and form vesicles by separating them from the plasma membrane with dynamin GTPase. After vesicle formation, the entire endocytic machinery, except the opioid receptor-β-arrestin complex, is disassembled, and the new vesicle fuses with an early endosome [33, 34]. From the early endosomes, opioid receptors can follow one of the three routes through vesicular trafficking: (1) recycling to the cell membrane, (2) transport to lysosomes for degradation, or (3) transport to the trans-Golgi compartment.

Opioid receptors' recycling can occur in either fast or slow recycling endosomes. For rapid recycling, receptors require dephosphorylation and β-arrestin uncoupling. Once opioid receptors return to the plasma membrane, a new ligand can activate them to restart G protein-dependent signaling. In contrast, the receptors that remain bound to β-arrestin are inefficiently recycled; thus, they are transported to the late endosomes and then the lysosomes for degradation or remain in the trans-Golgi compartment (Fig. 9.2).

The trans-Golgi compartment, also called the perinuclear cloud, is an intracellular region where receptors remain with limited mobility waiting for signals to initiate a new endosomal trafficking pathway [33, 35, 36]. The process by which

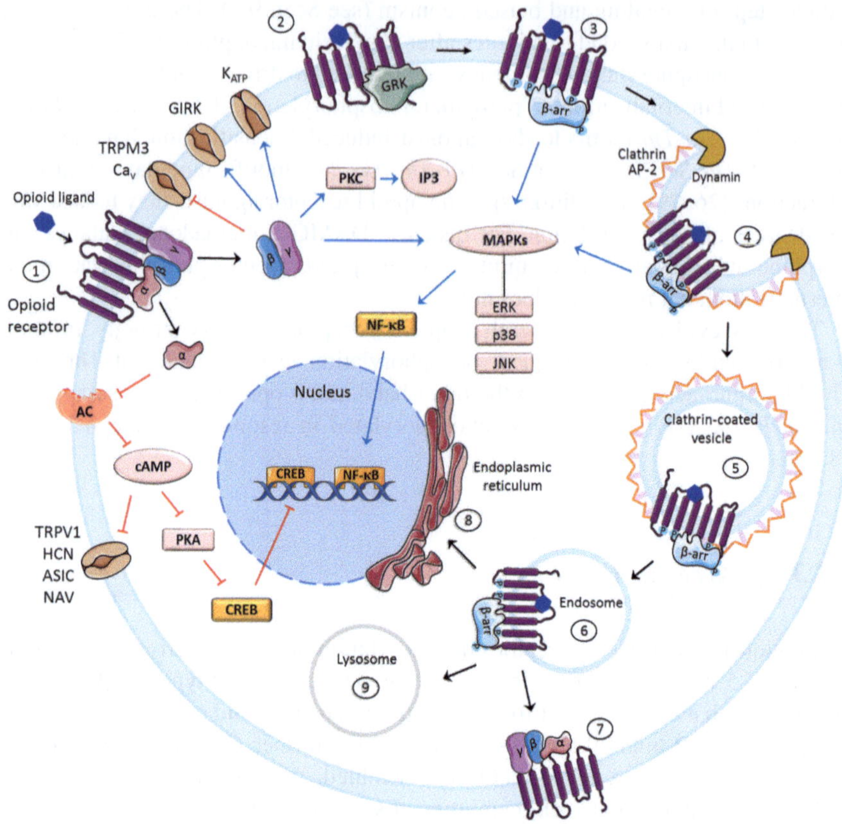

Fig. 9.2 Signaling pathways and vesicular traffic of opioid receptors. (1) Opioid receptor activation produces G protein dissociation: the α_i subunit inhibits the AC-cAMP-PKA-CREB pathway. A decrease in cAMP levels induces the blockade of the TRPV1, HCN, ASIC, and Na_v channels. The βγ complex increases the K^+ conductance of GIRK and K_{ATP} channels, activates the PLC-IP3 pathway, modulates the MAPK cascades, and induces NF-κB activation. (2) Once separated from the G protein, the opioid receptor is phosphorylated by GRKs. (3) Receptor phosphorylation induces the recruitment of β-arrestin. (4) The AP2 and clathrin proteins oligomerize to initiate the invagination of a membrane portion containing the opioid receptor. (5) The protein dynamin separates the clathrin-coated vesicle from the plasma membrane. (6) The endocytic machinery is disassembled, and the new vesicle fuses with an early endosome. From endosomes, the opioid receptors can follow different routes: (7) recycling to the cell membrane, (8) transport to the trans-Golgi compartment, or (9) transport to lysosomes for degradation. AC, adenylyl cyclase; cAMP, cyclic adenosine monophosphate; PKA, protein kinase A; CREB, cAMP response element-binding; TRPV1, transient receptor potential cation channel subfamily V member 1; HCN, hyperpolarization-activated cyclic nucleotide-gated channels; ASIC, acid-sensing ion channels; Na_v, voltage-gated sodium channel; GIRK, G protein-coupled inwardly rectifying potassium channel; K_{ATP}, ATP-sensitive potassium channel; PLC, phospholipase C; IP3, inositol trisphosphate; MAPK, mitogen-activated protein kinase; ERK, extracellular signal-regulated kinase; JNK, c-Jun N-terminal kinase; p38, p38 MAP kinase; NF-κB, nuclear transcription factor-κB; GRK, G protein-coupled receptor kinases; β-arr, β-arrestin; AP2, adaptor protein complex 2

cells determine receptors' fate is still unknown, but it requires phosphorylation and ubiquitination.

In addition to G protein-dependent signaling, internalization, and vesicular trafficking, opioid receptors can signal through β-arrestin proteins [37]. β-Arrestin activates MAPKs or PI3K/AKT transduction pathways. While MAPKs' signaling activates the nuclear transcription factor κB (NF-κB), PI3K/AKT proteins stimulate the mechanistic target of rapamycin (mTOR) signaling. The activation of these two pathways could mediate several effects of opioids such as cell growth, differentiation, motility, and survival (Fig. 9.2) [38–40].

9.3 Mechanisms that Modulate Ligand-Opioid Receptor Complexes' Effects

9.3.1 Biased Agonism

Biased agonism, also known as ligand-directed receptor signaling, is the ability of a ligand to preferentially stabilize the receptor in one or another active conformation to produce different trafficking and signaling pathways [4, 13]. The relevance of opioid-biased agonism emerged from numerous pharmacological studies on μ-opioid receptors. While several G protein-biased agonists produce analgesia with few side effects, ligands that preferentially activate β-arrestin signaling generate various adverse side effects such as analgesic tolerance, respiratory depression, and constipation [40]. Although biased agonism explains the differential effects of several μ-opioid ligands, recent studies have shown that even in models where β-arrestin signaling is impaired, activation of μ-opioid receptors produces some undesirable side effects [42–44]. In addition to biased agonism, efficacy and affinity are important in opioids' side effects [41].

Biased agonism of the δ- and κ-opioid receptors has been less explored. However, it appears that the activation of β-arrestin is not involved in some δ-opioid receptor-mediated side effects. In contrast, β-arrestin participates in κ-opioid receptor-induced aversion through p38 and mTOR activation [13].

9.3.2 Positive and Negative Allosteric Modulators

Allosteric modulators are small molecules that bind to pockets on receptors, separate from the orthosteric sites. There are three types of allosteric modulators: (a) positive allosteric modulators (PAMs), which enhance the affinity, potency, and maximal response of orthosteric ligands; (b) negative allosteric modulators (NAMs) that have PAMs' opposite effects; and (c) silent/neutral allosteric modulators (SAMs) that occupy the site of orthosteric agonists without interfering with their action but behave as competitive antagonists [4, 8, 44]. In the absence of an orthosteric agonist, allosteric modulators have little or no functional activity.

The sodium ion (Na^+) is a NAM of opioid receptors and other GPCRs. Na^+ binds at a conserved and well-described opioid receptor site, stabilizes the receptors to an inactive state (R), and reduces agonist binding to the orthosteric site. Sodium inhibits about 65% of the agonist binding and signaling in μ- and δ-opioid receptors, but only 20% in κ-opioid receptors [4, 17]. Other cations such as potassium (K^+) and lithium (Li^+) reduce agonist binding to the δ-opioid receptors.

In contrast, PAMs maintain opioid receptor configuration into the receptors' active state (R*) and can disrupt sodium binding [8]. PAMs can be attractive therapeutic tools because they improve desirable opioid effects without increasing the undesirable ones [45]. At a molecular level, PAMs avoid receptor downregulation and other compensatory mechanisms triggered by sustained receptor activation. Therefore, these molecules could produce less opioid tolerance and dependence. Manganese (Mn^{2+}) is one of the best-studied PAMs that restores full agonist binding in the presence of Na^+. Some other allosteric modulators are described in Table 9.3.

9.4 Structural and Molecular Aspects that Modify Opioid Receptor Signaling

Opioid signaling can vary depending on opioid receptors' structural modifications that result from a complex combination of genetic and molecular factors. On the one hand, genetic variations in opioid receptors' genes include alternative splicing and non-synonymous single-nucleotide polymorphisms. On the other hand, functional interactions occur when opioid receptors form dimers with equal or different GPCRs. These structural variations can affect ligands binding opioid receptors, alter the ligand-receptor complex conformation, change G protein coupling, and modify intracellular signaling.

Table 9.3 Positive and negative allosteric modulators of opioid receptors

Receptor	PAMs	NAMs	SAMs
MOR	BMS-986121 BMS-986122 MS1	Na^+ Cannabidiol Tetrahydrocannabinol (THC) Salvinorin A	BMS-986124
DOR	Mn^{2+} BMS-986187 Hexahydro-1H-xanthene-1,8(2H)-dione analogues	Na^+ K^+ Li^+ Cannabidiol THC	BMS-986122
KOR	Not described	Na^+	BMS-986122

Data taken from Refs. [8, 77–79]

9.4.1 Alternative Splicing

In most eukaryotes, deoxyribonucleic acid (DNA) consists of exons separated by introns. Exons are coding sections that contain the transcript sequences of messenger ribonucleic acid (mRNA). Introns are non-coding intervening sequences that are spliced out from the precursor mRNA to form a mature mRNA. The process of joining exons together is called splicing. Alternative splicing occurs when the precursor mRNA of a single gene produces diverse exon arrangements, encoding multiple protein isoforms that vary in their properties and activity [46].

The *Oprm1* gene codes for the μ-opioid receptor protein and has four major exons. Exon 1 codes for the extracellular amino terminus; the second and third exons code for the following six transmembrane segments, and exon 4 provides the amino-acid sequence for the intracellular carboxyl terminus. As previously mentioned, the highest homology of opioid receptors occurs in the amino-acid sequence of the seven transmembrane (7TM) domains; alternative splicing modifies mainly the genomic arrangement of regions outside of these domains. There are at least 29 functional splice variants of the human *Oprm1* gene, 2 of the human *Oprd1* gene, and 4 of the human *Oprk1* gene [47].

The splicing of the μ-opioid receptor gene generates different types of opioid receptors with structural and pharmacological differences (Fig. 9.3). The full-length splice variants containing exons 1, 2, and 3 encode the N-terminal region and the μ-opioid receptor 7TM domains. These proteins share identical binding pockets and recognize opioid ligands with similar affinities. However, structural variations at the C-terminal tail of the receptor affect the efficacy of some agonists for G protein coupling, modifying their capacity to inhibit cAMP production, and altering the signaling preference to G protein activation or β-arr2 recruitment. Also, 7TM splice variants show differential receptor phosphorylation, internalization, desensitization, and post-endocytic sorting [49–51].

Splice variants that lack exon 1 encode only six transmembrane domains (6TM) initially promoted by exon 11, which is located at 30 kb upstream of exon 1 [51]. The binding of morphine to these truncated 6TM μ-opioid receptors is similar to that produced with full-length opioid receptors. However, their activation increases the intracellular Ca^{2+} concentration, reduces K^+ conductance, and does not modify cAMP levels, suggesting an increase, not a decrease, in cellular excitability [52].

Single transmembrane (1TM) splice variants are encoded by exon 1 and contain the N-terminal domain, the first transmembrane segment, and a C-terminal tail with different splicing patterns [51]. 1TM splice variants may act as molecular chaperones of 7TM μ-opioid receptors in the endoplasmic reticulum to minimize their degradation and increase their expression at the cell membrane [53].

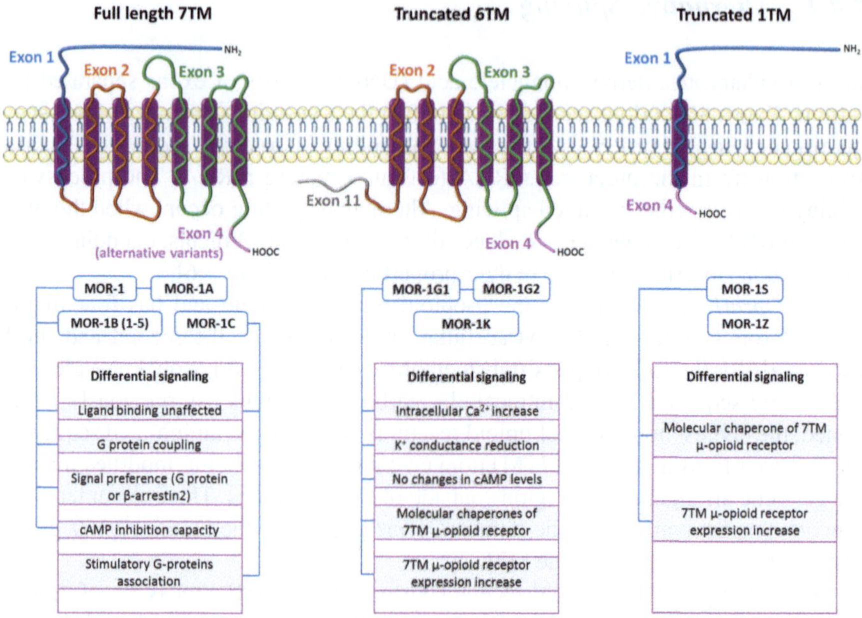

Fig. 9.3 μ-opioid receptor (MOR) splice variants and agonist-induced intracellular signaling. Different colors show μ-opioid receptor segments labeled by the exons that encode them. Representative splice variants of the human μ-opioid receptor are shown in rectangles and linked to the signaling variations they produce. Non-linked rectangles (MOR-1K and MOR-1Z) represent human splice variants with limited information regarding intracellular signaling

9.4.2 Single-Nucleotide Polymorphisms

Single-nucleotide polymorphisms (SNPs) are point mutations that produce single-base variations in specific genes. These polymorphisms have a frequency of approximately 1% in a population of random individuals. Synonymous SNPs come from different alleles that encode the same amino acid, whereas non-synonymous SNPs produce amino-acid substitutions [54].

Genome sequencing studies have identified more than 3324 polymorphisms in the human μ-opioid receptor gene, but almost 96% of these mutations rarely occur and have limited relevance at the population level. For example, of the 100 most studied SNPs of the *Oprm1* gene, only 24 are frequent in humans (>0.1%), and 10 are located in the first, second, and third exons (Fig. 9.4). Among these, only five appear to have significant clinical relevance because they share the following three criteria:

1. The SNP possesses a high allelic frequency reported in humans.
2. The mutation causes an amino-acid exchange that results in an altered protein.
3. The alteration of the protein has functional consequences.

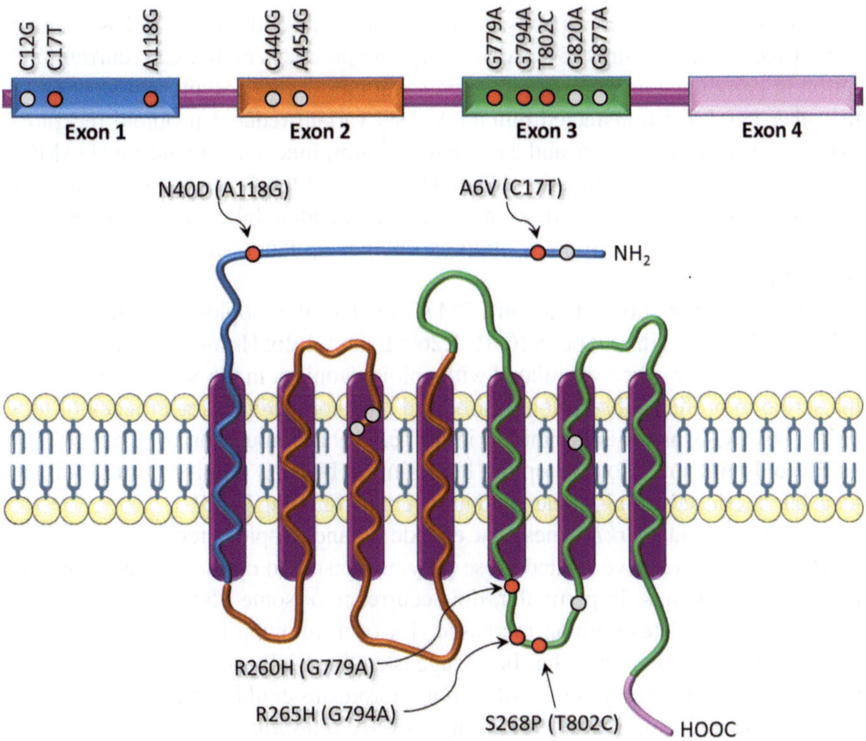

Fig. 9.4 Reported non-synonymous SNPs in the μ-opioid receptor. The upper panel shows frequent naturally occurring mutations in exons of the μ-opioid receptor gene. The lower panel shows the amino-acid arrangements of the μ-opioid receptor with colors that match their corresponding exons (blue, exon 1; orange, exon 2; green, exon 3; pink, exon 4). The red circles represent polymorphisms with functional consequences, and the gray circles represent polymorphisms with no reported information regarding intracellular signaling. Relevant polymorphisms are indicated with arrows and named by the amino acid exchanged and their corresponding nucleotides (between parentheses)

The most frequently occurring μ-opioid receptor gene mutation is located at the nucleotide position 118 in exon 1. This polymorphism is termed A118G or 118A<G due to the position and specific nucleotides exchanged, adenosine (A) to guanosine (G), whereas the encoded mutated protein is named Asn40Asp (N40D), referring to the replacement of asparagine (Asn; N) 40 by aspartic acid (Asp; D). Other SNPs located in exon 3 of the *Oprm1* gene follow the same nomenclature; such is the case of G779A (R260H), G794A (R265H), and T802C (S268P) [55, 56].

Non-synonymous SNPs produce signaling changes in μ-opioid receptors. In the case of A118G SNP, the encoded protein lacks an N-glycosylation site in the extracellular domain that alters the ability of the receptor to bind agonists. However, there is no consensus regarding the effects of this SNP. Some opioid agonists produce enhanced responses. For example, a μ-opioid receptor with the A118G variation has an increased capacity to bind β-endorphin and activate GIRK channels in transfected cell cultures [57]. Also, morphine produces higher Ca^{2+} current inhibition in ganglionic neurons of rats expressing the A118G polymorphism [58]. On the other hand, the transfected human A118G variant reduces μ-opioid receptors' expression at the cell surface and the potency of morphine, methadone, and DAMGO [59]. Another important mutation located at exon 1 of the *Oprm1* gene results in an alanine to valine (A6V) substitution and decreases adenylyl cyclase and ERK signaling in response to morphine, buprenorphine, fentanyl, and endogenous opioids [60].

Amino-acid variations at the third TM segment of the μ-opioid receptor can alter cell signaling in vitro. The S268P, R260H, and R265H mutant receptors are signaling-deficient when stimulated with opioid agonists. In the S268P variant case, the serine to proline substitution in μ-opioid receptors reduces the efficacy of morphine, DAMGO, and β-endorphin to inhibit cAMP accumulation [61]. The R260H and R265H mutant opioid receptors have similar binding affinities to opioid agonists but a reduced agonist-independent receptor signaling [62].

The *Oprd1* and *Oprk1* genes that encode δ- and κ-opioid receptors also have SNPs. Most studies have related these polymorphisms to opioid dependence, with inconclusive findings. In particular, the occurrence of some SNPs at exon 1 of the *Oprd1* gene correlates with an increased risk of developing heroin dependence in humans [63]. Additionally, it has been suggested that SNPs in *Oprk1* genes contribute to drug dependence, whereas silent polymorphisms could impact the function of the κ-opioid receptor without affecting mRNA transcription and stability [64]. Other signaling changes produced by these SNPs still remain to be understood. In any case, SNP occurrence can explain individual differences in the response to opioids.

9.4.3 Receptor Dimerization

Opioid receptor dimers are composed of two functional receptors with specific biochemical properties, different from individual components. Dimerization of the same opioid receptor produces a homodimer, whereas heterodimers involve interactions between two different opioid receptor subtypes or between opioid receptors and other GPCRs [65]. Although functional evidence for these receptor-receptor associations is controversial, bivalent agonists or different agonist co-administrations produce unique pharmacological effects absent in animals or cells where the heteromer is disrupted or blocked. Thus, the protein dimerization involving opioid receptors can be constitutive or driven by specific agonists, leading to different signaling pathways [12].

Various studies have documented μ-opioid receptor homodimer formation. Pharmacological approaches demonstrate that, by simultaneously targeting the μ-opioid receptor with the biased ligands DAMGO and morphine, it is possible to delay analgesic tolerance by promoting receptor desensitization and endocytosis [66]. A plausible explanation is that DAMGO, but not morphine, induces β-arrestin recruitment suggesting that pharmacological effects can be mediated by μ-opioid receptor homodimers at the cell membrane level. At the cytoplasmic level, the μ-opioid receptor homodimer alters transcriptional activity through ERK activation in primary dorsal root ganglion (DRG) neurons [67].

Opioid receptors also form heteromeric dimers (μ/δ, μ/κ, μ/NOP, δ/κ, δ/NOP, κ/NOP) with different pharmacological profiles and functional diversity. The μ/δ-opioid receptor heterodimer is highly co-expressed in the DRG and the spinal cord and has pharmacological relevance in pain control. In cells expressing μ/δ dimers, the δ-opioid receptor acts as an allosteric modulator that enhances the function of μ-opioid receptors for G protein coupling and cAMP inhibition. Moreover, the μ-opioid receptor agonists methadone and DAMGO internalize not only μ-opioid receptors but also μ/δ dimers. These structural variations lead to intracellular trafficking modifications and changes in the β-arrestin-ERK pathway after internalization [65, 68].

Other pairs of opioid receptor combinations also possess altered pharmacological properties. For example, the κ-opioid receptor allosterically modulates responses to δ agonists in heteromeric δ/κ dimers [69]; however, their specific intracellular signaling and functional implications are still poorly understood.

In neurons, some heterodimers consist of opioid and adrenergic receptors (μ/α_{2A}, μ/α_{2C}, δ/α_{2A}, δ/α_{2C}). In the μ/α_{2A} dimer case, the adrenoreceptor causes μ-opioid receptor inactivation and diminishes the downstream MAP kinase pathway. Additional heteromeric dimers form between opioid receptors and cannabinoid, chemokine, or glutamate receptors (e.g., μ/CB_1, δ/CB_1, δ/CXCR4, δ/CXCR2, μ/NMDA). In such cases, the activation of a single receptor is sufficient to initiate G protein signaling or produce agonist or antagonist interactions [65, 68]. Although the ligand binding to monomeric receptors can activate opioid receptor oligomers, some studies report that dimers have a restricted useful life but sufficient to change pharmacological outcomes in cellular and in vivo systems.

9.5 Genetic and Epigenetic Factors in Opioid Receptor Regulation

In addition to structural variation, proteins related to opioid receptors can have genetic alterations that modify the effects of specific ligands. These proteins include non-opioid receptors, channels, transporters, and signal transduction molecules [70, 71]. Moreover, polymorphisms for genes that encode adrenergic, glutamatergic, dopamine, and acetylcholine receptors also exist. Although some of these proteins form dimers with opioid receptors, their most studied implication focuses on genetic variations that produce a high vulnerability to opioid abuse in humans [70, 72].

At the cellular level, point mutations in genes that encode GIRK channels, voltage-gated calcium channels, β-arr2, adenylyl cyclase, GRK, ERK, and PLC modify intracellular signaling and the analgesic responses to opioids in animal models [71, 73].

Opioid receptor genes are rich in CpG islands. In these genomic regions, a cytosine (C) followed by a guanine (G) nucleotide frequently occurs. DNA methylation in cytosine nucleotides can modify gene expression without altering the nucleotide sequence. This process generally silences genes by physically blocking RNA polymerase II. The gene encoding μ-opioid receptor can be activated by decreasing the methyl-CpG-binding protein-2 (MECP2) expression or by adding artificial demethylation agents [74].

On the other side, gene expression of opioid receptors depends on the ability of the transcriptional machinery to access condensed genetic material. The DNA strands are wrapped around four histones (H2A, H2B, H3, and H4). The N-terminal tails of these proteins loosen or strengthen their grip on DNA. Histone acetylation reduces tightness between histones, opens chromatin, and facilitates gene transcription.

Prolonged opioid exposure can result in epigenetic modifications. Specifically, hyperacetylation in lysine residues (H3K9, H3K14, H3K18, and H3K27) has been observed in animals that self-administrate morphine and heroin. Also, a histone acetylation inducer promotes activation of the O*prm1* gene. Other modifications, such as histone methylation and phosphorylation, regulate the expression of opioid receptors but are less studied [75].

The epigenetic regulation of opioid receptors may alter gene transcription, activate or repress particular genes in response to drugs or other stimuli, and modulate splice variant expression. These regulatory mechanisms result in changes in signaling cascades, cellular structure, and synaptic activity of neurons expressing opioid receptors.

9.6 Final Considerations

Opioid agonists and partial agonists are essential tools for acute pain management and palliative care. In addition, opioid antagonists counteract opioid overdoses and are helpful in alcoholism treatment. Despite their benefits, opioids have caused numerous deaths among heroin users, and the increasing availability of potent synthetic drugs, such as fentanyl, has aggravated this situation (see Chap. 5). Understanding how opioids act is essential to take advantage of their desirable effects while minimizing their risks.

Numerous factors can modify beneficial or adverse opioid effects. The first variable is the ligand itself, which can act on several receptor types, with or without positive or negative allosteric modulators. Another variation source arises from the heterogeneous distribution of opioid receptor subtypes in the brain and peripheral tissues. As preclinical evidence has shown, the effects of a particular opioid vary on different experimental preparations depending on the basal activity of different

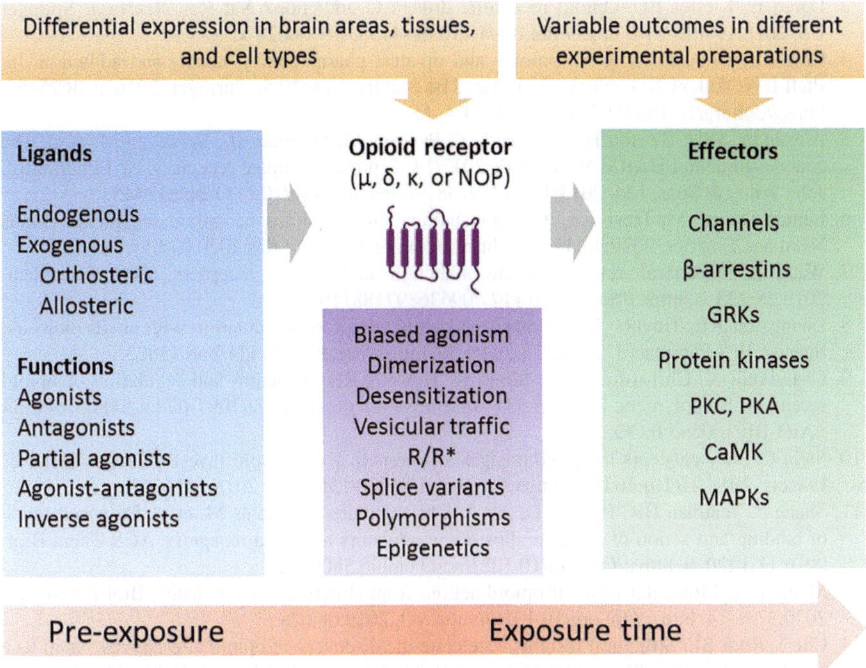

Fig. 9.5 Variables that modify opioid receptor signaling and effects. Opioid ligands produce outcomes that depend on the receptor type, receptor state and expression, signaling preference, structural modifications (dimerization, splice variants, polymorphisms, epigenetics), and intracellular traffic and signaling. Also, the responses can vary depending on the experimental preparations, pre-exposure to opioids, and exposure time

receptors, the presence and abundance of specific kinases, and other proteins needed for signal transduction. Similarly, at a clinical level, a great diversity of opioid users exists, from patients who use opioids for the first time to those who are dependent or become tolerant to opioids after prolonged medical treatment (Fig. 9.5). Variations in response can also occur depending on genetic factors. Consequently, the use of specific opioid agonists for particular conditions must be evaluated carefully by the physician to provide the best available treatment for patients in pain (Chaps. 10 and 11) or with substance use disorders (Chaps. 14 and 15).

References

1. Pert CB, Pasternak G, Snyder SH. Opiate agonists and antagonists discriminated by receptor binding in brain. Science (80-). 1973;182:1359–61. https://doi.org/10.1126/science.182.4119.1359.
2. Conibear AE, Kelly E. A biased view of μ-opioid receptors? Mol Pharmacol. 2019;96:542–9. https://doi.org/10.1124/mol.119.115956.

3. Darcq E, Kieffer BL. Opioid receptors: drivers to addiction? Nat Rev Neurosci. Springer US. 2018;19:499–514. https://doi.org/10.1038/s41583-018-0028-x.
4. Cruz SL, Granados-Soto V. Opioids and opiates: pharmacology, abuse, and addiction. In: Pfaff DW, Volkow ND, editors. Neurosci 21st century. New York: Springer; 2016. p. 3625–57. https://doi.org/10.1007/978-1-4939-3474-4_156.
5. Alexander SPH, Benson HE, Faccenda E, Pawson AJ, Sharman JL, Spedding M, et al. The concise guide to PHARMACOLOGY 2013/14: G protein-coupled receptors. Br J Pharmacol. John Wiley & Sons, Ltd. 2013;170:1459–581. https://doi.org/10.1111/bph.12445.
6. Lemos Duarte M, Devi LA. Post-translational modifications of opioid receptors. Trends Neurosci. Elsevier. 2020;43:417–32. https://doi.org/10.1016/j.tins.2020.03.011.
7. Wang S. Historical review: opiate addiction and opioid receptors. Cell Transplant. 2019;28:233–8. https://doi.org/10.1177/0963689718811060.
8. Livingston KE, Traynor JR. Allostery at opioid receptors: modulation with small molecule ligands. Br J Pharmacol. 2018;175:2846–56. https://doi.org/10.1111/bph.13823.
9. Chaturvedi K, Christoffers KH, Singh K, Howells RD. Structure and regulation of opioid receptors. Biopolymers. 2000;55:334–46. https://doi.org/10.1002/1097-0282(2000)55:4<334::AID-BIP1006>3.0.CO;2-S.
10. Stein C. New concepts in opioid analgesia [Internet]. Expert Opin Investig Drugs. Taylor & Francis. 2018;27(10):765–75. https://doi.org/10.1080/13543784.2018.1516204.
11. Shang Y, Yeatman HR, Provasi D, Alt A, Christopoulos A, Canals M, et al. Proposed mode of binding and action of positive allosteric modulators at opioid receptors. ACS Chem Biol. 2016;11:1220–9. https://doi.org/10.1021/acschembio.5b00712.
12. Manglik A. Molecular basis of opioid action: from structures to new leads. Biol Psychiatry. 2020;87:6–14. https://doi.org/10.1016/j.biopsych.2019.08.028.
13. Che T, Roth BL. Structural insights accelerate the discovery of opioid alternatives. Annu Rev Biochem. 2021;90:739–61. https://doi.org/10.1146/annurev-biochem-061620-044044.
14. Marino KA, Shang Y, Filizola M. Insights into the function of opioid receptors from molecular dynamics simulations of available crystal structures. Br J Pharmacol. 2018;175:2834–45. https://doi.org/10.1111/bph.13774.
15. Marmolejo-Valencia AF, Madariaga-Mazón A, Martinez-Mayorga K. Bias-inducing allosteric binding site in mu-opioid receptor signaling. SN Appl Sci. Springer International Publishing. 2021;3:566. https://doi.org/10.1007/s42452-021-04505-8.
16. Piltonen M, Krokhotin A, Parisien M, Bérubé P, Djambazian H, Sladek R, et al. Alternative splicing of opioid receptor genes shows a conserved pattern for 6TM receptor variants. Cell Mol Neurobiol. 2021;41:1039–55. https://doi.org/10.1007/s10571-020-00971-7.
17. Alexander SPH, Christopoulos A, Davenport AP, Kelly E, Mathie A, Peters JA, et al. THE CONCISE GUIDE TO PHARMACOLOGY 2019/20: G protein-coupled receptors. Br J Pharmacol. 2019;176(S1). https://doi.org/10.1111/bph.14748.
18. Remesic M, Hruby VJ, Porreca F, Lee YS. Recent advances in the realm of allosteric modulators for opioid receptors for future therapeutics. ACS Chem Neurosci. 2017;8:1147–58. https://doi.org/10.1021/acschemneuro.7b00090.
19. Quirion B, Bergeron F, Blais V, Gendron L. The delta-opioid receptor; a target for the treatment of pain. Front Mol Neurosci. 2020;13:52. https://doi.org/10.3389/fnmol.2020.00052.
20. Berthiaume S, Abdallah K, Blais V, Gendron L. Alleviating pain with delta opioid receptor agonists: evidence from experimental models. J Neural Transm. Springer Vienna. 2020;127:661–72. https://doi.org/10.1007/s00702-020-02172-4.
21. Margolis EB, Karkhanis AN. Dopaminergic cellular and circuit contributions to kappa opioid receptor mediated aversion. Neurochem Int. Elsevier. 2019;129:104504. https://doi.org/10.1016/j.neuint.2019.104504.
22. Faouzi A, Varga BR, Majumdar S. Biased opioid ligands. Molecules. 2020;25:4257. https://doi.org/10.3390/molecules25184257.
23. Palczewski K, Kumasaka T, Hori T, Behnke CA, Motoshima H, Fox BA, et al. Crystal structure of rhodopsin: a G protein-coupled receptor. Science. 2000;289:739–45. https://doi.org/10.1126/science.289.5480.739.

24. Koehl A, Hu H, Maeda S, Zhang Y, Qu Q, Paggi JM, et al. Structure of the μ-opioid receptor-Gi protein complex. Nature. 2018;558:547–52. https://doi.org/10.1038/s41586-018-0219-7.
25. Machelska H, Celik MÖ. Advances in achieving opioid analgesia without side effects. Front Pharmacol. 2018;9:1388. https://doi.org/10.3389/fphar.2018.01388.
26. Proft J, Weiss N. G protein regulation of neuronal calcium channels: back to the future. Mol Pharmacol. 2015;87:890–906. https://doi.org/10.1124/mol.114.096008.
27. Canals M. The complex roles of μ-opioid receptor phosphorylation: a key determinant in receptor signaling and regulation. Mol Pharmacol. 2015;88:814–5. https://doi.org/10.1124/mol.115.100180.
28. Williams JT, Ingram SL, Henderson G, Chavkin C, von Zastrow M, Schulz S, et al. Regulation of μ-opioid receptors: desensitization, phosphorylation, internalization, and tolerance. Dolphin AC, editor. Pharmacol Rev. 2013;65:223–54. https://doi.org/10.1124/pr.112.005942.
29. Allouche S, Noble F, Marie N. Opioid receptor desensitization: mechanisms and its link to tolerance. Front Pharmacol. 2014;5:280. https://doi.org/10.3389/fphar.2014.00280.
30. Lemel L, Lane JR, Canals M. GRKs as key modulators of opioid receptor function. Cell. 2020;9:2400. https://doi.org/10.3390/cells9112400.
31. Raehal KM, Schmid CL, Groer CE, Bohn LM. Functional selectivity at the μ-opioid receptor: implications for understanding opioid analgesia and tolerance. Sibley DR, editor. Pharmacol Rev. 2011;63:1001–19. https://doi.org/10.1124/pr.111.004598.
32. Bie B, Pan ZZ. Trafficking of central opioid receptors and descending pain inhibition. Mol Pain. 2007;3:37. https://doi.org/10.1186/1744-8069-3-37.
33. Shenoy SK, Lefkowitz RJ. β-Arrestin-mediated receptor trafficking and signal transduction. Trends Pharmacol Sci. Elsevier Ltd. 2011;32:521–33. https://doi.org/10.1016/j.tips.2011.05.002.
34. Delom F, Fessart D. Role of phosphorylation in the control of clathrin-mediated internalization of GPCR. Int J Cell Biol. 2011;2011:1–14. https://doi.org/10.1155/2011/246954.
35. Reiter E, Ahn S, Shukla AK, Lefkowitz RJ. Molecular mechanism of β-arrestin-biased agonism at seven-transmembrane receptors. Annu Rev Pharmacol Toxicol. 2012;52:179–97. https://doi.org/10.1146/annurev.pharmtox.010909.105800.
36. Magalhaes AC, Dunn H, Ferguson SSG. Regulation of GPCR activity, trafficking and localization by GPCR-interacting proteins. Br J Pharmacol. 2012;165:1717–36. https://doi.org/10.1111/j.1476-5381.2011.01552.x.
37. Dellibovi-Ragheb T, Altan-Bonnet N. Cloud storage for endosomes. EMBO J. 2016;35:1724–5. https://doi.org/10.15252/embj.201695080.
38. Che T, Dwivedi-Agnihotri H, Shukla AK, Roth BL. Biased ligands at opioid receptors: current status and future directions. Sci Signal. 2021;14:1–11. https://doi.org/10.1126/scisignal.aav0320.
39. Pineyro G, Nagi K. Signaling diversity of mu- and delta-opioid receptor ligands: re-evaluating the benefits of β-arrestin/G protein signaling bias. Cell Signal. Elsevier Inc. 2021;80:109906. https://doi.org/10.1016/j.cellsig.2020.109906.
40. Beaulieu J-M, Caron MG. β-Arrestin goes nuclear. Cell. 2005;123:755–7. https://doi.org/10.1016/j.cell.2005.11.010.
41. Kelly E. Efficacy and ligand bias at the μ-opioid receptor. Br J Pharmacol. John Wiley & Sons, Ltd. 2013;169:1430–46. https://doi.org/10.1111/bph.12222.
42. Kliewer A, Gillis A, Hill R, Schmiedel F, Bailey C, Kelly E, et al. Morphine-induced respiratory depression is independent of β-arrestin2 signalling. Br J Pharmacol. 2020;177:2923–31. https://doi.org/10.1111/bph.15004.
43. Kliewer A, Schmiedel F, Sianati S, Bailey A, Bateman JT, Levitt ES, et al. Phosphorylation-deficient G-protein-biased μ-opioid receptors improve analgesia and diminish tolerance but worsen opioid side effects. Nat Commun. Springer US. 2019;10:367. https://doi.org/10.1038/s41467-018-08162-1.

44. Imam MZ, Kuo A, Ghassabian S, Smith MT. Progress in understanding mechanisms of opioid-induced gastrointestinal adverse effects and respiratory depression. Neuropharmacology. Elsevier Ltd. 2018;131:238–55. https://doi.org/10.1016/j.neuropharm.2017.12.032.
45. Kaye AD, Cornett EM, Patil SS, Gennuso SA, Colontonio MM, Latimer DR, et al. New opioid receptor modulators and agonists. Best Pract Res Clin Anaesthesiol. Elsevier Ltd. 2018;32:125–36. https://doi.org/10.1016/j.bpa.2018.06.009.
46. Pryce KD, Kang HJ, Sakloth F, Liu Y, Khan S, Toth K, et al. A promising chemical series of positive allosteric modulators of the μ-opioid receptor that enhance the antinociceptive efficacy of opioids but not their adverse effects. Neuropharmacology. Elsevier Ltd. 2021;195:108673. https://doi.org/10.1016/j.neuropharm.2021.108673.
47. Gretton SK, Droney J. Splice variation of the mu-opioid receptor and its effect on the action of opioids. Br J Pain. 2014;8:133–8. https://doi.org/10.1177/2049463714547115.
48. Puig S, Gutstein HB. A “tail” of opioid receptor variants. J Clin Invest. 2017;127:1221–4. https://doi.org/10.1172/JCI93582.
49. Abbadie C, Pasternak GW. Differential in vivo internalization of MOR-1 and MOR-1C by morphine. Neuroreport. 2001;12:3069–72. https://doi.org/10.1097/00001756-200110080-00017.
50. Tanowitz M, Hislop JN, Von Zastrow M. Alternative splicing determines the post-endocytic sorting fate of G-protein-coupled receptors. J Biol Chem. Â© 2008 ASBMB. Currently published by Elsevier Inc; originally published by American Society for Biochemistry and Molecular Biology. 2008;283:35614–21. https://doi.org/10.1074/jbc.M806588200.
51. Narayan A, Hunkele A, Xu J, Bassoni DL, Pasternak GW, Pan YX. Mu opioids induce biased signaling at the full-length seven transmembrane C-terminal splice variants of the mu opioid receptor gene, OPRM1. Cell Mol Neurobiol. Springer US. 2021;41:1059–74. https://doi.org/10.1007/s10571-020-00973-5.
52. Pasternak GW, Childers SR, Pan Y-X. Emerging insights into mu opioid pharmacology. Handb Exp Pharmacol. 2020;258:89–125. https://doi.org/10.1007/164_2019_270.
53. Convertino M, Samoshkin A, Viet CT, Gauthier J, Li Fraine SP, Sharif-Naeini R, et al. Differential regulation of 6- and 7-transmembrane helix variants of μ-opioid receptor in response to morphine stimulation. PLoS One. 2015;10:e0142826. https://doi.org/10.1371/journal.pone.0142826.
54. Xu J, Xu M, Brown T, Rossi GC, Hurd YL, Inturrisi CE, et al. Stabilization of the μ-opioid receptor by truncated single transmembrane splice variants through a chaperone-like action. J Biol Chem. Â© 2013 ASBMB. Currently published by Elsevier Inc; originally published by American Society for Biochemistry and Molecular Biology. 2013;288:21211–27. https://doi.org/10.1074/jbc.M113.458687.
55. Robert F, Pelletier J. Exploring the impact of single-nucleotide polymorphisms on translation. Front Genet. 2018;9:507. https://doi.org/10.3389/fgene.2018.00507.
56. Lötsch J, Geisslinger G. Relevance of frequent μ-opioid receptor polymorphisms for opioid activity in healthy volunteers. Pharmacogenomics J. 2006;6:200–10. https://doi.org/10.1038/sj.tpj.6500362.
57. Spampinato SM. Overview of genetic analysis of human opioid receptors. Methods Mol Biol. 2021;2201:3–13. https://doi.org/10.1007/978-1-0716-0884-5_1.
58. Bond C, LaForge KS, Tian M, Melia D, Zhang S, Borg L, et al. Single-nucleotide polymorphism in the human mu opioid receptor gene alters β-endorphin binding and activity: possible implications for opiate addiction. Proc Natl Acad Sci. 1998;95:9608–13. https://doi.org/10.1073/pnas.95.16.9608.
59. Margas W, Zubkoff I, Schuler HG, Janicki PK, Ruiz-Velasco V. Modulation of Ca2+ channels by heterologously expressed wild-type and mutant human μ-opioid receptors (hMORs) containing the A118G single-nucleotide polymorphism. J Neurophysiol. 2007;97:1058–67. https://doi.org/10.1152/jn.01007.2006.
60. Kroslak T, Laforge KS, Gianotti RJ, Ho A, Nielsen DA, Kreek MJ. The single nucleotide polymorphism A118G alters functional properties of the human mu opioid receptor. J Neurochem. 2007;103:77–87. https://doi.org/10.1111/j.1471-4159.2007.04738.x.

61. Knapman A, Santiago M, Connor M. A6V polymorphism of the human μ-opioid receptor decreases signalling of morphine and endogenous opioids in vitro. Br J Pharmacol. 2015;172:2258–72. https://doi.org/10.1111/bph.13047.
62. Befort K, Filliol D, Décaillot FM, Gavériaux-Ruff C, Hoehe MR, Kieffer BL. A single nucleotide polymorphic mutation in the human μ-opioid receptor severely impairs receptor signaling. J Biol Chem. 2001;276:3130–7. https://doi.org/10.1074/jbc.M006352200.
63. Wang D, Quillan JM, Winans K, Lucas JL, Sadée W. Single nucleotide polymorphisms in the human μ opioid receptor gene alter basal G protein coupling and calmodulin binding. J Biol Chem. 2001;276:34624–30. https://doi.org/10.1074/jbc.M104083200.
64. Crist RC, Clarke T-K. OPRD1 genetic variation and human disease. In: Jutkiewicz EM, editor. Delta opioid receptor pharmacology and therapeutic applications. Cham: Springer International Publishing; 2018. p. 131–45. https://doi.org/10.1007/164_2016_112.
65. Gerra G, Leonardi C, Cortese E, D'Amore A, Lucchini A, Strepparola G, et al. Human kappa opioid receptor gene (OPRK1) polymorphism is associated with opiate addiction. Am J Med Genet B Neuropsychiatr Genet. 2007;144B:771–5. https://doi.org/10.1002/ajmg.b.30510.
66. Fujita W, Gomes I, Devi LA. Revolution in GPCR signalling: opioid receptor heteromers as novel therapeutic targets: IUPHAR review 10. Br J Pharmacol. 2014;171:4155–76. https://doi.org/10.1111/bph.12798.
67. He L, Fong J, von Zastrow M, Whistler JL. Regulation of opioid receptor trafficking and morphine tolerance by receptor oligomerization. Cell. 2002;108:271–82. https://doi.org/10.1016/S0092-8674(02)00613-X.
68. Rozenfeld R, Abul-Husn NS, Gomez I, Devi LA. An emerging role for the delta opioid receptor in the regulation of mu opioid receptor function. Sci World J. 2007;7:64–73. https://doi.org/10.1100/tsw.2007.219.
69. Zhang L, Zhang J, Hang L, Liu T. Mu opioid receptor heterodimers emerge as novel therapeutic targets: recent progress and future perspective. Front Pharmacol. 2020;11:1–11. https://doi.org/10.3389/fphar.2020.01078.
70. Massotte D. In vivo opioid receptor heteromerization: where do we stand? Br J Pharmacol. 2015;172:420–34. https://doi.org/10.1111/bph.12702.
71. Reed B, Butelman ER, Yuferov V, Randesi M, Kreek MJ. Genetics of opiate addiction. Curr Psychiatry Rep. 2014;16:504. https://doi.org/10.1007/s11920-014-0504-6.
72. Smith H. Variations in opioid responsiveness. Pain Physician. 2008;11(2):237–48. https://doi.org/10.36076/ppj.2008/11/237.
73. Mistry CJ, Bawor M, Desai D, Marsh DC, Samaan Z. Genetics of opioid dependence: a review of the genetic contribution to opioid dependence. Curr Psychiatr Rev. 2014;10:156–67. https://doi.org/10.2174/1573400510666140320000928.
74. Grim TW, Acevedo-Canabal A, Bohn LM. Toward directing opioid receptor signaling to refine opioid therapeutics. Biol Psychiatry. 2020;87:15–21. https://doi.org/10.1016/j.biopsych.2019.10.020.
75. Muñoa I, Urizar I, Casis L, Irazusta J, Subirán N. The epigenetic regulation of the opioid system: new individualized prompt prevention and treatment strategies. J Cell Biochem. 2015;116:2419–26. https://doi.org/10.1002/jcb.25222.
76. Browne CJ, Godino A, Salery M, Nestler EJ. Epigenetic mechanisms of opioid addiction. Biol Psychiatry. Elsevier Inc. 2020;87:22–33. https://doi.org/10.1016/j.biopsych.2019.06.027.
77. Burford NT, Traynor JR, Alt A. Positive allosteric modulators of the μ-opioid receptor: a novel approach for future pain medications. Br J Pharmacol. John Wiley & Sons, Ltd. 2015;172:277–86. https://doi.org/10.1111/bph.12599.
78. Burford NT, Clark MJ, Wehrman TS, Gerritz SW, Banks M, O'Connell J, et al. Discovery of positive allosteric modulators and silent allosteric modulators of the μ-opioid receptor. Proc Natl Acad Sci U S A. National Academy of Sciences. 2013;110:10830–5. https://doi.org/10.1073/pnas.1300393110.
79. Livingston KE, Stanczyk MA, Burford NT, Alt A, Canals M, Traynor JR. Pharmacologic evidence for a putative conserved allosteric site on opioid receptors. Mol Pharmacol. 2018;93:157–67. https://doi.org/10.1124/mol.117.109561.

Chapter 10
Opioids in Pain

Rocío Torres-Méndez, Rafael Acuña-Prats, Nayely V. Salazar-Trujillo, Silvia L. Cruz, and Juan C. Molina-Covarrubias

Abstract This chapter reviews pain definition and classification depending on location (somatic or visceral), duration (acute or chronic), and etiology (nociceptive, neuropathic, or nociplastic). It also reviews the different scales used to assess pain only (unidimensional) or pain and the emotional and cognitive changes associated with it (multidimensional pain assessing scales); the physiology of pain, including the ascending and descending pain pathways; and the analgesic ladder proposed by the World Health Organization (WHO) to treat pains of different severities. After this general review of pain essential aspects, this chapter addresses the role of opioids in acute pain treatment and the limitations of opioid use in chronic pain management.

Keywords Opioids · Pain · Nociception · Classification · Treatment · Opioid misuse

R. Torres-Méndez · R. Acuña-Prats
Hospital Ángeles Lomas, Huixquilucan, State of Mexico, Mexico

N. V. Salazar-Trujillo
Opioids Working Group, Global Studies Seminar, Faculty of Medicine, National University Autonomous of Mexico (UNAM), Mexico City, Mexico

S. L. Cruz
Opioids Working Group, Global Studies Seminar, Faculty of Medicine, National University Autonomous of Mexico (UNAM), Mexico City, Mexico

Department of Pharmacobiology, Cinvestav, Mexico City, Mexico

J. C. Molina-Covarrubias (✉)
Internal Medicine Physician, Private Practice, Mexico City, Mexico

S. L. Cruz (ed.), *Opioids*, https://doi.org/10.1007/978-3-031-09936-6_10

10.1 What Is Pain?

Pain is a physiological protective alert system, essential for detecting and minimizing contact with noxious or harmful stimuli. However, pain itself can become a disease when it becomes chronic due to the persistence or progression of the pathology that originated it or when the pathology has disappeared, but pain persists.

A taxonomy task force of the International Association for the Study of Pain (IASP) proposed a widely accepted definition of pain as "an unpleasant sensory and emotional experience associated with, or resembling that associated with, actual or potential tissue damage" [1]. A key feature of this definition is that pain is always a personal, subjective, and unpleasant experience. Indeed, pain is a biopsychosocial experience. Its perceived intensity does not depend solely on the extent of tissue damage but also on the psychological and adaptive factors of the individual who suffers from it.

Nociception (from the Latin *noci*, meaning to injure or damage) refers to the neural mechanisms by which noxious stimuli are perceived, transmitted, and modulated, independently of the emotional reaction to them. This term is preferred over pain when describing pathophysiological processes that do not consider the emotional aspects of pain experienced by the subject.

10.2 Types of Pain

10.2.1 General Classification

Different classifications of pain exist depending on the criterion used. If it is location, pain can be *somatic* or *visceral*. Somatic pain, also known as musculoskeletal pain, is originated in the skin, muscles, joints, and connective tissues and has a precise location. It is the kind of pain experienced when we cut our skin with a sharp knife or bump into a chair when shoeless. Visceral pain originates in internal organs in the chest, abdomen, or pelvis, is diffuse (i.e., difficult to locate with precision), and is usually associated with autonomic responses such as sweating, changes in blood pressure and heart rate, and gastrointestinal discomfort. Irritable bowel syndrome, endometriosis pain, and renal colic are examples of visceral pain [2].

A classification based on the etiology of pain recognizes three main categories:

- *Nociceptive* pain due to tissue injury [3, 4]
- *Neuropathic* pain, which is secondary to nerve damage [5]
- *Nociplastic* pain due to a sensitized nervous system [6]

Some overlapping can exist between these categories, as will be explained later. Table 10.1 summarizes the main characteristics and causes of these types of pain.

Depending on the duration, pain can be acute or chronic. *Acute pain* is the immediate consequence of intense nociceptive system activation by a thermal,

Table 10.1 Classification of pain according to its etiology

Type of pain	Cause	Example(s)
Nociceptive pain	Tissue injury, or local inflammation, and consequent activation of nociceptors innervating peripheral structures or internal organs	Somatic pain (e.g., acute trauma, arthritis, postoperative), visceral pain
Neuropathic pain	Lesion or disease affecting the somatosensory nervous system (direct toxic chemicals, metabolic insults, viral infections, autoimmune disorders)	Nerve compression, diabetic neuropathy, herpes zoster, chemotherapy
Nociplastic pain	Altered nociception function without clear evidence of actual or threatened tissue damage	Fibromyalgia, irritable bowel syndrome, and other pain disorders

mechanical, or chemical stimulus. It has an adaptive and protective function because it forces the individual to avoid situations that compromise its integrity. *Chronic pain* is a persistent or recurrent pain that lasts for at least 3 months [7].

Chronic pain is a disabling condition that significantly affects an individual's health status and life. It is frequently associated with various comorbidities such as depression, anxiety, fatigue, lack of energy, sleep disturbances, and neurocognitive changes, significantly affecting patients' quality of life [8].

The high prevalence of chronic pain represents a significant public health issue. Pain is common in chronic and progressive diseases. For example, approximately 70% of patients with advanced cancer suffer moderate to severe pain that may require opioid management [9]. This proportion is similar to that of various terminal illnesses.

The mechanisms by which acute pain becomes chronic and changes from physiological to pathological, or from protective to harmful, are still poorly understood. However, peripheral and central sensitizations appear to be critical elements in the development of pathological pain [10].

Peripheral sensitization refers to an increase in the excitability of peripheral nociceptors, which manifests as extreme sensitivity to pain (primary hyperalgesia). When these stimuli are prolonged over time, there is an increase and prolongation of excitability in the CNS (central sensitization), leading to secondary hyperalgesia and the development of chronic pain [11].

10.2.2 Classification of Chronic Pain

In 2015, the task force of the IASP for the classification of chronic pain proposed a new categorization of pathologic pain conditions for the 11th Revision of the International Classification of Diseases (ICD-11) [12]. This new classification aimed to harmonize with that of the World Health Organization by presenting a "classification system that is applicable in primary care and clinical settings for specialized pain management."

The definition of chronic pain as one lasting at least 3 months is the one used by ICD-11. In addition, chronic pain can be primary or secondary, depending on its cause, with several categories in each case [13].

Chronic primary pain is not better accounted for by another (secondary) chronic pain diagnosis and can therefore be considered as a disease in itself. This type of pain involves significant emotional distress or functional interference. There are five subtypes of chronic primary pain [14]:

- *Chronic widespread pain*: Diffuse musculoskeletal pain in at least four of five body regions and at least three or more body quadrants (as defined by upper-lower/left-right side of the body) and axial skeleton (neck, back, chest, and abdomen).
- *Chronic primary visceral pain*: Pain located in the head or neck, thoracic, abdominal, or pelvic region. The anatomical location is compatible with typical referral pain patterns from a specific internal organ. For example, pain resulting from a heart attack may produce pain in the arm in addition to the chest area.
- *Chronic primary musculoskeletal pain*: Pain located in the muscles, joints, bones, or tendons. Pain syndromes are related to their anatomical location (e.g., C.P. cervical pain, C.P. thoracic pain, C.P. low-back pain, C.P. limb pain).
- *Chronic primary headache or orofacial pain (OFP)*: Headache or OFP of unknown origin (idiopathic) occurs at least 15 days per month for longer than 3 months. Each episode of pain lasts at least 2 h (when untreated).
- *Complex regional pain syndrome*: Usually starts distally in an extremity after trauma and is disproportionate in magnitude or duration to the typical course (of pain) after similar tissue trauma, characterized by signs indicating autonomic and inflammatory changes in the affected region.

Chronic secondary pain syndromes are those that last or recur for periods longer than 3 months. Other diseases are the underlying cause of pain, for which pain may initially be regarded as a symptom. There are six subtypes of chronic secondary pain:

- *Chronic postsurgical or posttraumatic pain*: Pain that develops or increases in intensity after a surgical procedure or tissue injury and is sustained beyond the healing process, at least 3 months after surgery or tissue trauma [15].
- *Chronic secondary visceral pain*: Chronic pain secondary to an underlying condition originating from internal organs of the head or neck region or of the thoracic, abdominal, or pelvic areas. It can be caused by persistent inflammation, vascular mechanisms, or mechanical factors [16].
- *Chronic secondary musculoskeletal pain*: Chronic pain originating in musculoskeletal structures, such as bones or joints, and associated with an underlying disease [17].
- *Chronic secondary headache or orofacial pain OFP*: Headache or OFP that occurs at least 50% of days during at least 3 months and lasts at least 2 h per day. Headache and OFP are associated with the effects of a disease, infection, trauma, or other factors [18].

- *Chronic neuropathic pain*: Chronic pain secondary to peripheral or central nervous system (CNS) damage of different etiologies (e.g., metabolic, immunological, CNS disorders, viral infections) [19].
- *Chronic cancer-related pain*: Pain caused by primary cancer or metastases (chronic cancer pain) or by its treatment (chronic post-treatment cancer pain). It is important to differentiate chronic cancer-related pain from pain caused by a comorbid disease [20].

10.3 Pain Assessment

Because pain is a personal and subjective experience, it is not easy to measure. Moreover, patients suffering from pain cannot always give verbal or written responses to questionnaires to rate pain intensity. Consider, for example, the case of infants or people with severe intellectual disability. For this reason, several scales have been developed to use with different populations (e.g., pre-term and full-term infants, children, adults, elderly, and intensive care patients, among others).

Pain assessment can include verbal or numerical self-rating scales, behavioral observation scales, body movements, and physiological variables. Depending on the nature of such evaluations, scales have different limitations. There are two broad categories of scales for measuring pain: unidimensional and multidimensional [21]. The first type only evaluates pain intensity, while multidimensional scales evaluate other variables, such as emotional and cognitive impact.

10.3.1 Unidimensional and Multidimensional Scales of Pain

One of the best-known unidimensional scales is the faces pain scale. This self-assessment tool combines images (a face that does not reflect pain and others that represent discomfort of increasing intensity) and numbers to rate pain intensity. It is helpful for children over 3 years of age, people who have difficulty verbalizing emotions, and adults. Other unidimensional pain rating scales are the visual analog scale (VAS), the numeric rating scale (NRS), the verbal rating scale (VRS), and the verbal descriptor scale (VDS) (Table 10.2).

An example of a multidimensional pain assessment scale is the short form of the McGill Pain Questionnaire (SF-MPQ) which includes a list of 15 adjectives, 11 of which describe the type of pain (e.g., throbbing, stabbing, burning) and 4 adjectives describing affective qualities, such as the occurrence of tension, fear, and autonomic symptoms associated with pain. The patient can assign a value of 0 = none, 1 = mild, 2 = moderate, or 3 = severe to each word, which results in a final score; the higher the total, the more intense the pain. Other multidimensional pain scales are the extended version of the McGill Pain Questionnaire, the Brief Pain Inventory, and

Table 10.2 Unidimensional pain assessment scales

Scale name	Main features
Faces pain scale (FPS)	Self-assessment scale that combines images (faces that do not reflect pain to others that represent discomfort of increasing intensity) and numbers to rate pain. It can be used in children over 3 years of age and in adults
Numeric rating scale (NRS)	Patients select a value between 0 and 10, 0 and 20, or 0 and 100 that fits the intensity of their pain, where the number zero represents "no pain"and the upper limit means "worst possible pain"
Verbal rating scale (VRS)	Classifies pain as no pain, mild, moderate, or severe. As in the VAS, the values can be translated into a numerical value between 0 and 10
Verbal descriptor scale (VDS)	Vertical scale is also known as "pain thermometer." It ranges from "no pain"(at the bottom) to "most severe pain imaginable" (at the highest point)
Visual analog scale (VAS)	VAS is a widely used tool to assess pain intensity. The patient is asked to indicate perceived pain intensity along a 100 mm horizontal line

the Chronic Pain Grade Scale. Table 10.3 summarizes the main characteristics of these and other pain scales.

10.3.2 Pain Assessment Scales for Special Populations

Children under 3 years of age have very little ability for verbal communication [22]. The same is true for severely ill patients in the intensive care unit [23]. In these cases, pain evaluation implies observation of some specific variables. For example, the FLACC pain rating scale stands for the face (F), legs (L), activity (A), cry (F), and "consolability" (C). Each one of these variables can have a score of zero (face relaxed, normal position of legs, normal activity, no crying, and content) 1, or 2 (e.g., kicking, crying steadily, and difficult to console). Similar scales exist for neonates and patients with advanced dementia. Table 10.4 summarizes the main features of the most common scales to assess pain in special populations.

10.4 The Physiology of Pain

10.4.1 Nociceptors

Neuronal pathways that perceive, conduct, and process pain develop from embryonic stages and become fully integrated when the body reaches maturity. Although the main networks responsible for pain have already been identified, others are still unknown. The essential components of nociception are the nociceptors (or pain receptors), the chemical substances released, and the nervous pathways associated with the noxious stimulus [24].

Table 10.3 Multidimensional pain assessment scales

Scale	Main features
Brief Pain Inventory (BPI)	It is a 9-item self-administered questionnaire used to assess the severity of a patient's pain and its impact on daily functioning (difficulties with general activity, walking, mood, and sleep). It was developed by the Pain Research Group of the WHO Collaborating Centre for Symptom Evaluation in Cancer Care
Chronic Pain Grade Scale (CPGS)	Pain rating questionnaire that assesses pain intensity and associated disability. It classifies pain ranging from no pain (value of 0) to highly disability-severely limiting (IV). It has been used in clinical and epidemiological studies to assess and compare pain severity among groups
McGill Pain Questionnaire (MPQ)	Self-report questionnaire to evaluate different qualities of the subjective pain experience. It assesses three distinct components: sensory intensity, emotional impact, and cognitive evaluation. There are 20 subclasses, each containing two to six words and one pain intensity scale. Total scores range from 0 (no pain) to 78 (severe pain)
Short-Form McGill Pain Questionnaire (SF-MPQ)	This questionnaire consists of 15 words (11 sensory and 4 affective) which are rated on an intensity scale as 0 = none, 1 = mild, 2 = moderate, or 3 = severe. SF-MPQ is easier to use and takes less time to administer and complete than the original long form.
Short Form 36 Bodily Pain Scale (SF-36 BPS)	It is one of eight subscales of the SF-36, a generic measure of health status used in population surveys. A 6-point rating scale (ranging from "none" to very severe") is used to assess the intensity of bodily pain, and a 5-point scale (ranging from "none" to "extremely") is used to determine the impact of pain with everyday activities.
Measure of Intermittent and Constant Osteoarthritis Pain (ICOAP)	An 11-item questionnaire that assesses pain in individuals with osteoarthritis of the hip or knee; it considers both constant and intermittent pain experiences (pain intensity, frequency, mood, duality of life)
West Haven-Yale Multidimensional Pain Inventory (WHYMPI)	Tool to evaluate patients with chronic pain. The inventory consists of 3 parts, including 12 scales that assess the impact of pain on patients' lives, the responses of others to patients' pain communications, and the degree of patients' participation in various daily activities

Nociceptors are a subpopulation of peripheral free nerve terminals of sensory neurons capable of differentiating between innocuous and potentially harmful stimuli. These terminals are the final, non-specialized parts of nerve fibers located in the skin, bones, muscles, joints, and viscera. When tissue damage occurs, some chemicals that activate nociceptors are released. In particular, when cells rupture, potassium (K^+) concentration increases in the nociceptors' immediate environment, triggering the painful response. This occurs because K^+ is an abundant ion inside the cell but not on the outside. In addition, prostaglandins (related to inflammatory processes), bradykinin, histamine, and changes in the pH of the medium, among other factors, also activate nociceptors.

There are two main classes of nociceptors: the medium-diameter myelinated afferent fibers A-delta (Aδ), which transmit a painful stimulus rapidly, and the

Table 10.4 Pain assessment in critically ill patients and other special populations

Scale	Main features
Critically Ill Patients	
Behavioral Pain Scale (BPS)	Scale to assess pain in intensive care patients whose condition (e.g., sedation or mechanical ventilation) does not allow for the use of other tools. It consists of the sum of three subscales (facial expression, upper limb movements, and tolerance to mechanical ventilation), each of which has a specific evaluation
Critical-Care Pain Observation Tool (CPOT)	Pain assessment scale used in patients in the intensive care unit. It includes four behavioral domains: body movements, facial expression, muscle tension, and compliance with mechanical ventilation for intubated patients or vocalization for extubated patients
Special Populations	
Pain Assessment in Advanced Dementia Scale (PAINAD)	Tool to assess pain in adults with dementia or cognitive impairment. It can be used in both verbal and nonverbal patients. It includes the following items: breathing independent of vocalization, negative vocalization, facial expression, body language, and consolability
Pain Assessment in Seniors with Limited Ability to Communicate (PACSLAC)	A checklist used to assess pain in elderly/resident patients with dementia who cannot communicate verbally. Includes the following items: facial expressions, activity/body movement, social/personality/mood, and others (includes physiological changes, eating and sleeping, and vocal behaviors)
Pain Assessment in Neonates (PAIN) Scale	Scale for evaluation of pain in neonates that includes following items: facial expression, cry, breathing pattern, extremity movement, state of arousal, oxygen required for saturation >95%, and increased vital signs (heart rate)
Neonatal Infant Pain Scale (NIPS)	Behavioral scale for pain assessment that can be used in both pre-term and full-term infants. It includes six indicators: facial expression, crying, breathing patterns, arms, legs, and state of arousal
CRIES – Neonatal Pain Assessment Tool	It assesses crying, facial expression, insomnia, oxygenation level, and vital signs. It is applied in infants up to 6 months of age, mainly in neonatal intensive care units

small-diameter unmyelinated C fibers, which transmit pain stimulus slowly (Table 10.5). The conduction speed difference between these fibers explains why we experience an intense, localized, sharp pain immediately after an injury, followed by a more diffuse and dull pain a few seconds later [25].

10.4.2 Ascending Pain Pathways

Pain transmission involves first-, second-, and third-order neurons. This classification refers to the activation sequence from sensing the noxious stimuli to its integration (locating the sensation of pain and assigning it a negative emotional connotation) (Fig. 10.1).

Table 10.5 Differences between the two main types of nociceptive fibers

	Aδ fibers	C fibers
Distribution	Body surface, muscle, joints	Diverse tissues
Fiber diameter	Large (1–5 μm)	Small (<1.5 μm)
Myelinated	Yes	No
Conduction Velocity	Fast (5–30 m/s)	Slow (0.5–2.0 m/s)
Receptor type	Thermal and mechano-thermal	Polymodal (chemical, mechanical, and thermal)
Position of synapse	Laminae I and V (dorsal horn)	Lamina II (*substantia gelatinosa*)
Transmitter	Glutamate	Glutamate and substance P
Pain quality	Well-localized, rapid, sharp	Diffuse, slow, dull

The transmission of painful stimuli to the central nervous system involves four events [25]:

1. *Signal transduction* is when external stimuli (mechanical, thermal, or chemical) are converted into an electrical signal in nociceptive primary afferent neurons.
2. *Transmission* refers to the transfer of noxious impulses from nociceptors (Aδ and C fibers) to first-order neurons located in the dorsal horn of the spinal cord. Once the signal reaches the nociceptor terminals, depolarization causes calcium channel activation, precipitating calcium influx and glutamate and substance P release. Glutamate and substance P are excitatory neurotransmitters that activate postsynaptic receptors (AMPA/kainate and NK1, respectively), which generate excitatory postsynaptic currents (EPSCs) in second-order neurons. As a result, these second-order neurons conduct nociceptive stimuli across the spinal midline and up the spinothalamic tract into the thalamus. There, they synapse with third-order neurons that project to the cortex (pain, quality, and location recognition). In addition, some structures in the midbrain (e.g., reticular formation and periaqueductal gray matter) receive projections from the spinothalamic tract and mediate some pain-related behaviors, such as arousal and emotion.
3. *Modulation* covers neural activity modification (mainly pain-suppressive mechanisms) along the pain transmission pathway from the spinal cord to the higher brain stem and midbrain levels. Several neurotransmitters and neuropeptides modify the nerve transmission threshold, producing a higher or lower pain perception.
4. *Perception* includes processes occurring in the cortex. As previously mentioned, the third-order neurons project to the cortex, and it is here where pain sensations and their quality and location are integrated. At the same time, some structures in the midbrain (reticular formation and periaqueductal gray matter) receive projections from the spinothalamic tract and mediate pain-related behaviors, such as arousal and emotion.

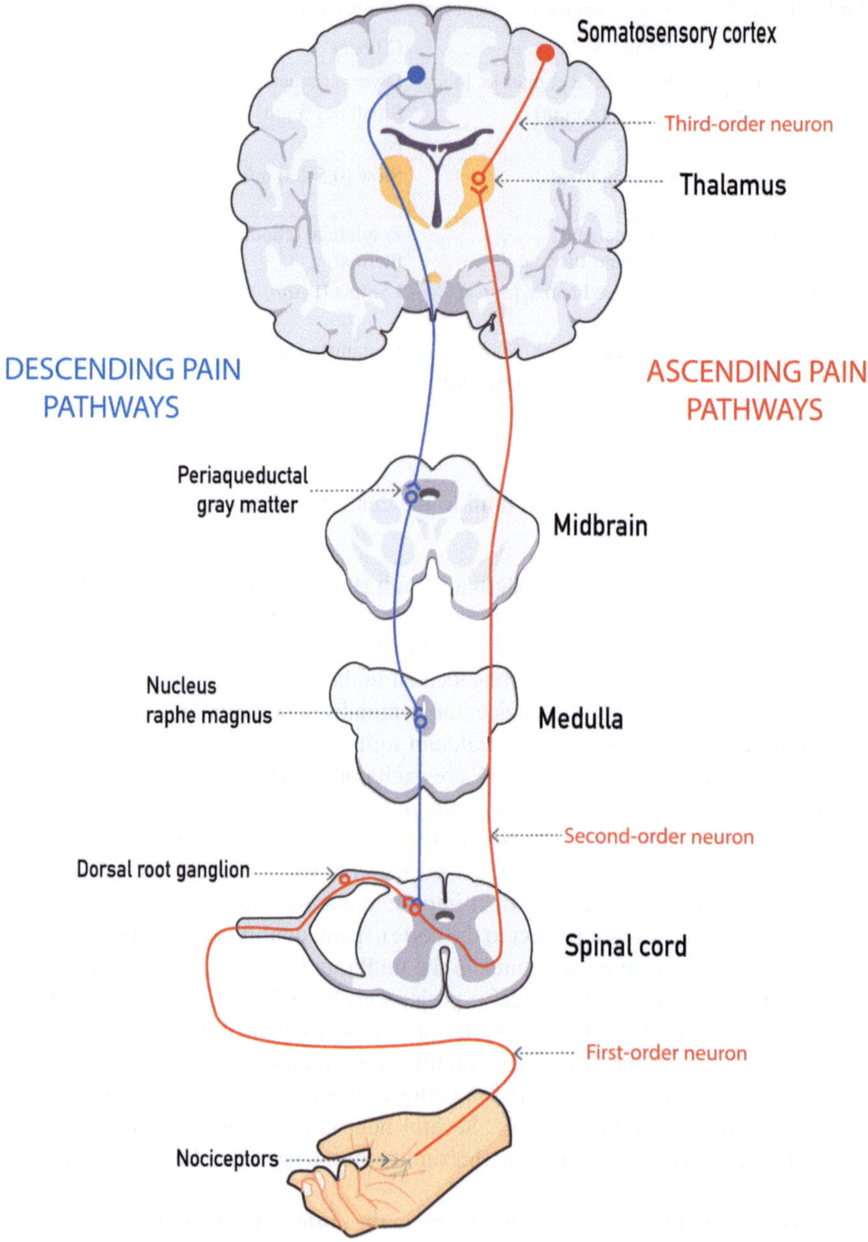

Fig. 10.1 Ascending and descending pain pathways

10.4.3 Descending Pain-Control Pathways

Descending pain pathways are top-down pain modulatory circuits that arise from the midbrain, inhibiting the ascending nociceptive information in the spinal cord (Fig. 10.1). The periaqueductal gray matter (PAG) is a crucial area in the descending modulatory pathways. Neurons from the PAG project down to the brain stem and activate the serotonergic neurons in the nucleus raphe magnus [26]. These neurons, in turn, release norepinephrine (NE) and serotonin (5-HT) in the spinal cord. NE and 5-HT activate small opioidergic interneurons located near the first-order and the second-order neuron's synapse. Activation of these interneurons results in the release of endogenous opioids (see Chap. 8), diminishing painful input. This may explain the efficacy of some antidepressants (e.g., tricyclics and selective serotonin reuptake inhibitors) that increase NE and 5-HT in treating chronic pain syndromes.

Morphine and other opioid agonists are excellent analgesics because they mimic the effects of endogenous opioids. Exogenous opioids inhibit pain transmission in the spinal cord and modulate the descending modulatory pain pathways.

10.5 The WHO Analgesic Ladder

The World Health Organization (WHO) proposed an analgesic ladder in 1986 as a therapeutic guide for pain management in adult cancer patients (WHO Cancer Pain and Palliative Care Program) [27]. This guide, developed by an international group of experts, has been modified over the years. It is a valuable tool for oncologic pain management and a reference for treating acute and chronic non-oncologic pain (e.g., neuropathic pain of various etiologies, musculoskeletal disorders, and visceral pain, among others).

The WHO analgesic ladder suggests starting pain treatment with non-opioid medications and increasing the potency of the medicines based on the patient's response.

The original ladder consisted of three main steps:

- *First step*. For mild pain: the first-choice analgesics are nonsteroidal anti-inflammatory drugs (NSAIDs) or acetaminophen with or without adjuvants.
- *Second step*. For moderate pain: weak opioids (e.g., hydrocodone, codeine, dihydrocodeine, hydrocodone, or tramadol) with or without NSAIDs and with or without adjuvants.
- *Third step*. For severe and persistent pain: potent opioids (e.g., morphine, methadone, fentanyl, oxycodone, buprenorphine, tapentadol, hydromorphone, or oxymorphone) with or without NSAIDs and with or without adjuvants.

Examples of NSAIDs are acetylsalicylic acid (aspirin), naproxen, ketorolac, phenacetin, and ibuprofen. Adjuvants are drugs that do not have significant analgesic effects by themselves but augment the analgesic efficacy of opioids. They are used to prevent or treat concurrent symptoms that exacerbate pain. Common adjuvants are anticonvulsants, antidepressants, anesthetics, and antiarrhythmics, among others.

A significant omission in the original WHO analgesic ladder was the lack of a criterion to integrate non-pharmacological treatments as therapeutic alternatives for pain management; hence a new step was later added. This *fourth step* includes invasive and minimally invasive procedures, which can be associated with strong opioids or other medications (intrathecal administration of analgesic and local anesthetic drugs with or without pumps, epidural analgesia, neurosurgical procedures, among others) [28].

Unlike chronic oncologic pain, acute pain tends to be more intense at the time of injury and to then decrease progressively. For this reason, the World Federation of Societies of Anesthesiologists proposed to use the WHO scale, but in reverse [29]. Thus, in acute pain, it is advisable to initiate treatment with potent analgesics, such as opioids, depending on the intensity of the pain, and consider switching to other agents as the pain subsides. This same therapeutic behavior can be applied in the case of chronic pain that tends to resolve (Fig. 10.2).

It is essential to keep in mind that to use the WHO ladder properly, the clinician must have the necessary skills to perform an adequate assessment of a patient with pain to establish the most appropriate pharmacological and non-pharmacological treatment in each case. If a physician does not have sufficient experience in diagnosing and managing pain, or if the patient does not respond to analgesics, it is advisable to refer the patient to a pain specialist [30].

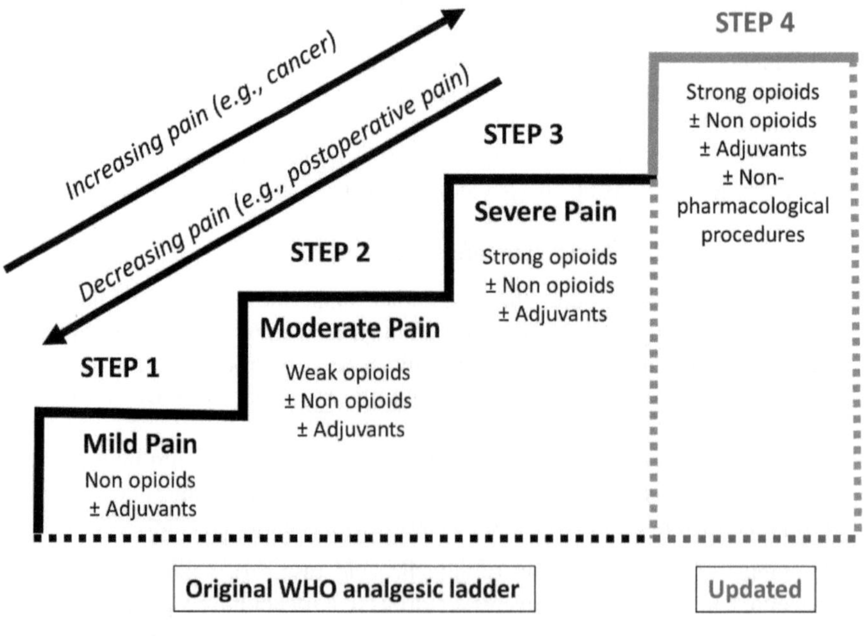

Fig. 10.2 The WHO analgesic ladder

10.6 Interdisciplinary Multimodal Pain Therapy Approach

The first pain clinics in the USA and Europe were established in the second half of the twentieth century and generally offered unimodal treatment approaches, such as pharmacotherapy (mainly with opioids) or nerve blockade [31]. However, this management approach aimed exclusively at treating pain as a symptom and proved to be ineffective. It is now widely recognized that chronic pain causes significant suffering to patients and has a relevant psychosocial and economic impact, making it necessary to establish multifaceted approaches such as interdisciplinary multimodal pain therapy (IMPT) [32].

Pain management practices are constantly being revised and updated. Although the physician's empathy for the patient is essential, it is insufficient for adequate pain treatment. The Institute of Medicine's Relieving Pain in America states that healthcare professionals "need and deserve greater knowledge and skills to be able to contribute to the necessary cultural transformation in the perception and treatment of people with pain" [33]. The World Health Organization has postulated that interprofessional education is a crucial aspect of collaborative medical activity to provide better healthcare for patients [34]. The University of Toronto Centre for the Study of Pain (UTCSP) developed the pain interprofessional curriculum design (PICD) model [35]; the core content on pain management was based on the IASP interprofessional pain curriculum domains and core content competencies related to pain management.

10.7 Relevant Aspects of Opioids in Clinical Practice

10.7.1 Opioid Pharmacokinetics

The existence of a wide variety of opioids with analgesic effects and different pharmacokinetic profiles gives physicians useful pharmacological tools to use the best one for each patient under specific conditions. Clinically used opioids differ in their absorption and distribution, which will determine the dose and administration route. In addition, each drug has a particular metabolism and elimination, which could be important to predict or avoid pharmacological interactions [36]. For example, some opioids suffer extensive first-pass metabolism and cannot be prescribed orally. In some cases, significant dose adjustments are required in case of switching from one administration route to another. The balance between absorption, distribution, metabolism, and excretion determines two important pharmacokinetic parameters: the time to peak, which is the time needed to reach the maximal concentration, and the elimination half-life ($t_{1/2}$), which is the time for concentration to decrease to 50%. The time to peak determines the onset of action, while the $t_{1/2}$ is related to the duration of drug effects. Table 10.6 provides information on administration routes, doses, metabolism, and $t_{1/2}$ of commonly prescribed opioids [37].

Table 10.6 Pharmacological data of clinically relevant opioids

Drug	Use	Administration route/dose[a]	Pharmacokinetic aspects	Comments
Morphine	Analgesic, antitussive	*p.o.*: 15–30 mg; *i.m., s.c.*: 5 mg	$t_{1/2}$: 2–3.5 h Active metabolite: M6G. First-pass metabolism	Morphine is the prototypical μ-opioid receptor preferring agonist
Heroin (diacetyl-morphine)	Analgesic, misused drug	*i.v., i.m.*: 5 mg *i.v.*: variable	$t_{1/2}$: 0.5 h First-pass metabolism	Morphine prodrug; more potent than morphine Deadly dose: 25 mg (i.v.)
Hydrocodone	Analgesic, antitussive	*p.o.*: 5–7.5 mg	$t_{1/2}$: 2–4.5 h Metabolism: CYP2D6 and CYP3A4	Prodrug opioid, the parent compound is a full μ-opioid receptor agonist. It is metabolized into its active moiety (hydromorphone) by CYP2D6. There is an extended-release presentation
Oxycodone	Analgesic	*p.o.*: 5–30 mg	$t_{1/2}$: 2.5–3 h Metabolism: CYP3A4	Frequently combined with non-opioids. A modified-release formulation was developed, intended to make the tablet difficult to manipulate for misuse
Fentanyl	Analgesic, preanesthetic medication	*p.o., s.l.*: 100–800 μg	$t_{1/2}$: 2–4 h It undergoes the first-pass metabolism via cytochrome CYP3A4	Immediate release is available in oral transmucosal (sublingual and buccal) and intranasal formulations. A transdermal patch is widely used in the treatment of both cancer and non-cancer chronic pain
Levorphanol	Analgesic	*p.o.*: 1–2 mg q.i.d or t.i.d	$t_{1/2}$: 11–16 h. Metabolism: glucuronide conjugation	μ-Opioid agonist with NMDA receptor antagonism and reuptake inhibition of both norepinephrine and serotonin
Methadone	Analgesic Assisted maintenance Tx	*p.o.*, parenteral: 2.5 mg *p.o.*: 20–40 mg	$t_{1/2}$: 20–37 h. Good absorption. Metanolism: CYP3A4 and CYP2B6	NMDA receptor antagonist activity. In most countries, it is available as a racemic mixture of two isomers (L and D methadone). Potential prologation of QTc interval

(continued)

Table 10.6 (continued)

Drug	Use	Administration route/dose[a]	Pharmacokinetic aspects	Comments
Bupre-norphine	Analgesic, assisted maintenance Tx	*i.v., i.m.:* 0.4 mg *s.l.:* 2–8 mg	$t_{1/2}$: Variable *i.v.:* 3 h *s.l.:* 30 h Main metabolism: CYP3A4	Partial (P) μ-receptor agonist. Buprenorphine transdermal patches are available in some regions in concentrations of 5, 7.5, 10, 15, and 20 μg per hour for a single weekly administration
Pentazocine	Analgesic, preanesthetic medication	*i.m.:* 30–50 mg *i.v.:* 20 mg	$t_{1/2}$: 2–3 h Hepatic metabolism	Agonist-antagonist. Available orally only in combination with naloxone
Butorphanol	Analgesic	*i.v.:* 0.5–2 mg *i.m.:* 1–4 mg	$t_{1/2}$: 4.5–5.6 h Extensive hepatic metabolism	Mixed agonist-antagonist. Available parenterally to treat moderate to severe pain and as a nasal spray to treat migraine
Nalbuphine	Analgesic Pre- and postoperative medication	*i.v., i.m.:* 10–20 mg	$t_{1/2}$: 3–5 h Extensive hepatic metabolism	Agonist-antagonist. It may be used to reduce some adverse effects from other opioid agonists (e.g., pruritus) without causing withdrawal or reducing the analgesic effect. It is not associated with biliary spasm or colic, hypotension, urinary retention, or prolonged QTc interval
Meperidine	Analgesic	*i.v., i.m.:* 50–100 mg	$t_{1/2}$: 2–4 h. First-pass metabolism (glucuronide conjugation)	Relatively weak opioid μ agonist with only approximately 10% effectiveness of morphine; significant anticholinergic and local anesthetic properties. Its metabolite normeperidine can accumulate after 2 days of administration causes adverse effects which are not reversible by naloxone
Tapentadol	Analgesic	Initial dose: 50 mg *p.o. b.i.d* (doses greater 600 mg daily are not recommended)	$t_{1/2}$: 4 h. Extensive first-pass metabolism (glucuronide conjugation)	Combines two analgesic mechanisms: a mu-opioid agonist and norepinephrine reuptake inhibitor. An extended-release presentation is available

(continued)

Table 10.6 (continued)

Drug	Use	Administration route/dose[a]	Pharmacokinetic aspects	Comments
Tilidine	Analgesic	(Combined with naloxone – Til/Nlx) p.o.,: 50/4–100/8 mg (q.i.d.); 50/4–150/12 mg retard formulation (b.i.d.), maximum (400–600) mg/day	Extensive first-pass metabolism (CYP3A4 and CYP2C19)	It is activated to active metabolite nortilidine. It is combined with an opioid antagonist to prevent its abuse
Tramadol	Analgesic	*p.o.:* 25–100 mg q.i.d (immediate-release); 100–300 mg once daily (extended-release)	$t_{1/2}$: 6 h Main metabolic pathways: CYP3A4 and CYP2D6	Prodrug mainly metabolized by CYP2D6 to O-desmethyltramadol (active metabolite) and then the analgesic activity is strongly modulated by such enzyme. M-Opioid receptor activity and also has weak effects on serotonin and norepinephrine reuptake inhibition
Naloxone	Opiate overdose	*p.o.:* 0.4–2 mg	$t_{1/2}$: 2–4 h	Receptor antagonist
Naltrexone	Assisted maintenance Tx. Opiate overdose	*p.o.:* 50 mg	$t_{1/2}$: 4–12 h	Receptor antagonist

[a]Doses are variable, depending on the intensity of pain and history of opioid use

10.7.2 *Opioid Rotation*

Opioid rotation refers to the practice of switching from one agent to another due to inadequate analgesia or the occurrence of unbearable adverse events (nausea, hallucinations, myoclonus, among others). Whenever there is a need to switch opioids, patients should be monitored closely to assess the efficacy and safety of the new drug. Dose adjustment with the new drug will most likely be necessary.

Calculation of total daily doses of opioids is an aspect that should always be considered when prescribing, maintaining, or reducing their administration. Various tables, like the one included in this chapter, can serve as guides for dose calculations (Table 10.7). However, clinicians should take all precautions and care when rotating

Table 10.7 Opioid conversion table – guidelines for equianalgesic dosing of opioid agonists

Opioid	Conversion factor (convert to MMEs)	Duration (h)	Dose equivalent morphine sulfate oral	Dose equivalent morphine sulfate (*i.m.*, *i.v.*, *s.c.*)	Dose equivalent morphine sulfate transdermal
Morphine	1	3–6	30 mg	10 mg	NA
Codeine	0.15	4–6	200 mg	125 mg	NA
Fentanyl (μg/h)	2.4	NA	NA	12.5 μg/h[a]	12.5 μg/h[a]
Hydrocodone	1	3–6	30–60 mg	NA	NA
Hydromorphone	4	4–5	7.5 mg	0.5–15	NA
Oxycodone	1.5	4–6	20 mg	NA	NA
Oxymorphone	3	3–6	10 mg	NA	NA
Tramadol	0.1	6–8	300 mg	NA	NA
Methadone	Variable[b]		2–20 mg	1–10 mg	NA

The dose conversions listed above are approximate and intended to serve only as an estimate of opioid requirements. Doses may vary, in part because of individual patient's genetics and pharmacokinetics

[a]Fentanyl is dosed in μg/h instead of mg/day, and absorption is affected by heat and other factors

[b]Methadone conversion factor increases as the dose is raised. *NA* not available.

opioids, considering the efficacy and safety profiles of the opioid agents involved. Treatment should be individualized and start with low doses and gradual increases to control pain [38].

When rotating opioids, morphine is the reference drug. The objective is to find how many milligrams of the new drug can produce the same analgesia as morphine. The following steps are suggested for calculating the daily dose of the new opioid: (1) determine the total daily doses of opioid medications received by the patient, (2) convert each dose to morphine milligram equivalents (MMEs = dose × conversion factor), (3) sum total doses if the patient is receiving more than one opioid, and (4) determine the equivalent daily dose of the new opioid. To do this, divide the calculated MMEs of the current opioid by the conversion factor of the new drug.

The conversion factors, shown in Table 10.7, consider several data, including the relative potency of opioids with respect to morphine (which is why morphine's correction factor is 1).

It is advisable to reduce the initial calculated opioid dose by 25–50% to avoid an unintentional overdose due to incomplete cross-tolerance and pharmacokinetic variations between individuals. If necessary, the clinician can increase the dose later according to the analgesic response obtained.

It is beyond the scope of this chapter to review all the details involved in opioid rotation, but specialized sources on this topic may be consulted for further information [39–42].

10.8 Opioids in Acute Pain Treatment

10.8.1 Opioids in Postoperative Pain

Following extensive surgery, most patients experience moderate to severe acute pain that can be difficult to control. In addition, inadequate pain management negatively affects the rehabilitation and recovery process. Some of the procedures most frequently associated with severe pain from the early postoperative period are oncologic, orthopedic, spinal, and gynecologic surgeries. Unfortunately, surgery for a pathology that causes pain does not always eliminate or reduce it. Sometimes, the surgical trauma itself can lead to its evolution toward chronicity.

The most important predictors of the occurrence of severe perioperative pain are preoperative pain, anxiety, surgery in young patients, and the type of surgical procedure performed [43].

Perioperative pain management includes measures taken before hospital discharge to reduce or eliminate pain before, during, and after a surgical procedure. Effective and safe management of perioperative pain should consist of a care plan tailored to the patient's underlying disease(s), the presence of a chronic pain-causing condition, prior opioid use, and the surgical procedure to be performed. In most cases, multimodal analgesic regimens based on scientific evidence are applied [44].

Perioperative techniques for postoperative pain management include central regional (neuraxial) analgesia with opioids, patient-controlled analgesia with systemic opioids, and peripheral regional analgesic techniques. Opioids are commonly used to treat moderate to severe acute pain, and the physician's choice depends mainly on preferred strategy and the availability of drugs and administration equipment [45].

The main challenges in postoperative pain management with opioids are providing adequate analgesia while preventing and minimizing side effects, the risks of abuse, overdose, and addiction.

Treatment of acute pain using a multimodal approach is recommended whenever possible. Multimodal (or "balanced") analgesia refers to the combined use of analgesics with different modes or sites of action to improve analgesic efficacy by reducing opioid use ("opioid-sparing effect") and the risks associated with their use during the postoperative period [46]. Some of these agents include acetaminophen, gabapentinoids (gabapentin and pregabalin), dexamethasone, local anesthetics (e.g., bupivacaine and lidocaine), NMDA receptor antagonists, and opioids (e.g., morphine, hydrocodone, oxycodone, codeine, tramadol) [47].

10.8.2 Opioids in the Emergency Department

Pain is one of the most frequent symptoms of patients seeking care in the emergency department, and its management presents significant challenges. Although the precept of "treat first what kills first" is valid in managing patients with a medical emergency, patients should not unnecessarily suffer pain and its physiological and psychological effects.

In medical emergencies, pain is often not adequately treated due to lack of knowledge about the physiological relationship between trauma and pain, the negative consequences of inadequate pain treatment, and lack of training on how to use opioids or the beneficial effects of pain management in the recovery and healing of the patient [48, 49].

Pain management relies on the appropriate assessment of each case based on clinical experience, not on pre-established protocols. Fear of adverse events associated with opioid administration should not play a role, but on the other hand, clinicians should be knowledgeable about the specific requirements to provide adequate pain management to patients with drug use disorders. Particular attention is needed to possible cross-tolerance in patients with opioid misuse or opioid stabilization therapy (see Chap. 15). There can also be barriers on the part of patients, such as cultural factors, reluctance to inform the presence of pain, refusal of pharmacological treatment, or opioid misuse [50].

Patients suffering from renal colic or abdominal pain require opioid management less frequently than those suffering from a fracture or pain due to severe musculoskeletal trauma. In addition, renal colic shows a characteristic pattern of severe episodic pain that disappears once the stone causing the ureteral obstruction is expelled [51]. In these cases, the physician should consider the pros and the cons of using opioids.

Opioids are not the first option for biliary colic because they produce spams of the Oddi sphincter [52]; however, they can be used in certain circumstances while waiting for admission to the hospital.

The pain care standards require continuous recording and evaluation and establish guidelines for the monitoring and administering of analgesic drugs. They should also provide information to monitor the appropriateness and efficacy of the analgesic therapies used.

10.8.3 Opioids in the Intensive Care Unit

Critically ill patients have acute physiologic disturbances and organ failure. Nearly half of intensive care unit (ICU) patients present moderate to severe pain due to a variety of causes, including underlying illness or injury, pre-existing chronic pain syndrome, recent surgery, and invasive or other ICU procedures (e.g., tracheal intubation, nasogastric tubes, mechanical ventilation, routine nursing care such as postural changes) [53]. In addition, pain at the site of central venous catheter placement or other central access routes is often a cause of long-term complications and pain [54].

Pharmacological treatment of pain in the ICU often focuses on opioids, which, after a prolonged stay and repeated administration, can lead to accumulation and risk of dependence at discharge. Therefore, analgesia in the ICU aims to provide the patient with a state of well-being, considering the clinical situation, individual pain tolerance, and analgesic treatments' side effects [55].

To attenuate adverse physiologic responses associated with pain and prevent the development of chronic pain syndromes, analgesic management strategies should include multimodal analgesia (e.g., non-opioid analgesics, regional anesthesia, or nonpharmacologic therapy), opioid tapering, and switching from parenteral opioids such as fentanyl (i.v.) to other extended-release oral agents such as methadone, oxycodone, or morphine sulfate after several days of treatment. Once the traumatic or postoperative pain subsides, it is possible to progressively reduce the opioid doses, but this could take longer than usual in patients with multiple pain sites [56].

10.9 Opioid Use in Non-cancer Chronic Pain (NCCP)

Approximately 50% of the US adult population has at least one chronic disease, and more than 25% has at least two [57]. Multiple chronic diseases (MCDs) cause individuals to have significant physical and emotional challenges leading to severe deterioration of their health and quality of life.

The use of prescription opioids for non-cancer pain management (NCCP) increased significantly in recent decades in the USA. This practice contributed to an increase in the frequency of opioid use disorders and overdose deaths [58]. This crisis has led to a critical evaluation of the role of opioids in the treatment of NCCP. The results of a meta-analysis published in 2018 [59] comparing opioid versus placebo use in NCCP concluded that opioid use (1) produced a slight improvement in pain relief, sleep quality, and physical functioning, (2) was associated with marginal modification in social functioning, (3) did not improve emotional functioning, and (4) more frequently resulted in various adverse events (nausea, vomiting, constipation, dry mouth, dizziness, drowsiness, and pruritus). When opioids were compared with other agents used in various types of pain (e.g., nonsteroidal anti-inflammatory drugs and antidepressants), all drugs produced similar improvements in pain and physical functioning. Only anticonvulsants were marginally less effective as analgesics. Therefore, opioids are not the first-choice drugs for chronic pain, and, if used, they require careful medical supervision to avoid the development of opioid use disorders. For now, it remains valid to perform an individual assessment on each patient suffering from pain and choose the best alternative in each case based on personal clinical experience, the advice of medical pain specialists, and the recommendations of major local and international pain organizations [60].

10.9.1 Guidelines for Opioid Prescription in NCCP

Some general approaches to the management of opioids in NCPN during the initiation, continuation, and discontinuation phases of treatment have been proposed and are summarized below [61, 62].

Aspects to Consider Before Use of Opioids for NCCP

- Establish the rationale for the use of opioids in NCCP (non-opioid pharmacologic therapy always should be preferred).
- Have the support and guidance of a pain specialist and a pediatrician in the case of patients in childhood or adolescence.
- Explain to the patient the realistic benefits of opioid therapy, as well as the risks and frequency of adverse events associated with their use.
- Establish a management agreement with the patient that includes informed consent.
- When starting opioid therapy, define treatment goals and consider discontinuation strategies.
- Initiate opioid administration (lowest effective dose) with immediate-release rather than extended-release dosage forms.
- Consider weaning or discontinuing other agents that may result in interactions or increased risk from concomitant use (e.g., CNS depressants such as benzodiazepines).
- Evaluate efficacy and safety within 1–4 weeks of onset of opioid treatment.

Issues in Continuation Treatment with Opioids

If the dose being used is equal to or less than 50 mg oral morphine equivalent daily dose (oMEDD):

- Monitor efficacy, adverse effects, and patient adherence to the administration schedule within 1–4 weeks of onset of treatment and at least every 3 months thereafter.
- Follow applicable medical-legal guidelines and safe prescribing practices for drugs and controlled substances.
- Proceed with downward adjustments to opioid doses according to the occurrence of adverse events.
- Before increasing the opioid dose to more than 50 mg of MEDD, obtain permission from the institution's pain management service.
- If opioid treatment is expected to last longer than 12 months, it is essential to seek for advice from a pain medicine specialist.

If the dose is greater than 50 mg of oMEDD:

- Inform the patient, family, and caregivers about the possibility of lack of efficacy with opioids and their associated risks of toxicity and adverse events, especially at higher doses (above 50 mg oMEDD).
- Seek the support of medical pain management groups for regular guidance and direction on behaviors to be followed during continued opioid treatment and the weaning phase.
- Monitor and be alert to the risk of opioid misuse.
- Have a plan that includes measures to be followed in case of overdose or opioid misuse (use of antagonists, substitution therapy; see Chaps. 14 and 15).

Steps to Be Taken When Discontinuing Opioid Treatment

- Explain to the patient the need to stop opioid therapy due to its limited usefulness, risk of adverse events, and/or potential harms, even at low doses.
- Conduct an opioid tapering protocol.
- Opioid weaning is usually necessary after the patient leaves the intensive care unit. It is advisable to time the tapering of opioid administration to coincide with the progressive resolution of the pathologies that gave rise to the pain.

10.9.2 Opioid Use in Various Medical Conditions

10.9.2.1 Opioids in Rheumatologic Pain

Pain is a critical symptom in most rheumatic diseases, but initially, the mechanisms involved in its origin, amplification, and evolution toward chronicity may vary according to the specific condition. For example, in systemic lupus erythematosus (SLE), pain is associated with other symptoms such as fatigue, sleep disorders, and memory disturbances due to inflammatory, neuropathic, and central components. Administration of acetaminophen and tramadol may be useful in certain cases of patients with widespread pain [63].

Management with mild opioids (codeine or tramadol) could be considered when first-line non-opioid analgesics do not produce a good response in patients with osteoarthritis (O.A.) pain, but poor tolerability and restricted efficacy limit their use [64]. In any case, any analgesic intervention with opioids should always be considered a short-term therapeutic trial, limited to between 4 and 12 weeks. A long-term opioid treatment (≥26 weeks) in patients with NCCP increases the risk of developing opioid use disorders and only has a good analgesic response in about 25% of patients [65].

10.9.2.2 Opioids for Pain Following Spinal Cord Injury

There is a significant association between spinal cord injury (SCI) and chronic pain. Overall, four out of five patients complain of the presence of ongoing pain of musculoskeletal or neuropathic origin. SCI-associated pain is accompanied by memory impairment and anxiety symptoms. This central pain syndrome can precipitate a chronic, disabling condition with significant impairment of quality of life.

The efficacy of treatment for pain secondary to SCI is often suboptimal and requires extended periods of use. There are data indicating that the use of high doses of opioids in the first 24 h post spinal cord injury correlates with increased pain during the chronic phase [66]. There are still no studies on the analgesic efficacy of long-term opioid monotherapy and the impact on patients' quality of life [67]. Therefore, whenever it is necessary to manage pain in patients with SCI, it is crucial to consolidate an interdisciplinary medical group that includes a pain management specialist to evaluate the most appropriate treatment scheme and avoid the risk of developing an opioid use disorder [68, 69].

10.9.2.3 Opioids and COVID-19

COVID-19 is a multiorgan disease characterized by multiple clinical manifestations [70]. Although it mainly affects the respiratory and cardiovascular systems, involvement of the nervous system (central and peripheral) and the musculoskeletal system have also been found [71]. In addition, COVID-19 causes pain through multiple mechanisms, including direct and indirect effects on nociceptors, which can give rise to or aggravate a painful syndrome [72].

The pathophysiology of neurological damage is associated with hypercoagulability and hyper-inflammation triggered directly by a viral infection and postinfectious immune mechanisms. These alterations can lead to neuropathic pain or intracranial nerve involvement (cranial polyneuritis) [73].

In some patients, neurological manifestations persist up to several months after infection and could potentially give rise to chronic pain. Therefore, clinicians should always be alert to the onset of pain in patients with COVID-19 and monitor its evolution, even after the acute phase of the infection is overcome [74].

In the COVID-19 pandemic, opioids have demonstrated their usefulness in facilitating patient intubation and as adjuvants in the treatment of cases of acute pain. Some negative points identified have been their potential immunosuppressive effects and possible overuse in the clinic. Although undoubtedly valuable for COVID-19, opioids should be used with caution in patients, considering that these agents can cause respiratory depression and worsen respiratory symptoms [75].

References

1. Raja SN, Carr DB, Cohen M, Finnerup NB, Flor H, Gibson S, et al. The revised International Association for the Study of Pain definition of pain: concepts, challenges, and compromises. Pain. 2020;161(9):1976–82.
2. Sikandar S, Dickenson AH. Visceral pain: the ins and outs, the ups and downs. Curr Opin Support Palliat Care. 2012;6(1):17–26.
3. Shimoji K, Kurokawa S. Anatomical pahysiology of pain. In: Shimoji K, Nader A, Hamann W, editors. Chronic pain management in general and hospital. Singapore: Springer; 2021. p. 21–42.
4. Cheng J. Overview of pain states. In: Cheng J, Rosenquist RW, editors. Fundamentals of pain medicine. Cham: Springer; 2018. p. 1–6.
5. Cohen SP, Mao J. Neuropathic pain: mechanisms and their clinical implications. BMJ (Clinical research ed). 2014;5(348):f7656.
6. Popkirov S, Enax-Krumova EK, Mainka T, Hoheisel M, Hausteiner-Wiehle C. Functional pain disorders – more than nociplastic pain. NeuroRehabilitation. 2020;47(3):343–53.
7. Treede RD, Rief W, Barke A, Aziz Q, Bennett MI, Benoliel R, et al. Chronic pain as a symptom or a disease: the IASP Classification of Chronic Pain for the International Classification of Diseases (ICD-11). Pain. Lippincott Williams and Wilkins. 2019;160:19–27.

8. Dale R, Stacey B. Multimodal treatment of chronic pain. Med Clin N Am. 2016;100(1):55–64.
9. Neufeld NJ, Elnahal SM, Alvarez RH. Cancer pain: a review of epidemiology, clinical quality and value impact. Future Oncol. 2017;13(9):833–41.
10. Nijs J, Leysen L, Vanlauwe J, Logghe T, Ickmans K, Polli A, et al. Treatment of central sensitization in patients with chronic pain: time for change? Expert Opin Pharmacother. 2019;20(16):1961–70.
11. Gebhart GF, Schmidt RF. Primary and secondary hyperalgesia. In: Gebhart GF, Schmidt RF, editors. Encyclopedia of pain [Internet]. Berlin, Heidelberg: Springer Berlin Heidelberg; 2013. p. 3174. Available from: https://doi.org/10.1007/978-3-642-28753-4_201764.
12. Treede R-D, Rief W, Barke A, Aziz Q, Bennett MI, Benoliel R, et al. A classification of chronic pain for ICD-11. Pain. 2015;156(6):1003–7.
13. Treede R-D, Rief W, Barke A, Aziz Q, Bennett MI, Benoliel R, et al. Chronic pain as a symptom or a disease: the IASP Classification of Chronic Pain for the International Classification of Diseases (ICD-11). Pain. 2019;160(1):19–27.
14. Nicholas MK, Blyth FM. Are self-management strategies effective in chronic pain treatment? Pain Manag. 2016;6(1):75–88.
15. Schug SA, Lavand'homme P, Barke A, Korwisi B, Rief W, Treede R-D. The IASP classification of chronic pain for ICD-11: chronic postsurgical or posttraumatic pain. Pain. 2019;160(1):45–52.
16. Aziz Q, Giamberardino MA, Barke A, Korwisi B, Baranowski AP, Wesselmann U, et al. The IASP classification of chronic pain for ICD-11: chronic secondary visceral pain. Pain. Lippincott Williams and Wilkins. 2019;160:69–76.
17. Perrot S, Cohen M, Barke A, Korwisi B, Rief W, Treede R-D. The IASP classification of chronic pain for ICD-11: chronic secondary musculoskeletal pain. Pain. 2019 Jan;160(1):77–82.
18. Benoliel R, Svensson P, Evers S, Wang SJ, Barke A, Korwisi B, et al. The IASP classification of chronic pain for ICD-11: chronic secondary headache or orofacial pain. Pain. Lippincott Williams and Wilkins. 2019;160:60–8.
19. Scholz J, Finnerup NB, Attal N, Aziz Q, Baron R, Bennett MI, et al. The IASP classification of chronic pain for ICD-11: chronic neuropathic pain. Pain. Lippincott Williams and Wilkins. 2019;160:53–9.
20. Bennett MI, Kaasa S, Barke A, Korwisi B, Rief W, Treede RD. The IASP classification of chronic pain for ICD-11: chronic cancer-related pain. Pain. Lippincott Williams and Wilkins. 2019;160:38–44.
21. Shimoji K, Aida S. Pain Measurements. In: Shimoji K, Nader A, Hamann W, editors. Chronic pain management in general and hospital. Singapore: Springer; 2021. p. 173–200.
22. Franck LS, Greenberg CS, Stevens B. Pain assessment in infants and children. Pediatr Clin N Am. 2000;47(3):487–512.
23. Bantel C, Laycock H, Ward S, Halmshaw C, Nagy I. Pain in intensive care: a personalised healthcare approach. J Intensive Care Soc. 2013;14(4):312–8.
24. Leung E. Physiology of pain. In: Sackheim KA, editor. Pain management and palliative care. New York: Springer Science+Business Media; 2015.
25. Zaky S. Anatomy of the nervous system. In: Abd-Elsayed A, editor. Pain. Springer Nature Switzerland AG; 2019. p. 3–7.
26. Ossipov MH, Morimura K, Porreca F. Descending pain modulation and chronification of pain. Curr Opin Support Palliat Care. 2014;8(2):143–51.
27. Jadad AR, Browman GP. The WHO analgesic ladder for cancer pain management. Stepping up the quality of its evaluation. JAMA. 1995;274(23):1870–3.
28. Anekar AA, Cascella M. WHO analgesic ladder [Internet]. In: StatPearls. StatPearls Publishing; 2022 [cited 2022 Jan 15]. Available from: https://www.ncbi.nlm.nih.gov/books/NBK554435/#!po=50.0000.
29. Charlton E. The management of postoperative pain [Internet]. In: Update in anaesthesia; 1997 [cited 2022 Jan 18]. p. 2–17. Available from: https://e-safe-anaesthesia.org/e_library/10/Management_of_postoperative_pain_Update_1997.pdf.

30. National Institute for Health and Care Excellence (NICE). Medicines optimisation in chronic pain [Internet]. 2017 [cited 2022 Jan 18]. Available from: https://www.nice.org.uk/advice/ktt21/resources/medicines-optimisation-in-chronic-pain-pdf-58758008162245.
31. Sabatowski R, Schafer D, Kasper S, Brunsch H, Radbruch L. Pain treatment: a historical overview. Curr Pharm Des. 2004;10(7):701–16.
32. Pfingsten M, Arnold B, Böger A, Brinkschmidt T, Casser H-R, Irnich D, et al. Sektorenübergreifende interdisziplinäre multimodale Schmerztherapie [Cross-sectoral interdisciplinary multimodal pain therapy: recommendations on structural and process parameters of the ad hoc commission "Interdisciplinary Multimodal Pain Therapy" of the German Pain Society (Deutsche Schmerzgesellschaft e. V.)]. Der Schmerz. 2019;33(3):191–203.
33. Loeser JD. Relieving pain in America. Clin J Pain. 2012;28(3):185–6.
34. World Health Organization (WHO). Framework for action on interprofessional education & collaborative practice [Internet]. 2010 [cited 2022 Jan 24]. Available from: https://www.who.int/publications/i/item/framework-for-action-on-interprofessional-education-collaborative-practice.
35. Watt-Watson J, Lax L, Davies R, Langlois S, Oskarsson J, Raman-Wilms L. The pain interprofessional curriculum design model. Pain Med. 2017;18(6):1040–8.
36. Zöllner C, Stein C. Opioids. In: Stein C, editor. Analgesia. Springer-Verlag Berlin Heidelberg; 2007. p. 32–54.
37. Cruz SL, Granados-Soto V. Opioids and opiates: pharmacology, abuse, and addiction. In: Pfaff DW, Volkow ND, editors. Neuroscience in the 21st century. 2nd ed. New York: Springer Science+Business Media, LLC; 2016. p. 3625–58.
38. Liao S, Skavinski J, Capasso J, Pirrello RD. Pain management: a practical approach for hospital clinicians. In: Pantilat S, Anderson W, Gonzales M, Widera E, editors. Hospital-based palliative medicine: a practical, evidence-based approach. New Jersey: Wiley; 2015. p. 11–36.
39. Fine PG, Portenoy RK. Establishing "best practices" for opioid rotation: conclusions of an expert panel. J Pain Symptom Manag. 2009;38(3):418–25.
40. Nalamachu SR. Opioid rotation in clinical practice. Adv Ther. 2012;29(10):849–63.
41. Peppin J, Smith HS. Toward a systematic approach to opioid rotation. J Pain Res. 2014;2014:589.
42. Treillet E, Laurent S, Hadjiat Y. Practical management of opioid rotation and equianalgesia. J Pain Res. 2018;11:2587–601.
43. Ip HYV, Abrishami A, Peng PWH, Wong J, Chung F. Predictors of postoperative pain and analgesic consumption. Anesthesiology. 2009;111(3):657–77.
44. Vetter TR, Kain ZN. Role of the perioperative surgical home in optimizing the perioperative use of opioids. Anesth Analg. 2017;125(5):1653–7.
45. Shanthanna H, Ladha KS, Kehlet H, Joshi GP. Perioperative opioid administration. Anesthesiology. 2021;134(4):645–59.
46. Polomano RC, Fillman M, Giordano NA, Vallerand AH, Nicely KLW, Jungquist CR. Multimodal analgesia for acute postoperative and trauma-related pain. Am J Nurs. 2017;117(3):S12–26.
47. Schwenk ES, Mariano ER. Designing the ideal perioperative pain management plan starts with multimodal analgesia. Korean J Anesthesiol. 2018;71(5):345–52.
48. Samcam I, Papa L. Acute pain management in the emergency department. In: Pain management. InTech; 2016.
49. Sinnenberg LE, Wanner KJ, Perrone J, Barg FK, Rhodes KV, Meisel ZF. What factors affect physicians' decisions to prescribe opioids in emergency departments? MDM Policy Pract. 2017;2(1):238146831668100.
50. Kant J, Dombagolla M, Lai F, Hendarto A, Taylor DM. Analgesia in the emergency department: why is it not administered? Emerg Med J. 2018;30(1):12–7;emermed-2018-207629.
51. Hosseininejad SM, Jahanian F, Erfanian Irankar S, Moosazadeh M, Hosseini SA, Shahbakhti N, et al. Comparing the analgesic efficacy of morphine plus ketamine versus morphine plus placebo in patients with acute renal colic: a double-blinded randomized controlled trial. Am J Emerg Med. 2019;37(6):1118–23.

52. Wu S-D. Effects of narcotic analgesic drugs on human Oddi's sphincter motility. World J Gastroenterol. 2004;10(19):2901.
53. Azzam PN, Alam A. Pain in the ICU. J Intensive Care Med. 2013;28(3):140–50.
54. Rao MH. Effects of fentanyl on procedural pain and discomfort associated with central venous catheter insertion: a prospective, randomized, double-blind, placebo controlled trial. Indian J Crit Care Med. 2014;18(7):421–6.
55. Stamenkovic DM, Laycock H, Karanikolas M, Ladjevic NG, Neskovic V, Bantel C. Chronic pain and chronic opioid use after intensive care discharge – is it time to change practice? Front Pharmacol. 2019;22:10.
56. de Pinto M, Dagal A, Naidu RK. Management of pain in the critically ill. In: Irwin RS, Lilly CM, Mayo PH, Rippe JM, editors. Intensive care medicine. 8th ed. Lippincott Williams & Wilkins; 2018. p. 9–21.
57. Boersma P, Black LI, Ward BW. Prevalence of multiple chronic conditions among US adults, 2018. Prev Chronic Dis. 2020;17(17):200130.
58. Jennings WG, Perez N, Delcher C, Wang Y. Introduction to the opioid crisis. In: Opioid prescribing rates and criminal justice and health outcomes. SpringerBriefs in criminology. Cham: Springer International Publishing; 2020. p. 1–3.
59. Busse JW, Wang L, Kamaleldin M, Craigie S, Riva JJ, Montoya L, et al. Opioids for chronic noncancer pain. JAMA. 2018;320(23):2448.
60. Volkow N, Benveniste H, McLellan AT. Use and misuse of opioids in chronic pain. Annu Rev Med. 2018;69(1):451–65.
61. CDC guideline for prescribing opioids for chronic pain — United States, 2016 [Internet]. [cited 2022 Jan 22]. Available from: https://www.cdc.gov/mmwr/volumes/65/rr/rr6501e1.htm.
62. ACI pain management network – agency of clinical innovation AU. Quick steps through opioid managment [Internet]. [cited 2022 Jan 23]. Available from: https://aci.health.nsw.gov.au/chronic-pain/health-professionals/quick_steps_through_opioid_management.
63. di Franco M, Guzzo MP, Spinelli FR, Atzeni F, Sarzi-Puttini P, Conti F, et al. Pain and systemic lupus erythematosus. Reumatismo. 2014;66(1):33–8.
64. Lang LJ, Pierer M, Stein C, Baerwald C. Opioids in rheumatic diseases. Ann N Y Acad Sci. 2010;1193(1):111–6.
65. Dowell D, Haegerich TM, Chou R. CDC guideline for prescribing opioids for chronic pain-United States, 2016. JAMA. 2016;315(15):1624–45.
66. Stampas A, Pedroza C, Bush JN, Ferguson AR, Kramer JLK, Hook M. The first 24 h: opioid administration in people with spinal cord injury and neurologic recovery. Spinal Cord. 2020;58(10):1080–9.
67. Guy SD, Mehta S, Casalino A, Côté I, Kras-Dupuis A, Moulin DE, et al. The CanPain SCI clinical practice guidelines for rehabilitation management of neuropathic pain after spinal cord: recommendations for treatment. Spinal Cord. 2016;54(S1):S14–23.
68. DiPiro ND, Murday D, Corley EH, DiPiro TV, Krause JS. Opioid use among individuals with spinal cord injury: prevalence estimates based on state prescription drug monitoring program data. Arch Phys Med Rehabil. 2021;102(5):828–34.
69. Bryce TN. Opioids should not be prescribed for chronic pain after spinal cord injury. Spinal Cord Ser Cases. 2018;4(1):66.
70. Bittencourt JV, Reis FJJ, Nogueira LAC. Pain in COVID-19 patients: a call to action for physical therapists to provide pain management after an episode of COVID-19. Braz J Phys Ther. 2021;25(4):367–8.
71. Fernández-de-las-Peñas C, Rodríguez-Jiménez J, Fuensalida-Novo S, Palacios-Ceña M, Gómez-Mayordomo V, Florencio LL, et al. Myalgia as a symptom at hospital admission by severe acute respiratory syndrome coronavirus 2 infection is associated with persistent musculoskeletal pain as long-term post-COVID sequelae: a case-control study. Pain. 2021;162(12):2832–40.

72. McFarland AJ, Yousuf MS, Shiers S, Price TJ. Neurobiology of SARS-CoV-2 interactions with the peripheral nervous system: implications for COVID-19 and pain. Pain Rep. 2021;6(1):e885.
73. Román GC, Spencer PS, Reis J, Buguet A, Faris MEA, Katrak SM, et al. The neurology of COVID-19 revisited: a proposal from the Environmental Neurology Specialty Group of the World Federation of Neurology to implement international neurological registries. J Neurol Sci. 2020;414:116884.
74. Clauw DJ, Häuser W, Cohen SP, Fitzcharles M-A. Considering the potential for an increase in chronic pain after the COVID-19 pandemic. Pain. 2020;161(8):1694–7.
75. Palamim CVC, Boschiero MN, Faria AG, Valencise FE, Marson FAL. Opioids in COVID-19: two sides of a coin. Front Pharmacol. 2022;6:12.

Chapter 11
Opioids in Palliative Care

Nayely V. Salazar-Trujillo, Rocío Torres-Méndez, and Juan C. Molina-Covarrubias

Abstract Palliative care is an essential part of the right to health of all human beings. This medical specialty aims to improve the quality of life of patients with chronic or life-threatening conditions, their families, and caregivers. This chapter presents some conditions that benefit from palliative care and the role of opioids as pain killers and agents used to alleviate symptoms in patients with cystic fibrosis, end-of-life breathlessness, cardiovascular diseases, sickle cell disease, chronic kidney disease, and advanced chronic liver disease. It also mentions specific opioids that can be used or avoided in patients with impaired kidney or liver function, reviews opioid use for pain management in children in palliative care and at the end of life, and provides references to main opioid prescribing guidelines.

Keywords Palliative care · Opioids · Cancer pain · Noncancer pain

11.1 Introduction

Palliative care aims primarily to improve the quality of life of patients facing difficult situations related to chronic and life-threatening illnesses. It also seeks to prevent and alleviate suffering through the early identification, assessment, and treatment of pain, addressing its physical, psychosocial, and spiritual aspects [1]. Measures provided by palliative care are directed at patients, their families, and caregivers [2].

N. V. Salazar-Trujillo
Opioids Working Group, Global Studies Seminar, Faculty of Medicine, National University Autonomous of Mexico (UNAM), Mexico City, Mexico

R. Torres-Méndez
Hospital Ángeles Lomas, Huixquilucan, Mexico

J. C. Molina-Covarrubias (✉)
Internal Medicine Physician, Private Practice, Mexico City, Mexico

S. L. Cruz (ed.), *Opioids*, https://doi.org/10.1007/978-3-031-09936-6_11

Treating patients with chronic or life-threatening illnesses presents challenges regarding the safe and effective use of drugs. Factors related to each individual, such as age, sex, and comorbidities, must be considered in selecting the best drugs, doses, and routes of administration. Moreover, because many patients with chronic conditions receive multiple medications, the physician should always be alert to the possibility of drug-drug interactions. Given the risks of adverse effects associated with polypharmacy, it is necessary to consider nonpharmacological options in patients suffering pain. These may include physiotherapy, cognitive behavioral therapy, massages, ice, heat, rest, biofeedback, and transcutaneous electric nerve stimulation. Therefore, a holistic approach should always be considered, recognizing that pharmacotherapy is only one of several strategies in patient management [3].

11.1.1 The Origin of Palliative Care

More than a century ago, religious institutions were the first to provide palliative care as a charitable act. Later, in 1948, Cicely Saunders, a British nurse, social worker, and physician, opened the first hospice to provide care and support to the terminally ill to make the end-of-life period as comfortable as possible. The success of her initiative gave rise to what is known today as palliative medical care [4].

In 1964, Saunders introduced the concept of "total pain" to recognize that pain in dying patients is an overwhelming experience comprising physical, emotional, social, and spiritual dimensions. Using a novel multidimensional approach, which included special attention to the patient's narratives and advocacy for emotionally engaged terminal care, she established St. Christopher's Hospice in London in 1967, thus setting an important precedent worldwide [5]. By 1990, the World Health Organization (WHO) recognized palliative care as a formal medical specialty to alleviate suffering and improve the quality of life of patients with life-limiting illnesses or serious injuries [6].

Palliative care is an essential part of the health rights of all human beings. Therefore, integrated person-centered health services must provide it while paying particular attention to the specific needs and preferences of each individual.

Most adults who come to require palliative care management have chronic conditions such as cardiovascular disease (38.5%), cancer (34%), chronic respiratory disease (10.3%), HIV/AIDS (5.7%), or diabetes (4.6%). Patients with chronic liver disease, renal failure, multiple sclerosis, rheumatoid arthritis, Parkinson's disease, various neurological diseases, dementia, congenital anomalies, and antibiotic-resistant tuberculosis may also need it [7]. In addition, approximately 80% of patients with AIDS or cancer and two thirds of those with cardiovascular disease or chronic obstructive pulmonary disease (COPD) will experience moderate to severe pain at the end of their lives, requiring opioid management.

Opioids can help control pain and other types of physical symptoms that cause distress, such as shortness of breath (dyspnea). Controlling these manifestations early is an ethical duty to alleviate patients' suffering; however, opioid use is not devoid of adverse effects.

11.1.2 Pathophysiology of Pain in Terminal Illness

Facing a potentially terminal disease is always challenging. Although pain may not be the most frequent symptom at this stage, it is the most feared; therefore, its adequate management represents a physician's professional, moral, and ethical obligation. Pain can strongly decrease the quality and satisfaction of the remaining life; contribute to anxiety, depression, despair, and loss of self-efficacy; and can interfere with medical decision-making [8].

According to a systematic meta-analysis conducted in 2007 of studies conducted between 1965 and 2006, 64% of patients with advanced cancer had pain [9]. In addition, pain in various late-stage noncancer diseases, including congestive heart failure (CHF), end-stage renal disease (ESRD), and chronic obstructive pulmonary disease (COPD), ranged from 34% to 77% [10].

Cancer pain often results from a mix of mechanisms. For example, the etiology of pain related to tissue lesions or involvement of neural structures may be due to the tumor itself, its treatment (e.g., chemotherapy), or the presence of comorbidities. For this reason, some mechanisms are nociceptive, while others are inflammatory or neuropathic (see Chap. 10) [11]. This complexity, which can also occur with other types of advanced illnesses, such as HIV/AIDS, may complicate treatment.

Recent animal studies are shedding light on some of the specific mechanisms that may result in cancer pain. Notably, whereas nociceptive pain results from activation of sensory afferents, neuropathic pain originates from damage to nerves and may prominently include disturbances in ion channels. Therefore, treatments are different for these two types of pain [12].

11.1.3 Causes and Mechanisms of Pain in Palliative Care Patients

As previously mentioned, pain due to cancer is one of the most common symptoms experienced by palliative care patients. The negative consequences of pain undertreatment are significant, yet pain treatment can be inadequate for many reasons.

In cancer patients, tumor growth can cause pain. Other reasons include treatment side effects and coexisting pain conditions (Box 11.1) [13]. In noncancer situations, chronic pain may result from chronic degenerative disorders of the spine and joints. Central pain can occur because of spinal cord injury and neuropathic pain due to metabolic disorders (diabetes), viral infections (shingles), and other causes [14].

Box 11.1 Common Pain Syndromes in Palliative Medicine

Cancer Pain Syndromes

- Pain due to tumors, cancer invasion, or metastasis (e.g., bone pain, abdominal pain, plexopathy pain, chest wall pain)
- Postsurgical pain (e.g., mastectomy, radical neck dissection, limb pain syndrome)
- Chemotherapy-induced pain (e.g., vinca alkaloids, platinum-based agents, taxanes)
- Radiation-therapy induced pain (e.g., mucositis, neuropathies, lymphedema, pain secondary to radiopharmaceuticals)

Noncancer Pain

- Neuropathic pain (e.g., peripheral neuropathy, postherpetic neuralgia central pain, complex regional pain syndrome, trigeminal neuralgia, phantom limb syndrome)
- Headaches (e.g., cluster-type, tension-type, chronic migraine, medication overuse, and miscellaneous)
- Musculoskeletal pain (e.g., osteoarthritis, rheumatoid arthritis, chronic low back pain, chronic shoulder pain)
- Myofascial pain (e.g., lateral epicondylitis, quadratus lumborum pain)
- Fibromyalgia (i.e., chronic widespread pain)
- Medical disease pain (e.g., sickle cell disease, multiple sclerosis, chronic fatigue syndrome, HIV pain other)

11.2 Opioids for Pain Management in Palliative Care

During 2018, 18 million new cases of cancer were diagnosed worldwide. This figure puts the need of efficient pain management into perspective because up to two thirds of people with cancer that experience pain need a strong opioid. This proportion is similar or even higher in many other advanced and progressive conditions, such as respiratory, cardiovascular, renal, liver, and neurodegenerative diseases [15].

Since 2006, the World Cancer Declaration, amended in 2008 and 2013, included as one of its targets, "to make effective pain control and distress management services universally available" [16]. There are several key documents that also highlight the importance of adequate pain control and provide practical guidelines, including: "Improving palliative and supportive care for adults with cancer" (NICE cancer service guide 2004) [17], "Control of pain in adults with cancer" (Scottish Intercollegiate Guidelines Network guideline 106) [18], "A Strategic Direction for Palliative Care Services in Wales" (Government of the Welsh Assembly 2005) [19], and "End of Life Care Strategy" (Department of Health 2008) [20]. These documents are constantly updated and are available in websites.

Strong opioids, especially morphine, are the main medications for treating pain related to advanced and progressive diseases; however, significant differences exist among individual drugs in their bioavailability, metabolism, and response in patients. As a result, each patient needs an individualized evaluation to determine the best opioid for them. In addition, since drug doses cannot be estimated or calculated in advance, they must be adjusted and carefully titrated according to each scenario, providing patients comfort and safety.

In palliative medicine, drugs are seldom used to cure or modify underlying diseases but rather to improve symptoms. Consequently, patients often use these drugs until death. Knowledge of pharmacokinetic variation between patients and across time as the condition worsens, combined with an understanding of the primary modes of drug action, underpins the logical selection and use of the most appropriate treatment [21].

For safe opioid analgesic prescription, it is necessary to recognize the different types of pain (nociceptive, neuropathic, or nociplastic) and use more effective drugs for each case. As a guide for using opioids to relieve cancer pain, the World Health Organization (WHO) developed a 3-step ladder scale, with strong opioids on top of the ladder (see Chap. 10) [22].

A critical aspect to consider in the pharmacological management of pain is monitoring the appearance of manifestations derived from potential drug-drug interactions resulting from the combination of the agents considered in guidelines, including the WHO analgesic scale (see Chap. 10).

11.2.1 Opioids in the Control of Symptoms in Progressive Illnesses

Main medical entities which opioid treatment can play a significant role include cystic fibrosis, dyspnea, various cardiovascular diseases, post-stroke pain, and pain associated with sickle cell disease. Use of opioids in patients with chronic kidney disease or advanced chronic liver disease requires specific treatments as described in the following sections.

11.2.1.1 Cystic Fibrosis

Cystic fibrosis (CF) is a genetic disease with a torpid and progressive course that affects about 70,000 individuals worldwide. It is associated with frequent lung infections, gastrointestinal disorders, and various comorbidities that decrease the quality of life and life expectancy [23].

CF patients show a reduced function of an anion channel protein called cystic fibrosis transmembrane regulator (CFTR). This protein facilitates the absorption/secretion of fluids and electrolytes in epithelia of various organs (e.g., lungs and intestine). A deficient functioning of CFTR causes changes in acidity, which in turn modify the physiological microbial barrier and mucus fluidity, leading to the obstructive pulmonary disease characteristic of this condition [24].

Understanding of this disease has increased in recent years, allowing for the development of novel therapeutic alternatives aimed at correcting and substituting the abnormal protein function. This can improve the patient's evolution and, in some cases, prolong life expectancy [25].

In the advanced stage of CF, patients can present cough, fatigue, and dyspnea. In addition, approximately 80% will suffer from pain at some point, even those with stable lung disease. This condition worsens the quality of life, elevates the risk of illness exacerbations, and impairs physical function. General guidelines for opioid use in adult patients with CF are to:

1. Use low doses and adjust upward according to response
2. Start with immediate-release opioids and not with controlled-release formulations
3. Reevaluate the evidence of risk and benefit before increasing the dose above 50 mg morphine equivalent per day (MME/day) (see Chap. 10)
4. If possible, avoid the use of doses higher than 90 mg morphine equivalents

These recommendations also apply to other pathologies with chronic pain [26].

There is evidence that pain is significantly more intense and lasts longer in adults with CF than in the pediatric population, but the incidence of this condition is similar in both groups [27]. Despite this, opioids tend to be used less frequently in centers caring for children with CF than in institutions dedicated to the treatment of the adult population.

11.2.1.2 Dyspnea

Dyspnea, also known as breathlessness, is a common symptom at the end-of-life stage and affects people's daily functioning. In a cohort study that included a large group of patients with various types of cancer, about 50% reported moderate to severe respiratory distress [28]. Despite this, clinicians often fail to recognize this symptom in people with cancer or other terminal diseases.

It is essential to keep in mind that not all patients with dyspnea have hypoxemia and, therefore, will not benefit from oxygen therapy. As occurs with other conditions, treatment of the patient with dyspnea requires individualized evaluation.

Several systematic reviews have examined the efficacy of oral or parenteral opioids in treating dyspnea in cancer patients and other groups. Overall, some beneficial effects exist for orally and parenterally administered opioids. This effect is counterintuitive because opioids, at high doses, produce respiratory depression. However, the doses of morphine useful to reduce breathlessness are lower than those used for pain treatment. Most patients respond well to 10 mg over 24 h (extended-release formulation), and the titration rate, if needed, is much lower compared to analgesic treatments [29].

Several systematic reviews have examined the efficacy of oral or parenteral opioids in treating dyspnea in cancer patients and other groups. Overall, some beneficial effects exist for orally and parenterally administered opioids [29]. This effect is counterintuitive because opioids, at high doses, produce respiratory depression. However, the doses of morphine useful to reduce breathlessness are lower than those used for pain treatment; most patients respond well to 10 mg over 24 h (extended-release formulation), and the titration rate, if needed, is much lower compared to analgesic treatments [30]. A review of trials involving patients with chronic obstructive pulmonary disease (COPD) and associated dyspnea showed that morphine had a good safety profile, without risk of respiratory depression or other related complications [31]. Thus, opioids, at the doses recommended, can be a safe and predictable alternative to reduce chronic dyspnea in patients with COPD or cancer.

11.2.1.3 Cardiovascular Diseases

Cardiovascular diseases are one of the leading causes of morbidity and mortality worldwide. Opioids, and morphine in particular, have been used in several conditions for their analgesic, anxiolytic, and hemodynamic effects [32].

Acute heart failure (AHF) Along with diuretics and nitrate vasodilators, morphine is one of the most frequently prescribed drugs in AHF patients. This is based on morphine's beneficial hemodynamic effects because it reduces preload and vascular resistance, diminishes heart rate, and decreases the agitation and anxiety accompanying AHF. Despite this, the actual benefit of opioids on patients with this cardiovascular condition is still controversial. According to recent systematic reviews, opioid use appears to be associated with an increased risk of short-term mortality; however, the results are not conclusive, and further research is needed [33]. In any case, when using opioids in AHF, it is crucial to weigh their potential deleterious effects, such as decreased respiratory drive (hypoxic and hypercapnic), which occurs at high doses and is directly proportional to the analgesic drug potency [34].

Chronic heart failure (CHF) affects about 18 million people worldwide. It is characterized by an impaired heart pump function that causes blood congestion in various vascular territories such as the lungs, liver, abdomen, and lower extremities [35]. CHF is a terminal disease, and opioids may be necessary to treat chronic pain in patients and alleviate some symptoms, such as exercise-associated dyspnea. Approximately 88% of patients with chronic heart failure have dyspnea, and opioids have been used as an option to control this symptom [36] (see above); however, the use of opioids in CHF is still controversial because their action on the pathophysiology of CHF is not well understood [37].

Acute coronary syndrome (ACS)/myocardial infarction (MI) The main treatment goals in a cardiac ischemic event (as occurring in ACS and MI) are to stabilize the patient and limit myocardial damage as much as possible. Morphine administration is an option for patients with recurrent or persistent symptoms associated with ACS after management with vasodilators (such as nitroglycerin) because it reduces chest pain and the associated anxiety symptoms [38]. According to various clinical guidelines, opioid management should begin with intravenous boluses of morphine (1–5 mg), repeated until achieving pain relief. Throughout this procedure, it is essential to closely monitor vital signs, mainly blood pressure. On the other hand, it is worth noting that a delay in the onset of action of some antiplatelet agents (e.g., P2Y12-type inhibitors such as clopidogrel and ticlopidine) can occur if these agents are administered shortly after intravenous morphine [39].

11.2.1.4 Post-Stroke Pain

According to recent data, 13 million new strokes occur every year worldwide [40]. Post-stroke pain treatment represents a clinical challenge considering its multiple potential origins and the different pharmacological and non-pharmacological therapeutic approaches required to treat it depending on the type of pain. The most common is musculoskeletal pain (70%), which can be managed with non-opioid analgesics; local neuromuscular blockade, if pain results from muscle spasticity; or other non-pharmacological methods, such as physiotherapy. Approximately 1–12% of patients who have a stroke subsequently suffer from chronic pain of central origin [41]. In these cases, anticonvulsants and antidepressants (e.g., pregabalin, gabapentin, and tricyclic antidepressants) are the first-line drugs. If pain persists, opioids could be adjuvants for these agents.

Paradoxically, both opioid receptor agonists (e.g., morphine and levorphanol) and the opioid receptor antagonist naloxone have been used in post-stroke pain. The possible mechanisms of action could involve anti-inflammatory effects which are mediated by non-classical opioid receptors (e.g., TLR4; see Chap. 12). However, despite this possibility, a recent meta-analysis indicates that there is still little information to support the actual beneficial effects of opioids in post-stroke management [42].

Pain after suffering a stroke is a common symptom that the clinician can easily overlook because of impaired communication, cognition problems, or other concomitant illnesses that are common in these patients. Unfortunately, it often develops into chronic, difficult-to-control pain that can significantly reduce patients' rehabilitation and quality of life. Other relatively frequent complications are peripheral neuropathic pain because of spasticity, joint subluxation, and complex regional pain syndrome. In addition, several types of post-stroke pain syndromes can occur with both nociceptive and neuropathic components. The most common are central post-stroke pain (CPSP), the complex regional pain syndrome (CRPS), pain secondary to spasticity, shoulder pain, and headache [43].

Central post-stroke pain (formerly known as thalamic syndrome) refers to pain occurring after a stroke. It can be challenging to characterize because the patient can describe it in many ways (e.g., sharp, burning, or shooting). Most of the time, the pain appears 1, 3, or 6 months after the event, and it can occur concomitantly with other pathologies that cause pain. CPSP results from damage to the pain pathways. It can occur with or without peripheral receptors' stimulation, with local allodynia or hyperalgesia [44]. In addition, spontaneous pain (not awakened by any stimulus) of a continuous or paroxysmal type may appear. The diagnosis of CPSP is by exclusion, i.e., when there is no other cause explaining its origin. The pathophysiology of CPSP remains unclear, although some conditions may favor its appearance, such as stroke severity, smoking, and a history of depression, among others [45].

Complex regional pain syndrome (CRPS) was first described more than a century ago and remains one of the most misunderstood pathological. CRPS usually appears as severe chronic pain, generally affecting a single limb, and is accompanied by various sensory, motor, cutaneous, and autonomic disturbances. In addition, allodynia and hyperalgesia occur in all patients suffering from this pathological entity. There is evidence of local sympathetic nervous system hyperactivity in CRPS that results in altered temperature regulation in both affected and unaffected extremities, which sometimes improves with a sympathetic blockade; however, this does not fully explain CRPS pathophysiology. In addition, although other triggering mechanisms for this type of pain have been proposed (such as tissue hypoxia, magnification of the inflammatory response, or somatic hypersensitization), there are not enough data to confirm this [46].

Pain secondary to spasticity. Spasticity is an involuntary and often painful contraction of several muscle groups caused by central nervous system's damage secondary to a stroke. It usually appears 3–6 months after the event and remains as a long-term sequela that can lead to loss of function. In addition, muscle contractures can cause intense painful episodes [47].

Shoulder pain usually occurs on the same side of the body affected by the stroke. It can be of two types: shoulder subluxation or frozen shoulder. The first is due to the weakness of muscles that provide stability to the joint, while the second one results from an inflammatory process in the joint capsule that limits shoulder movements. Shoulder pain occurs in up to half of the patients who have had a major stroke, accompanied by significant spastic paralysis with sensory disturbances [48].

Headache after stroke (ischemic or hemorrhagic) affects approximately 6–44% of patients during the acute post-stroke phase. It is like a tension-type headache and can be treated as such [49].

11.2.1.5 Sickle Cell Disease

Sickle cell disease (SCD) is a disorder caused by a genetic mutation that results in the sickle hemoglobin (Hb) variant HbS. This disease is most common in individuals of African descent, and there are approximately 330,000 new cases every year worldwide [50]. HbS impacts the morphological characteristics of erythrocytes, which can become trapped in the microcirculation, leading to ischemia, myocardial infarction, reperfusion, inflammation, and tissue damage. These vaso-occlusive crises (VOCs) are accompanied by painful episodes that are a common cause of admission to emergency units [51].

Acute painful crises (APC) management in SCD remains a clinical challenge despite advances in understanding and managing acute pain in other clinical conditions. There is no specific indication for opioid use in this condition. Pethidine was used once, but it has been suspended because of its short duration of action and possible accumulation of norpethidine, a toxic metabolite [52].

Morphine could be the opioid of choice in APC; however, its use has also raised concerns, including the risk of (a) increased chest crises (characterized by fever, respiratory difficulties, and pulmonary infiltrate) due to increased vascular permeability associated with alterations in endothelial function and (b) the risk of toxicity caused by the active metabolite morphine 6-glucuronide (M6G) in patients with impaired kidney function (see below).

In addition to morphine, various specialized hospital settings use diamorphine (clinical heroin) and hydromorphone. Oxycodone is usually prescribed for home use. Intravenous or transmucosal administration of fentanyl has also been used in APC, but the literature concerning its efficacy in this disease is scarce, promising information on the combined use of morphine with other analgesics such as ibuprofen and arginine in APC, apparently with good clinical response [53].

11.2.1.6 Chronic Kidney Disease (CKD)

Renal disease, frequently associated with various chronic pathologies such as diabetes and hypertension, is becoming increasingly common. In addition, some chemotherapeutic agents can be potentially nephrotoxic (e.g., cisplatin, ifosfamide, permetrexed, methotrexate, vincristine, and daunorubicin, among others). Approximately half of the patients undergoing long-term dialysis have chronic pain; however, it is often unrecognized and undertreated [54].

It should always be kept in mind that patients with cancer and renal insufficiency may have higher concentrations of drugs or their metabolites. For example, morphine's metabolite M6G is eliminated through kidneys and can accumulate in patients with impaired kidney function, causing toxic effects such as myoclonus, impaired cognitive function, and excessive sedation [55].

The recommended opioids for renal insufficiency are those that, unlike morphine, do not have hydrophilic metabolites. Only highly lipophilic opioids meet these characteristics (e.g., fentanyl, alfentanil, sufentanyl, and methadone). As a brief guideline, Table 11.1 summarizes several considerations that must be considered when using opioids in patients with impaired renal function.

Table 11.1 Opioids in patients with advanced chronic kidney disease (CKD)

Opioid	Recommendations	Comments
Morphine	Do not use in patients with severe chronic kidney disease	Morphine's M3G metabolite produces neuroexcitatory effects and may cause seizures and myoclonus. Morphine's active metabolite M6G is slowly dialyzed
Hydromorphone	Recommended for CKD. Extensive hepatic metabolism with a minimal amount of unchanged drug in urine	The metabolite of hydromorphone derived from conjugation with glucuronide acid (hydromorphone-3-glucuronide) is dialyzable
Hydrocodone	There is limited information on its safety in renal disease. Approximately one quarter of this drug is excreted in the urine (unchanged or as a metabolite)	The role of hydrocodone in the management of pain for patients with advanced CKD remains unclear
Fentanyl (IV and transmucosal)	It is recommended in CKD, except in opioid-naïve patients. Safe when used in a single dose. Use lower doses and more extended periods between doses with repeated use	No active metabolites. Not significantly removed by dialysis. Possibility of adsorption to CT190 dialysis membranes
Buprenorphine	Recommended. Hepatic metabolism	Dialysis does not appear to alter buprenorphine plasma levels significantly
Methadone	Recommended. No dose adjustment is required in most cases	No active metabolites. In patients with anuria, methadone is excreted in the feces, so it does not accumulate in the plasma. It is not removed by dialysis
Pethidine	Should be avoided	
Oxycodone	Safe when used as a single dose. Use lower doses and more extended periods between doses with repeated use	Less than 10% is excreted unchanged in the urine; however, its active metabolites may accumulate with prolonged administration and cause CNS toxicity even in dialysis patients
Codeine	Its use is not recommended in patients with severe renal impairment	Accumulation of metabolites derived from glucuronide conjugation can modify efficacy and adverse events profiles
Meperidine	It is not recommended in renal failure due to the accumulation of normeperidine	Seizures may occur
Tramadol	Use with caution in patients with severe renal disease	Tramadol levels are significantly reduced with hemodialysis, so it is recommended to administer it after this procedure

11.2.1.7 Advanced Chronic Liver Disease

Impaired liver function is a condition frequently observed in the clinic, so it is always advisable to explore this possibility in patients presenting pain to adopt adequate strategies for management. Liver disease can be primary (e.g., viral hepatitis or alcoholic liver cirrhosis) or secondary (e.g., metastatic).

The main metabolic pathway of most opioids is through oxidative processes via various cytochrome P450 liver enzymes (e.g., CYP2D6 and CYP3A4) [56]. There is therefore a risk of accumulation and toxic effects in patients with hepatic impairment. Because the metabolism of morphine, oxymorphone, hydromorphone, and buprenorphine is through conjugation with glucuronide acid, they may be the opioids of choice in this context. Nevertheless, when using morphine, hydromorphone, or oxycodone, it is necessary to keep low doses and prolonged administration intervals. Tramadol may be safe, but the experience with this drug is limited, so it should be used with caution in this population. Fentanyl appears to have a good safety profile in patients with moderate liver dysfunction.

After opioid administration to patients with impaired hepatic function, clinicians should monitor excessive sedation, cognitive impairment, or encephalopathy manifestations. The latter condition can be precipitated by poor elimination of ammonia (which is highly toxic to the brain) due to constipation associated with opioid use and requires opioid discontinuation [57].

Table 11.2 summarizes recommendations to use opioids for patients with chronic liver disease.

Table 11.2 Opioid use in patients with impaired liver function

Opioid	Recommendation	Comments
Morphine, hydromorphone	Use with caution	Advisable to increase the interval between doses. Monitor appearance of adverse events
Oxycodone	Use with caution	Advisable to increase the interval between doses. Variations in analgesic response may occur
Fentanyl	Use with caution	Decreased hepatic blood flow may affect clearance. Monitor for adverse events (mainly with patch formulation)
Tramadol	Use with caution. Avoid if possible	Hepatic biotransformation to active metabolites; unpredictable analgesic effect and adverse events
Codeine	Use with caution. Avoid if possible	Hepatic biotransformation to active metabolites; efficacy and adverse events could change
Methadone	Not recommended	Risk of accumulation and toxic metabolites

11.3 Opioids for Pain in Children in Palliative Care

11.3.1 Pediatric Palliative Care

In 2007, the European Association of Palliative Care (EAPC) and the Fondazione Maruzza Lefebvre D'Ovidio Onlus published a series of documents that currently serve as standard guidelines for pediatric palliative care. Among them, the document "IMPaCCT: Standards for pediatric palliative care in Europe" resulted from a Conference in 2006, held by a group of health professionals from Europe, Canada, Lebanon, and the United States that met to discuss the situation of Pediatric Palliative Care, the guidelines established, and the need for their implementation [58]. These recommendations focus on:

- The assessment and alleviation of physical, psychological, and social suffering of sick children by health professionals
- Providing effective palliative care with a broad and multidisciplinary approach, including families and communities despite limited resources
- Providing palliative care at all levels of care, such as hospitals, health centers, and children's homes

The IMPaCCT document defines a life-limiting illness as a condition, such as Duchenne muscular dystrophy, in which premature death is expected. A life-threatening illness is one with a high probability of premature death due to severe disease, but with a possibility of long-term survival in adulthood, such as when children are treated for cancer.

The use of the term "terminal illness" can be confusing; it often refers to all children with disabling conditions, but also to those with life-threatening illnesses when death becomes unavoidable, while some others use it for children with terminal disease. From these latter perspectives, it would be difficult to consider cystic fibrosis or Batten disease as terminal illnesses, even when they require palliative care. To avoid confusion, the Guidelines for the Development of Paediatric Palliative Care Services, published by the Association for Children with Life-threatening or Terminal Conditions and their Families (ACT) and the Royal College of Paediatrics and Child Health in the United Kingdom have categorized disorders for children who should receive palliative care into four blocks [59]:

Group 1: Children with life-threatening illnesses where curative treatment is possible but may fail and for whom access to palliative services may be necessary alongside curative treatment or in case of failure to curative treatment

Group 2: Children with inevitable premature death, but who can undergo long periods of intensive treatment to prolong life, and who can participate in everyday activities (e.g., Duchenne muscular dystrophy and cystic fibrosis)

Group 3: Children with the progression of the disease, with no options for curative treatment, only palliative care, which can last for several years, for example, Batten disease and muscular dystrophy

Group 4: Children with an irreversible but non-progressive disease with complex health needs that often lead to complications and increase the likelihood of premature death Examples include severe cerebral palsy and other disabilities caused by brain or spinal cord injury.

In the context of pediatric palliative care, pain is the main symptom, both in patients with oncological pathology and in children with non-malignant but disabling or life-threatening conditions.

11.3.2 Incidence of Pain and Pain Management in Pediatric Palliative Care

Recent data estimate that between four and eight million children worldwide require palliative care [60, 61]. Thus, children and adolescents between 0 and 19 years old represent 7% of the global palliative care needs.

Main attention needs correspond to Africa and Southwest Asia (51.8% and 19.5%, respectively); followed by the regions of the oriental Mediterranean (12%), Eastern Pacific (7.7%), and the region of the Americas (6.2%), the European area represents the 2.8% of the total.

Adequate pain treatment in children is complex not only because of the broad etiology of pain but also due to the difficulties in recognizing pain in patients with neurologic conditions or problems with communication, i.e., newborns and children. In addition, there is little clinical and laboratory evidence for the use of opioids in this population. Moreover, there are significant differences in the pharmacokinetics of opioids in children compared to the adult population.

The WHO (2020) recommends using morphine at the end of life in pediatric palliative care in particular scenarios [62], considering that:

1. Morphine should not be an independent treatment and opioid prescriptions should depend on the context of the biopsychosocial model, balancing benefits and risks.
2. Only trained physicians should prescribe opioids and be held responsible for the follow-up and continuous assessment of the child and treatment adjustments.
3. There are significant differences in the pharmacokinetics of morphine in children, as well as variations in individual sensitivity and pain perception.
4. Children and their families should receive information on the risk of developing dependence and tolerance, as well as adverse effects, and how to treat them.
5. Although morphine is an excellent option for breakthrough pain and works for chronic pain in palliative care, the dose and time of administration require periodic adjustments.

The administration of morphine and other opioids requires an individual titration of the dose using scales for pain assessment already validated for children, as there is no predetermined maximum dose limit for each patient. The main principle is to increase the doses until the maximum analgesia with the minimum adverse effects is obtained. Other potent opioids, such as fentanyl (transdermal or transmucosal) or buprenorphine (transdermal) can be alternatives if morphine produces intolerable side effects [63].

Although there is little evidence on the ideal dose to treat pediatric acute pain, children, in general, need less than 50% of the opioid doses prescribed to adults.

11.4 Evaluating the Risk of Opioid Misuse in Palliative Care Patients

Chemical coping and addiction can occur in patients under prolonged opioid treatment. Chemical coping refers to the intake of increasing doses of medications as an inappropriate method to cope with psychological, emotional, or spiritual distress [64]. Not all those with chemical coping are dependent on opioids, but this behavior frequently precedes addiction. Cancer patients receiving opioid treatment who have a history of alcohol dependence or drug misuse, are under 65 years old, and have a psychiatric disorder, high emotional stress, or limited behavioral coping mechanisms are considered at risk of becoming chemical copers or persons with opioid dependence.

It is now recognized that chemical coping can decrease some total pain components because physical pain and emotional pain share neurobiological mechanisms and are integrated in the same brain area, the cingulate cortex. However, the risk of developing dependence, neurotoxicity, and overdose with prolonged opioid treatments should not be underestimated [65].

Addiction (dependence) can be diagnosed using the criteria contained in the *Diagnostic and Statistical Manual of Mental Disorders*, version 5 (DSM-5.0), or the International Classification of Diseases, version 11 (ICD-11). These criteria include the desire to reduce or stop drug intake, using drugs despite being aware of the negative consequences associated with their use, tolerance to the desirable drug effects, the occurrence of an abstinence syndrome when reducing or stopping the drug intake, and repeated attempts to quit drug use without success (see Chap. 13).

Specific behaviors and screening tests help identify at-risk patients. Several aberrant behaviors on the part of the patient suggest opioid misuse. Examples of these behaviors are using opioid analgesics to experience euphoric effects (to "get high"), attempts to forge prescriptions to get opioids, licking or chewing fentanyl patches or dissolving their content for intravenous injection, and using opioid analgesics by routes other than the oral or transdermal administration (i.e., snorting, smoking, or injecting crushed opioids). Other minor aberrant behaviors include buying medicines from another person, lying about losing prescriptions, or exaggerating pain to get more opioids. Additional warning signs are social isolation and concerns on the part of family members [66].

The CAGE-AID questionnaire is a drug use screening test. Its name is an acronym for *c*utting down on drinking or drug use, being *a*nnoyed when people criticize someone's drinking or drug use, feeling *g*uilty about alcohol or drug use, and drinking alcohol or using drugs first thing in the morning as an *e*ye-opener. Developed initially to detect alcohol use disorders, the CAGE test was later *a*dapted to *i*nclude other *d*rugs. This test indicates risky consumption when the patient fulfills at least two of these criteria [67]. There is a wide variability in the prevalence of substance use disorders detected with the CAGE questionnaire in cancer patients, from 4% to 38%, depending on the population, type of cancer, and other factors [68].

The Opioid Risk Tool (ORT) is another screening test that patients should answer before receiving opioids for pain management. It assigns a numerical value to different features related to gender, age, family, and personal history of substance misuse, history of sexual abuse in childhood, and psychological disease [69].

Adolescents and young adults are particularly susceptible to the nonmedical use of controlled medications, especially the 12- to 15-year-old population, which requires close monitoring because substance misuse predisposes them to intravenous use of other substances. Information on the prevalence of substance use disorders in the pediatric population in need of palliative care is scarce because the traditional assessment tools for identifying risk factors for substance use disorders do not apply to pediatric populations. However, the lack of reliable data on the magnitude of this situation should not be a reason to ignore the problem or prevent inappropriate opioid misuse.

The National Institute on Drug Abuse recently launched two online brief drug use screening tools validated for adolescents 12–17 years old. One of them, the "Screening to Brief Intervention" (S2BI), asks how many times in the past year the adolescent has used tobacco, alcohol, marijuana, prescription drugs, illegal drugs, inhalants, herbs, or synthetic drugs. The test provides four possible responses: never, once, or twice, monthly, or weekly or more. It also gives examples of prescription drugs (such as pain medications or Adderall not prescribed by a physician), illegal drugs (such as cocaine or ecstasy), inhalants (such as nitrous oxide), herbs, or synthetic drugs (such as salvia, "K2," or bath salts) [70].

Another screening test for adolescents is the Brief Screener for Tobacco, Alcohol, and Other Drugs (BSTAD), which asks simple questions regarding what drugs and how many days the adolescent used in the past year [71].

Although opioid misuse can occur in the pediatric population, few references exist on the risk of addiction or aberrant behaviors in children. It is important to consider safe prescription guidelines with the pediatric population given the well-recognized biological vulnerability, health determinants, and risks of this age group, but not to the point of avoid limiting access to pain medications to children in need of care.

Regardless of whether patients needing palliative care are adults, adolescents, or children, a multidisciplinary team must provide pain treatment and symptom management following precise indications on opioid use when treating persons at risk of chemical coping or addiction. Also, patients need a close follow-up to ensure safe access to these drugs, without stigmatizing their use or restricting access when needed, favoring psychosocial interventions in conjunction with pharmacological treatment, and keeping in mind that inadequate pain management can generate behavioral disorders.

11.5 End-of-Life Care

End-of-life care aims to provide comfort to the patient through minimally invasive procedures. However, during the final days of end-stage patients, oral administration often becomes problematic. As swallowing becomes complex, the clinician should use other routes of administration (i.e., sublingual, transdermal, rectal). In addition to difficulty drinking and eating, terminally ill patients may also have drowsiness, immobility, and significant cognitive function impairment. The physician must manage these symptoms and provide support to family members and caregivers in this challenging and stressful period [72].

A general recommendation at this end-of-life stage is to review all the medications used by the patient, keeping only those that are strictly indispensable while establishing effective pain management treatment tailored to the patient's needs.

Delirium and agitation, also common at this stage, can be due to hydroelectrolyte imbalance, metabolic disorders, or other conditions that require diagnosis and treatment. If pharmacological management is required, the physician can prescribe various antipsychotics (e.g., haloperidol, olanzapine, risperidone) [73].

As previously mentioned, opioids are highly effective drugs for pain treatment and have recognized benefits in dyspnea, a common condition in the dying process. However, in terminally ill patients, it is essential to monitor for constipation, which may occur due to opioid use and insufficient fluid intake; in both cases, laxatives or stool softeners should be part of the treatment plan [72].

Despite their ease of administration, extended-release opioids, such as morphine, oxycodone, and fentanyl patches, are not recommended as initial treatment for pain management or dyspnea due to their unpredictable pharmacokinetics and the difficulty of their titration. In these cases, continuous subcutaneous infusion of opiates (CSIO) is a helpful alternative [74], in addition to the fact that the technique for its installation is simple and allows for easy dose adjustment. The most common opioids administered subcutaneously are morphine, fentanyl, hydromorphone, and methadone [75]. Table 11.3 shows the suggested starting doses of the main opioids used to treat pain and dyspnea in terminally ill patients.

Although restrictions and recommendations on the use of opioids in both acute and chronic pain have been tightened in recent years, these agents remain an essential alternative option in patients at the end of life. Effective management of each of

Table 11.3 Suggested starting opioid doses in terminally ill opioid-naïve patients with moderate to severe pain/dyspnea

Opioid	Oral dose	Intravenous or subcutaneous dose	Number of administrations over 24 h
Morphine	2.5–10 mg	2–10 mg	6–8
Fentanyl	Not available	25–100 μg	8–12
Hydromorphone	2–4 mg	0.5–2 mg	6–8
Oxycodone	2.5–10 mg	Not available	6–8

the disorders that appear at this stage, including pain, can enable patients, their families, and caregivers to go through the dying process in the most comfortable and dignified manner [76].

11.6 The Prescription Drug Monitoring Program (PDMP)

Several countries worldwide, but especially the United States, are experiencing an opioid overdose crisis (see Chap. 5). In recent years, a significant increase in the number of opioid overdose deaths has put health authorities on alert to push for initiatives that would lead to greater vigilance and safety of patients receiving these drugs.

The prescription drug monitoring program (PDMP) is an electronic database used in the United States that stores information on drugs dispensed by pharmacies and other services for tracking prescribing behavior and the use of controlled drugs, like opioids [77]. In addition, PDMP provides health authorities with timely information on opioid use by region or state, enabling them to deliver agile and targeted responses. The PDMP is also a promising tool for health-care providers to see patients' histories to inform their prescribing decisions. Some states have implemented policies that require providers to check the local PDMP before prescribing certain controlled substances under specific circumstances.

According to the cooperative agreement "Overdose Data to Action," administered by the Centers for Disease Control and Prevention (CDC) in the United States, "public health departments need accurate and timely data to save lives and prevent opioid misuse, opioid use disorder, and overdose." This document further adds that "accurate and timely information on the dispensing of controlled substances is a critical piece of data for these purposes." The goal is to apply this control and monitoring tool in all the states and territories. The PDMP was initially considered a regulatory and compliance monitoring tool; therefore, state laws did not allow sharing data from this electronic tool with public health personnel. However, this situation has changed, and pharmacists and physicians can now use the information collected to improve prescribing habits and reduce the risks associated with opioid administration. In addition, the PDMP databases allow for the dissemination of measures and interventions related to opioid management for the benefit of public health [78]. Application of programs like the PDMP in other countries could promote international collaboration and save lives.

11.7 Guidelines for the Safe Prescriptions of Opioids in Palliative Care

The abuse of prescription or illegal opioids represents a growing public health problem in the United States and other countries. According to data obtained by the Centers for Disease Control and Prevention (CDC), about 20% of individuals

Table 11.4 Opioid prescribing guidelines

Guideline	Description
Guideline for Prescribing Opioids for Chronic Pain – United States, 2016. https://stacks.cdc.gov/view/cdc/38440	Includes recommendations for first-contact health-care professionals who prescribe opioids for various chronic pain types that do not include oncologic pain, palliative care, or end-of-life care. This guideline is intended to encourage communication between physicians and patients about the benefits and risks of opioid administration in chronic pain, such as the risks associated with long-term use, including opioid use disorders
NICE clinical guidelines for opioid use in palliative care, 2012. https://www.nice.org.uk/guidance/cg140	The updated guidance developed by the National Institute for Health and Care Excellence (NICE) targets non-specialists initiating treatment with strong opioids for adults with advanced and progressive disease
Adult Cancer Pain, Version 3.2019, NCCN Clinical Practice Guidelines in Oncology. https://www.nccn.org/guidelines/recently-published-guidelines	The complete version of the National Comprehensive Cancer Network (NCCN) Guidelines for Adult Cancer Pain addresses additional aspects of this topic, including pathophysiologic classification of cancer pain syndromes, comprehensive pain assessment, management of pain crisis, ongoing care for cancer pain, pain in cancer survivors, and specialty consultations

receiving prescription opioids misuse these agents. Based on the above, several guidelines have been implemented to help clinicians make decisions in primary care and various specialties on the rational use of opioids. Table 11.4 summarizes the main current guidelines.

Palliative care prevents unnecessary suffering and improves the quality of life of patients, family members, and caregivers. The use of opioids requires a thorough assessment of the patient's needs, careful supervision, and sound knowledge of pain assessment and treatment.

References

1. World Health Organization. Palliative care [Internet]. 2020 [cited 2021 Nov 3]. Available from: www.who.int/news-room/fact-sheets/detail/palliative.
2. von Gunten CF. Interventions to manage symptoms at the end of life. J Palliat Med. 2005;8:S-88.
3. Cherny NI, Fallon M, Kaasa S, Portenoy R. Oxford textbook of palliative medicine. Oxford University Press; 2015.
4. Clark D. Cicely Saunders and her associates: a kaleidoscope of effects. In: Clark D, editor. To comfort always – a history of palliative medicine since nineteenth century. Nerw York: Oxford University Press; 2016. p. 85–116.
5. Wood J. Cicely Saunders, "Total Pain" and emotional evidence at the end of life. Med Humanit. 2021. https://doi.org/10.1136/medhum-2020-012107.
6. World Health Organization. Cancer pain relief and palliative care. Technical Report Series 804. Geneva: WHO; 1990.
7. World Health Organization. Palliative care [Internet]. 2020. Available from: https://www.who.int/news-room/fact-sheets/detail/palliative-care.
8. Hagarty AM, Bush SH, Talarico R, Lapenskie J, Tanuseputro P. Severe pain at the end of life: a population-level observational study. BMC Palliat Care. 2020;19(1):60.

9. van den Beuken-van Everdingen MHJ, de Rijke JM, Kessels AG, Schouten HC, van Kleef M, Patijn J. Prevalence of pain in patients with cancer: a systematic review of the past 40 years. Ann Oncol. 2007;18(9):1437–49.
10. Solano JP, Gomes B, Higginson IJ. A comparison of symptom prevalence in far advanced cancer, AIDS, heart disease, chronic obstructive pulmonary disease and renal disease. J Pain Symptom Manag. 2006;31(1):58–69.
11. Falk S, Dickenson AH. Pain and nociception: mechanisms of cancer-induced bone pain. J Clin Oncol. 2014;32(16):1647–54.
12. Pineda-Farias JB, Saloman JL, Scheff NN. Animal models of cancer-related pain: current perspectives in translation. Front Pharmacol. 2020;11:610894.
13. Kubiak M, Vidal M, Reddy SK. Causes and mechanisms of pain in palliative care patients. In: Bruera E, Higginson IJ, von Gunten CF, Morita T, editors. Textbook of palliative medicine and supportive care. 3rd ed. Boca Raton: CRC Press; 2021. p. 207–19.
14. Marcus DA. Chronic pain. Totowa: Humana Press; 2009.
15. National Institute for Health and Care Excellence (UK). Palliative care for adults: strong opioids for pain relief, vol. 140. NICE Clinical Guidelines; 2016.
16. Union for International Cancer Control (UICC). World Cancer Declaration – UICC [Internet]. 2013 [cited 2021 Nov 7]. Available from: www.uicc.org/sites/main/files/atoms/files/131119_UICC_WorldCancerDeclaration_2013_1.pdf.
17. NICE National Institute for Health and Care Excellence. Palliative care for adults: strong opioids for pain relief – Clinical guideline CG140 [Internet]. Last updated: 3 Aug 2016. 2012 [cited 2021 Nov 3]. Available from: www.nice.org.uk/guidance/cg140.
18. Scottish Intercollegiate Guidelines Network (SIGN). Control of pain in adults with cancer [Internet]. 2008 [cited 2022 Jan 7]. Available from: https://www.palliativedrugs.com/download/081114_SIGN106_quick_ref_guide.pdf.
19. Welsh Assembly Government. A strategic direction for palliative care services in Wales [Internet]. 2003 [cited 2022 Jan 7]. Available from: http://www.wales.nhs.uk/documents/pall-care-final-e.pdf.
20. Department of Health (DH). End of life care strategy: promoting high quality care for adults at the end of their life [Internet]. 2008 [cited 2021 Dec 2]. Available from: https://assets.publishing.service.gov.uk/government/uploads/system/uploads/attachment_data/file/136431/End_of_life_strategy.pdf.
21. Opioid Prescribing Work Group – Minnesota U.S. Chronic pain opioid prescribing recommendations [Internet]. [cited 2022 Jan 7]. Available from: https://mn.gov/dhs/opip/opioid-guidelines/pain-phase/chronic-pain.jsp.
22. Jadad AR, Browman GP. The WHO analgesic ladder for cancer pain management. Stepping up the quality of its evaluation. JAMA. 1995;274(23):1870–3.
23. Elborn JS. Cystic fibrosis. Lancet (London, England). 2016;388(10059):2519–31.
24. VanDevanter DR, Kahle JS, O'Sullivan AK, Sikirica S, Hodgkins PS. Cystic fibrosis in young children: a review of disease manifestation, progression, and response to early treatment. J Cyst Fibros. Elsevier B.V. 2016;15:147–57.
25. Polineni D, Nichols DAH. Inflammation in CF: key characteristics and therapeutic discovery. In: Davis SD, Rosenfeld M, Chmiel J, editors. Cystic fibrosis. Springer Nature Switzerland AG; 2020. p. 161–80.
26. Yaoli Y, Trina H, Zubin M, Anita L, Catherine D, Andrew TB. Attitudes of pain and opioids prescription practices in U.S. cystic fibrosis centers. J Cyst Fibros. 2021;20(1):127–32.
27. Mherekumombe M, Collins JJ. Persisting pain in childhood medical illness. In: Stevens BJ, Hathway G, Zempsky WT, editors. Oxford textbook of pediatric pain. 2nd ed. Oxford: Oxford University Press; 2021. p. 343–50.
28. Dudgeon DJ, Kristjanson L, Sloan JA, Lertzman M, Clement K. Dyspnea in cancer patients. J Pain Symptom Manag. 2001;21(2):95–102.
29. Peel ET, Burns GP. Breathlessness. In: Burke SJ, Peel ET, editors. Integrated palliative care of respiratory disease. Cham: Springer Nature; 2019. p. 19–35.

30. Bausewein C, Currow DC, Johnson MJ. Palliative care in respiratory disease. Sheffield: ERS; 2016.
31. Verberkt CA, van den Beuken-Van Everdingen MHJ, Schols JMGA, Hameleers N, Wouters EFM, Janssen DJA. Effect of sustained-release morphine for refractory breathlessness in chronic obstructive pulmonary disease on health status: a randomized clinical trial. JAMA Intern Med. 2020;180(10):1306–14.
32. Zelis R, Mansour EJ, Capone RJ, Mason DT. The cardiovascular effects of morphine. J Clin Invest. 1974;54:1247–58.
33. Miró Ò, Gil V, Peacock WF. Morphine in acute heart failure: good in relieving symptoms, bad in improving outcomes. J Thorac Dis. 2017;9(9):E871–4.
34. Verbraecken J. Opiates in acute heart failure: more harmful than beneficial. Ann Am Thorac Soc. 2019;16(9):1121–3.
35. Ziaeian B, Fonarow GC. Epidemiology and aetiology of heart failure. Nat Rev Cardiol. 2016;13(6):368–78.
36. Lowey SE. Palliative care in the management of patients with advanced heart failure. Adv Exp Med Biol. 2018;1067:295–311.
37. Johnson MJ, Clark AL. The mechanisms of breathlessness in heart failure as the basis of therapy. Curr Opin Support Palliat Care. 2016;10(1):32–5.
38. Ibanez B, James S, Agewall S, Antunes MJ, Bucciarelli-Ducci C, Bueno H, et al. 2017 ESC guidelines for the management of acute myocardial infarction in patients presenting with ST-segment elevation. Eur Heart J. 2018;39(2):119–77.
39. Duarte GS, Nunes-Ferreira A, Rodrigues FB, Pinto FJ, Ferreira JJ, Costa J, et al. Morphine in acute coronary syndrome: systematic review and meta-analysis. BMJ Open. 2019;9(3):e025232.
40. World Stroke Organization. Learn about stroke [Internet]. [cited 2021 Dec 7]. Available from: https://www.world-stroke.org/world-stroke-day-campaign/why-stroke-matters/learn-about-stroke.
41. Klit H, Finnerup NB, Jensen TS. Central post-stroke pain: clinical characteristics, pathophysiology, and management. Lancet Neurol. 2009;8(9):857–68.
42. Scuteri D, Mantovani E, Tamburin S, Sandrini G, Corasaniti MT, Bagetta G, et al. Opioids in post-stroke pain: a systematic review and meta-analysis. Front Pharmacol. 2020;11:587050.
43. Payton H, Soundy A. The experience of post-stroke pain and the impact on quality of life: an integrative review. Behav Sci. 2020;10(8):128.
44. Treister AK, Hatch MN, Cramer SC, Chang EY. Demystifying poststroke pain: from etiology to treatment. PM R. 2017;9(1):63–75.
45. Harrison RA, Field TS. Post stroke pain: identification, assessment, and therapy. Cerebrovasc Dis. 2015;39(3–4):190–201.
46. Chae J. Poststroke complex regional pain syndrome. Top Stroke Rehabil. 2010;17(3):151–62.
47. Yang S, Chang MC. Poststroke pain. Semin Neurol. 2021;41(1):67–74.
48. Dyer S, Mordaunt DA, Adey-Wakeling Z. Interventions for post-stroke shoulder pain: an overview of systematic reviews. Int J Gen Med. 2020;13:1411–26.
49. Harriott AM, Karakaya F, Ayata C. Headache after ischemic stroke: a systematic review and meta-analysis. Neurology. 2020;94(1):e75–86.
50. Aygun B, Odame I. A global perspective on sickle cell disease. Pediatr Blood Cancer. 2012;59(2):386–90.
51. Darbari DS, Sheehan VA, Ballas SK. The vaso-occlusive pain crisis in sickle cell disease: definition, pathophysiology, and management. Eur J Haematol. 2020;105(3):237–46.
52. Howland MA, Goldfrank LR. Why meperidine should not make a comeback in treating patients with sickle cell disease. Ann Emerg Med. 2008;51(2):203–5.
53. Puri L, Nottage KA, Hankins JS, Anghelescu DL. State of the art management of acute vaso-occlusive pain in sickle cell disease. Paediatr Drugs. 2018;20(1):29–42.
54. Nicolaysen A. Nephrotoxic chemotherapy agents: old and new. Adv Chronic Kidney Dis. 2020;27(1):38–49.

55. Dean M. Opioids in renal failure and dialysis patients. J Pain Symptom Manag. 2004;28(5):497–504.
56. Cruz SL, Granados-Soto V. Opioids and opiates: pharmacology, abuse, and addiction. In: Pfaff DW, Volkow ND, editors. Neuroscience in the 21st century. 2nd ed. New York: Springer Science+Business Media, LLC; 2016. p. 3625–58.
57. Soleimanpour H, Safari S, Shahsavari Nia K, Sanaie S, Alavian SM. Opioid drugs in patients with liver disease: a systematic review. Hepat Mon. 2016;16(4):e32636.
58. EAPC Taskforce on Palliative Care for Children and Adolescents. IMPaCCT: standards for paediatric palliative care in Europe. Eur J Palliat Care. 2007;14(3):109–14.
59. Royal College of Paediatrics and Child Health. A guide to children's palliative care. 4th ed. [Internet]. 2018 [cited 2021 Dec 9]. Available from: https://www.togetherforshortlives.org.uk/wp-content/uploads/2018/03/TfSL-A-Guide-to-Children's-Palliative-Care-Fourth-Edition-5.pdf.
60. International Association for the Study of Pain (IASP). Palliative pain care in children and adolescents [Internet]. 2019 [cited 2021 Dec 9]. Available from: https://www.iasp-pain.org/resources/fact-sheets/palliative-pain-care-in-children-and-adolescents/.
61. Worldwide Hospice Palliative Care Alliance (whpca). Global Atlas of Palliative Care, 2nd Ed 2020 [Internet]. 2020 [cited 2021 Dec 9]. Available from: http://www.thewhpca.org/resources/global-atlas-on-end-of-life-care.
62. World Health Organization. Guidelines on the management of chronic pain in children. Geneva: World Health Organization; 2020.
63. Rigal M, Martino R, Hain R. Opioids and the World Health Organization pain ladder. In: Hain R, Meiring M, Rapaport A, Goldman A, editors. Oxford textbook of palliative care for children. 3rd ed. Oxford: Oxford University Press; 2021. p. 176–87.
64. Bruera E, Moyano J, Seifert L, Fainsinger RL, Hanson J, Suarez-Almazor M. The frequency of alcoholism among patients with pain due to terminal cancer. J Pain Symptom Manag. 1995;10(8):599–603.
65. del Fabbro E. Assessment and management of chemical coping in patients with cancer. J Clin Oncol. 2014;32(16):1734–8.
66. Passik SD, Kirsh KL, Donaghy KB, Portenoy RK. Pain and aberrant drug-related behaviors in medically ill patients with and without histories of substance abuse. Clin J Pain. 2006;22(2):173–81.
67. Aertgeerts B, Buntinx F, Kester A. The value of the CAGE in screening for alcohol abuse and alcohol dependence in general clinical populations: a diagnostic meta-analysis. J Clin Epidemiol. 2004;57(1):30–9.
68. Webber K, Davies AN, Leach C, Bradley A. Alcohol and drug use disorders in patients with cancer and caregivers: effects on caregiver burden. BMJ Support Palliat Care. 2020;10(2):242–7.
69. Webster LR, Webster RM. Predicting aberrant behaviors in opioid-treated patients: preliminary validation of the opioid risk tool. Pain Med. 2005;6(6):432–42.
70. National Institute on Drug Abuse (NIDA) USA.gov. Screening to Brief Intervention (S2BI) [Internet]. [cited 2022 Jan 11]. Available from: https://www.drugabuse.gov/ast/s2bi/#/.
71. National Institute on Drug Abuse (NIDA). Screening tools for adolescent substance use [Internet]. [cited 2021 Dec 10]. Available from: https://www.drugabuse.gov/nidamed-medical-health-professionals/screening-tools-resources/screening-tools-adolescent-substance-use.
72. Albert RH. End-of-life care: managing common symptoms. Am Fam Physician. 2017;95(6):356–61.
73. Agar MR. Delirium at the end of life. Age Ageing. 2020;49(3):337–40.
74. Anderson SL, Shreve ST. Continuous subcutaneous infusion of opiates at end-of-life. Ann Pharmacother. 2004;38(6):1015–23.
75. Clinical Excellence Commission SHPN. Management of PAIN in the last days of life – ADULT [Internet]. 2017 [cited 2021 Dec 12]. Available from: https://www.cec.health.nsw.gov.au/__data/assets/pdf_file/0008/359342/LDOL-Flowchart-PAIN-2017.pdf.

76. Paes P. End-of-life care. In: Bourke SJ, Peel T, editors. Integrated palliative care of respiratory disease. 2nd ed. Cham: Springer; 2019. p. 225–52.
77. Rutkow L, Smith KC, Lai AY, Vernick JS, Davis CS, Alexander GC. Prescription drug monitoring program design and function: a qualitative analysis. Drug Alcohol Depend. 2017;180:395–400.
78. Centers for Disease Control and Prevention (CDC). Leveraging prescription drug monitoring program (PDMP) data in overdose prevention and response [Internet]. 2021 [cited 2021 Dec 12]. Available from: https://www.cdc.gov/drugoverdose/pdf/Leveraging-PDMPs-508.pdf.

Chapter 12
Opioids and the Immune System

Claudia Gonzalez-Espinosa, Iris K. Madera-Salcedo, Luz M. Molina-Martínez, and Frida L. Martínez-Cuevas

Abstract Opioids have multiple effects on the immune system (IS). Experimentally, the effects of opioid administration range from a severe inhibition to strong activation of immune responses, depending on the compound, schedule administration, experimental model, or clinical condition. On the other hand, endogenous opioids play a central role in the complex circuitry that mediates the IS and nervous system (NS) communication, tuning the intensity of reactions such as inflammation and pain or the mechanisms for sensing tissue damage and triggering a stress response. This chapter reviews studies showing increased susceptibility to infections and altered immune parameters produced by opioids and some mechanisms involved in direct and indirect actions of opioids on innate and adaptive immunity, the influence on genetic factors and aging on opioid effects, and pathologies where opioids exert immunomodulatory actions, including current information about COVID-19.

Keywords Opioids · Inflammation · Cytokines · Immunosuppression · Infection · Pattern-recognition receptors

C. Gonzalez-Espinosa (✉) · F. L. Martínez-Cuevas
Departamento de Farmacobiología, Centro de Investigación y de Estudios Avanzados (Cinvestav), Mexico City, Mexico
e-mail: cgonzal@cinvestav.mx

I. K. Madera-Salcedo
Departamento de Inmunología y Reumatología, Instituto Nacional de Ciencias Médicas y Nutrición Salvador Zubirán, Mexico City, Mexico

L. M. Molina-Martínez
Laboratorio de Farmacología y Conducta, Instituto de Neurociencias, CUCBA, Universidad de Guadalajara, Guadalajara, Jalisco, Mexico

S. L. Cruz (ed.), *Opioids*, https://doi.org/10.1007/978-3-031-09936-6_12

12.1 Introduction

To understand the effects of opioids on immune responses, it is necessary to consider the complexity of the two highly organized systems involved: (1) the opioidergic system with all its ligands, receptors, signaling cascades, and final effects on distinct cell types (Chaps. 8 and 9) and (2) the immune system (IS) with its primary and secondary lymphoid organs, particular responding cells and active molecules produced against damage.

The major design principles of the IS are:

- Layering: new processes are built on top of initial, more general processes.
- Scaffolding: early steps in the immune system provide the conditions needed for the later steps.
- Parallel processing: several events occur at the same time, not always synchronized, in distinct parts of the body.
- Dynamic engagement: cells act briefly and are then replaced by other cells.
- Variable network connectivity: mediators and cells acting in one immune process can be incorporated into another ongoing response, thus altering the outcome [1].

Opioids can modify the intensity of processes involved in all the organizational levels of the IS, both directly and indirectly, altering the direction and final consequences of a given immune response.

Classical μ-, δ-, and κ-opioid receptors (ORs), as well as non-classical ORs expressed in immune cells, mediate the direct actions of opioids on the IS. Indirect actions are mainly related to opioid effects on the hypothalamus-pituitary-adrenal (HPA) axis and the sympathetic nervous system (SNS), as explained in the following sections. Also, conditions involving changes in the IS functioning, such as aging, modify opioid effects. Altogether, evidence shows that opioids alter the individual response to infection and tissue damage, as supported by numerous studies documenting increased susceptibility of opioid users to infections and chronic diseases.

12.2 Altered Immune Parameters Associated with Opioid Administration

Initial observations of immunosuppressive opioid actions in humans showed that people dependent on heroin or morphine and methadone-treated patients had a high incidence of infectious diseases of viral, bacterial, and fungal etiology [2, 3]. It was initially thought that such infections resulted from transmission of microorganisms through fomites (passive vectors), including the drug paraphernalia used for injection, shared hypodermic needles, or contaminated syringes. It was believed that non-diagnosed acquired immunodeficiency syndrome (AIDS) and lifestyle factors such as malnutrition influenced opioid users' immunosuppression. However, further controlled studies clearly showed an altered IS function in opioid users [4, 5], and comparisons between healthy and non-dependent subjects who received opioids further supported the notion that these drugs might have immunosuppressive or immunomodulatory effects on their own [6].

Opioids appear to be immunosuppressive regardless of the reasons for use (pain relief, anesthetic, psychoactive effects) (Table 12.1). Recent studies show that prescribed opioids increase the incidence of Gram-positive infections and death [7]. Opioid use also contributes to human immunodeficiency virus (HIV) neuropathogenesis by acting on immune cells such as monocytes, macrophages (MΦs), microglia, and T cells [11, 19, 20]. In addition, patients with rheumatoid arthritis or burn injuries (conditions that occur with local and systemic inflammation) have a higher infection risk after opioid treatment [18, 29]. Activation of μ-OR affects cell differentiation, migration, and cytokine synthesis in the human epidermis, creating a niche that favors microbial survival in patients with burns [33].

High doses of opioid analgesics in trauma patients are also associated with infectious complications related to pneumonia, bacteremia, urinary tract infection, and wound infection [23]. Other studies have shown a higher incidence of diseases associated with repeated parenteral opioid administration, such as viral infections and infective endocarditis (Table 12.1) [34].

Clinical use of opioids has been related to poor surgical outcomes and the appearance of other diseases, such as infections and cancer [35]. In humans, immunosuppression after surgery has traditionally been associated with factors such as hypothermia, poor lung ventilation, or pre-existing health conditions [36], but evidence indicates that fentanyl, remifentanil, and morphine, can produce deleterious effects because of their immunosuppressive actions [37–39].

Fukada et al., in 2016, showed that paradoxical effects of opioids on the IS can be related to different administration schedules because the timing of administration can determine whether morphine exacerbates or inhibits an infection. For example, morphine treatment before a lipopolysaccharide (LPS) challenge (which mimics an infection with Gram-negative bacteria) suppresses lethal endotoxic shock in mice, but exacerbates it when administered after LPS [40].

Morphine and other μ-OR agonists do not produce similar immunosuppressive effects. Recent clinical trials suggest that there are two distinct groups of opioids: those with significant immunosuppressive effects (i.e., codeine, dihydrocodeine, methadone, morphine, or fentanyl) and those with less immunosuppressive effects (i.e., buprenorphine, hydromorphone, oxycodone, or tramadol) [41]. Moreover, the administration scheme and individuals' condition and health status modify the final effect of a specific opioid on selected immune responses.

The mechanisms proposed for opioid-induced immunosuppression are diverse. Some involve direct activation of ORs on immune cells; others require stress response activation. A few mechanisms are related to alterations in gut microbiota (dysbiosis), which compromise the intestinal barrier and allow changes in gut permeability, leading to systemic deleterious inflammation. The following section presents some general aspects of the IS function to understand the consequences of opioid actions on immune responses.

Table 12.1 Common diseases associated with the use of opiates and effects of opioid treatment on patients with distinct pathologies

Evidence of alterations on immune system in opioid users			
Users	**Opioid(s)**	**Common diseases**	**References**
Occasional users surgery, effects of opioid anesthetics	Alfentanil, butorphanol, codeine, dezocine, dihydrocodeine, fentanyl, hydrocodone, hydromorphone, oxycodone, levorphanol, meperidine, methadone, morphine, nalbuphine, oxymorphone, pentazocine, propoxyphene, remifentanil, sufentanil and buprenorphine	Gram-negative bacterial and fungal infections. Sepsis	[7]
Post-cesarean delivery	Morphine	Increases the risk for herpes type I reactivation	[8]
Dependent users	Heroin, morphine	Infectious endocarditis	[9, 10]
	Heroin, morphine and fentanyl	Increases viral replication and immunopathogenesis of HIV, HCV, HBV (in vitro)	[11–13]
	Heroin, morphine, fentanyl	Abscess, septic arthritis, phlebitis, cellulitis, osteomyelitis	[14, 15]
Users under assisted therapy	Buprenorphine/naloxone maintenance	High incidence of dental caries	[16]
	Methadone, buprenorphine, slow-release oral morphine	Increases asymptomatic vaginal infections and candidiasis in pregnant women	[17]
		Pneumonia, cellulitis, bacteremia without pneumonia, pyelonephritis and septic	[18]
Effects of opioid treatment on immunologic status of patients with distinct pathologies			
Pathology	**Opioid treatment**	**Side effects/complications**	**References**
HIV	Morphine	High HIV replication (in vitro)	[19]
		Increased development of neuropathologies (in vitro)	[20, 21]
		High incidence of developing pneumonia	[22]
Trauma	Morphine, fentanyl, hydrocodone, hydromorphone, and tramadol	High incidence to present: pneumonia, bacteremia, urinary tract infection, and wound infection	[23]
Burns	Morphine, hydromorphone, and oxycodone	Infectious complications: cellulitis, pneumonia, fungal sepsis, necrotizing fasciitis, sepsis-like syndrome, septicemia, sinusitis, wound infection, yeast infection, urinary tract infection, and graft infection*	[24]

(continued)

Table 12.1 (continued)

Primary and advanced cancer	Morphine, oxycodone, and fentanyl	High incidence to acquire bacterial infections	[25, 26]
Cirrhosis	Hydromorphone, fentanyl, methadone, morphine sulfate, oxycodone, and tramadol	Altered gut microbiota and increased inflammation	[27]
Pancreatitis	Prescribed opioids	Small intestinal bacteria overgrowth	[28]
Rheumatoid arthritis	Codeine, morphine, and methadone	Infectious complications: pneumonia, meningitis, encephalitis, septicemia, cellulitis, soft tissue infections, endocarditis, pyelonephritis, infective arthritis, and osteomyelitis*	[29]
Chronic pain	Morphine, fentanyl, and methadone	High risk of infections: pneumonia, cellulitis, bacteremia without pneumonia, pyelonephritis, and septic arthritis/osteomyelitis	[18]
Pneumonia	Prescribed opioids: codeine, morphine sulfate, fentanyl, tramadol, and methadone, among others	Increased risk of invasive pneumococcal disease	[30]
Alzheimer	Morphine, hydromorphone, oxycodone, and fentanyl	Increased risk of pneumonia	[31]
Other diseases present in older adults	Methadone, morphine, fentanyl, and codeine	High rate of developing pneumonia	[32]

*Patients can develop one or more complications; however, in this study were eligible those patients with only the first infection manifested

12.3 Overview of the Immune System

The immune system is organized based on the type and time course of reactions elicited by infection or tissue damage.

The innate IS (also called the innate immunity responses) constitutes the first line of defense against germs or damage in the body. In general, it is nonspecific, acts rapidly, and has limited efficiency. It comprises physical barriers for the entrance of external substances and specialized secretions (such as tears, mucus, and saliva) and involves the activation of tissue-resident mast cells (MCs), macrophages (MΦs), and natural killer (NK) cells, among others.

The adaptive immune system orchestrates adaptive immunity responses, activating T and B lymphocytes which proliferate and produce cytokines or differentiate and produce antibodies, respectively. Adaptive immunity responses via T cells initiate after antigen presentation by dendritic cells to T-cell receptor (TCR), whereas B cell antibody production starts after antigen recognition via the B-cell receptor

(BCR). Alteration of the early and efficient innate immune response against pathogenic insults promotes an impaired adaptive response that results in uncontrolled inflammation and host tissue damage (Fig. 12.1).

Effective connection between innate and adaptive immune responses is mainly orchestrated by cytokines. Cytokines are low molecular weight, signaling proteins that are released in response to diverse stimuli. They have specific effects on the activation, proliferation, and differentiation of immune cells. Chemokines (chemotactic cytokines) are also proteins produced by immune cells, but their main action is to induce the chemotaxis needed for the recruitment of cells to sites of infection and/or tissue damage. By doing this, cytokine and chemokine production modifies the interaction between the cellular elements of a given immune response.

Pro-inflammatory cytokines alert the IS against invading pathogens. Some pro-inflammatory cytokines are interleukin (IL)-1β, tumor necrosis factor (TNF)-α, IL-2, IL-6, and chemokines such as IL-8, CCL2, and CCL5 (also known as RANTES). Anti-inflammatory cytokines limit the actions of pro-inflammatory ones. Major anti-inflammatory cytokines include IL-10, IL-4, transforming growth factor (TGF)-β, and soluble cytokines receptors [42].

Opioids can alter the activity of distinct immune cells and modify the course of an immune reaction by direct interaction with immune cells or by activation of neuroendocrine mechanisms that modulate global immune responses. The following sections present current knowledge on that matter.

12.4 Direct Effects of Opioids on the IS

12.4.1 Direct Actions of Opioids on Innate Immune Cells: Focus on Inflammation

Immune cells express all described ORs. As mentioned in Chap. 9, ORs are G protein-coupled receptors (GPCR). To date, five ORs have been identified; three classical, μ-, κ-, and δ-OR; and two non-classical receptors, the nociceptin/orphanin FQ (ORL-1 or NOP) receptor and the novel atypical Mas-related G protein-coupled receptor X2 (MRGPRX2), expressed in MCs, basophils, eosinophils, and other immune cells [43–45].

Opioid receptors activate similar signal transduction cascades in neurons and immune cells, but with some relevant particularities, since the latter ones are not considered excitable cells. As mentioned in Chap. 9, activation of classical ORs and NOP promotes $G_{i/o}$ dissociation into $G\alpha_{i/o}$ and $G\beta\gamma$ subunits. $G\alpha_{i/o}$ inhibits adenylyl cyclase and PKA activity, modulating ion channels activated by capsaicin (TRVP1) and voltage-gated sodium channels. Meanwhile, the $G\beta\gamma$ complex blocks Ca^{2+} channels (Ca_v^{2+}), increases K^+ channels conductance (GIRK and K_{ATP}), and activates PKC and mitogen-activated protein kinases (MAPKs) [46]. ORs are desensitized by phosphorylation and then internalized by clathrin-dependent pathways, a

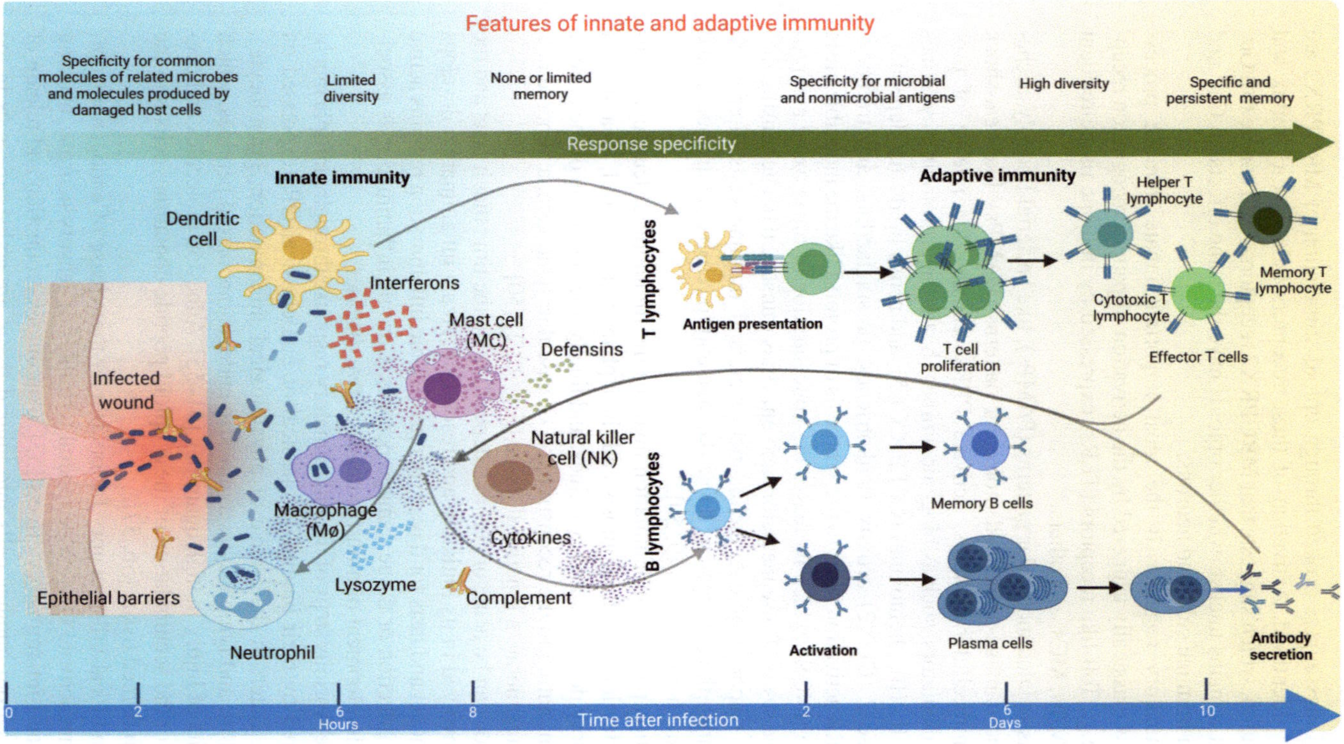

Fig. 12.1 Features of innate and adaptive immunity. Innate immunity constitutes the early line of defense that is in place even before infection. It is composed by various cell types, such as mast cells (MCs), monocytes, macrophages (MΦ), dendritic cells (DCs), and natural killer cells (NKs), and active molecules, such as proteins of the complement system, cytokines, and enzymes. Structures that are common to groups of related microbes or are released by damaged cells activate innate immune responses, which shape adaptive immunity trough the interaction of phagocytic cells with T cells (among other mechanisms). Those events result in the production of antigen-specific T and B lymphocytes. Adaptive immunity is based on lymphocytes properties; they can respond selectively to a huge number of different antigens, leading to specific memory and efficient effector response. Created with BioRender.com

mechanism initially controlled by GPCR kinases (GRKs) and followed by β-arrestin binding [47].

The second messengers evoked through opioid stimulation of MRGPRX2 are not completely identified. Activation of these receptors induces Ca^{2+} influx and MAPKs activation, which suggests that MRGPRX2 receptors could couple to a $G\alpha_q$ protein [48]. It remains unclear whether opioids have the same effects on Ca^{2+} and K^+ channels in immune cells and neurons.

The inflammatory response is the most important innate immune process (Fig. 12.2). It is classically triggered by the activation of pattern recognition receptors (PRRs), e.g., Toll-like receptors (TLRs) expressed by immune tissue-resident cells, such as MΦs, MCs, and others.

Pathogen-associated molecular patterns (PAMPs), like bacterial lipopolysaccharide (LPS), and damage-associated molecular patterns (DAMPs), like heat-shock proteins or intracellular proteins, activate PRRs, causing the release of pro-inflammatory mediators, such as biogenic amines (histamine, serotonin), cytokines (TNF-α, IL-6, IL-1β), chemokines (CCL2, CCL5), and active lipids that depend on cyclooxygenase 2 (COX2) activity (leukotrienes, prostaglandins, and tromboxanes). Also, PRR activation leads to the synthesis of proteins like the inducible nitric oxide synthase (iNOS), which increases the concentration of nitric oxide and promotes oxidative stress. These mediators and enzymes translate damage sensing into a coordinated response of endothelial, neuronal, and immune cells directed to remove or contain the pathogen or tissue damage and to resolve inflammation and restore the tissue homeostasis [49, 50].

The resolution phase of inflammation involves the limitation of leukocyte infiltration, the induction of cell apoptosis and their phagocytic removal, clearance of pro-inflammatory dead cells and cytokines, and finally, the beginning of the healing processes, culminating in the reconstruction of vascular, lymphatic, and nerve networks, which regenerate a functional tissue (Fig. 12.2) [50].

Inflammation can be acute or chronic, depending on the time that it lasts and the mechanisms involved in its initiation and resolution. Acute inflammation develops in minutes to hours and lasts for days, whereas chronic inflammation remains active for as long as infection or tissue damage persists. Chronic inflammation is associated with the development of multiple degenerative diseases.

Opioids alter various steps of the inflammatory response through multiple mechanisms triggered by their binding to classical or non-classical ORs expressed on immune cells. Later, in the course of physiological inflammation, granulocytes, monocytes, MΦs, and lymphocytes synthesize endogenous opioid peptides, which infiltrate the site of injury in high quantities [51–55]. Adhesion molecules and chemokines control the accumulation of these opioid peptide-containing cells into the extravascular inflamed tissue [56, 57].

Stressors and local-inflammatory factors trigger the release of endogenous opioid peptides that activate ORs on peripheral terminals of sensory neurons, evoking analgesia and anti-inflammatory effects [53]. Continuous production and release of opioid peptides from immune cells into injured tissue are active processes in the

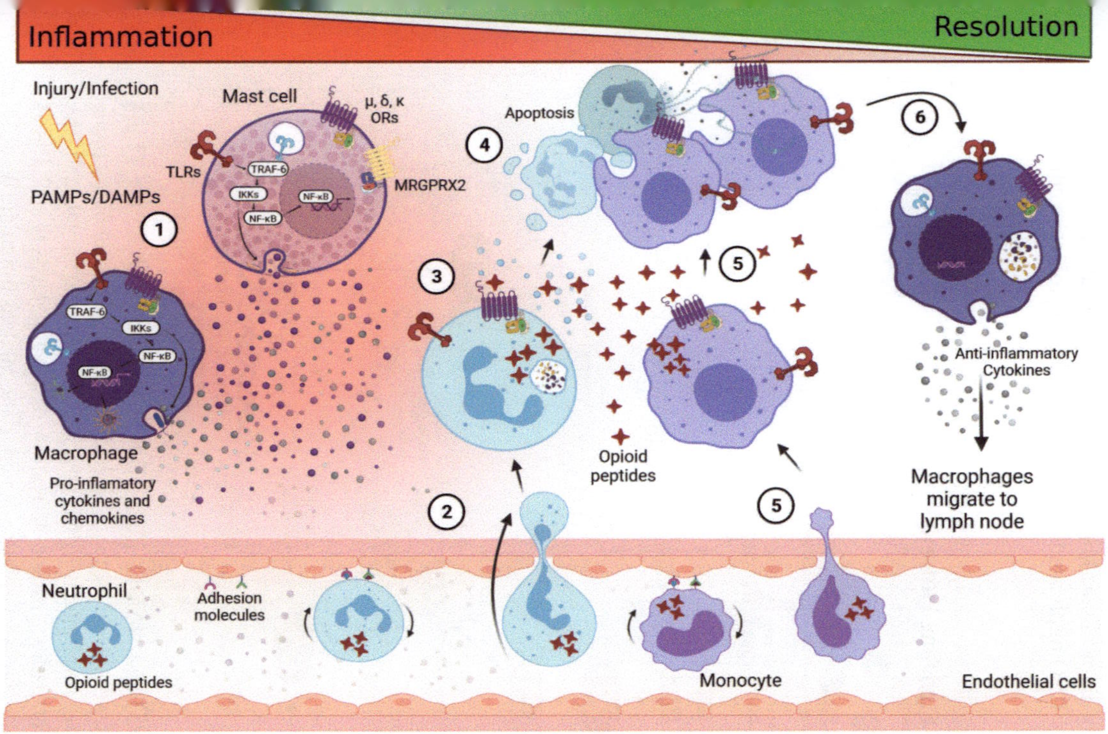

Fig. 12.2 The inflammatory response. (1) After a harmful stimulus or infection, PAMPs or DAMPs bind to PRRs and activate tissue-resident MCs and MΦs. For example, triggering of TLRs promotes the release of pro-inflammatory mediators (e.g., TNF-α, IL-1β, and IL-12, among others), favoring neutrophil migration toward the inflamed tissue. Nod-like receptor (NLRP) activation leads to the formation of inflammasomes that results in IL-1β secretion. (2) Neutrophils travel from the circulation to the damaged tissues through blood vessels with the participation of adhesion molecules highly expressed in endothelial cells. (3) Neutrophils secrete mediators to kill pathogens or dead cells into the compromised tissue and phagocyte them. (4) In addition, neutrophils release chemical mediators and extracellular DNA traps by the induction of programmed cell death (NETosis and apoptosis) capturing and killing pathogens to prevent them from spreading. (5) In parallel, non-inflammatory monocytes arrive and eliminate the pro-inflammatory molecules, pathogens, and cellular debris, promoting its reprogramming to mature MΦs. Both leukocytes, neutrophils and monocytes, release opioid peptides into the inflamed tissue supporting the beginning of the resolution steps of the inflammatory response. (6) Reprogrammed macrophages release anti-inflammatory cytokines to stop leukocyte infiltration and attract immune cells that will support the tissue remodeling and homeostasis recovery. Created with BioRender.com

resolution phase of inflammation. Due to those actions, opioids can suppress this first defense mechanism, promoting infection or damage progression (Fig. 12.2).

Studies on the molecular mechanism of opioids' inhibitory actions on inflammation triggered by PRRs can help identify molecules that could be targets to prevent opioid-induced immunosuppression. For example, the TLR4 signaling system involves activation of intracellular kinases, ubiquitin ligases, and the NF-κB transcription factor (Box 12.1) implicated in pro-inflammatory cytokine synthesis.

Box 12.1 TLR4 Signaling Pathways

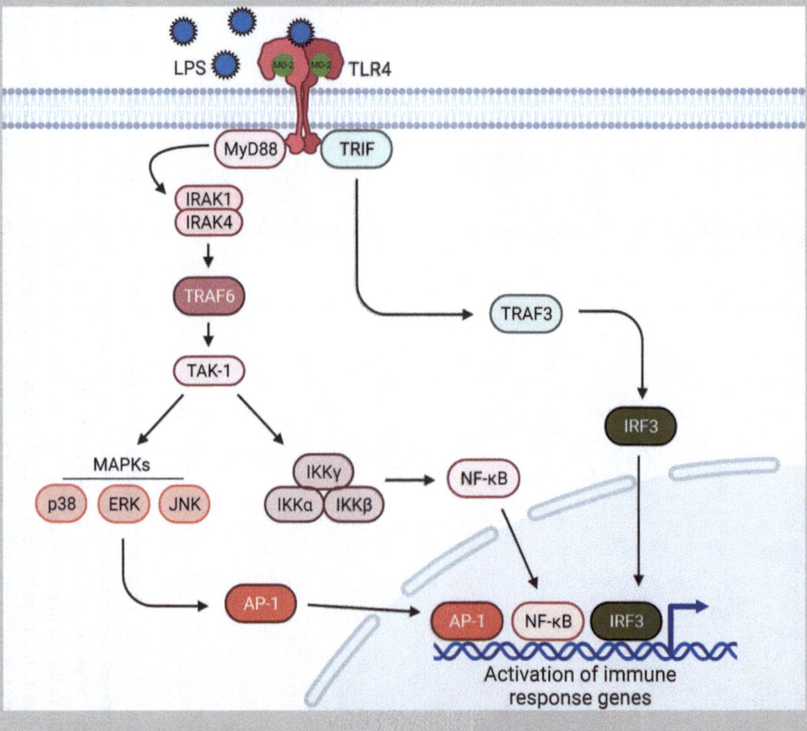

PAMPs (e.g., LPS) and DAMPs (e.g., HMGB proteins) activate the TLR4 receptor complex, composed by the TLR4/MD-2 dimer [58]. Signaling pathways include those coordinated by MyD88 or TRIF adapter proteins. In the MyD88-dependent pathway, after activation of several kinases, AP-1 and NF-κB transcription factors are translocated to the nucleus to initiate de novo mRNA synthesis for distinct pro-inflammatory cytokines. In the TRIF-dependent pathway, TRIF adapter binds to the ubiquitin ligase TRAF3,

(continued)

Box 12.1 (continued)

promoting the activation of the interferon regulatory factor 3 (IRF3) and stimulating the transcription of type I interferon genes (IFNs) [42, 59]

PAMPs pathogen-associated molecular patterns, LPS lipopolysaccharide, DAMPs damage-associated molecular patterns, HMGB high mobility group box 1 proteins, MyD88 myeloid differentiation primary response gene 88), IRAK interleukin-1 receptor-associated kinases, TRAF6 TNF receptor-associated factor 6, TAK1 TGF-β-activated kinase 1, MAPKs mitogen-activated protein kinases, IKK the IκB kinase complex, AP-1 activating protein 1, NF-κB nuclear factor κ-light chain enhancer of activated B cells. Created with BioRender.com

Opioids can inhibit several steps in the TLR4 signaling cascades (Fig. 12.3). For example, in tissue-resident MCs, μ-OR activation by morphine and fentanyl and δ-OR activation by morphine prevent the early secretion of TNF-α induced by TLR4 stimulation through a process that involves the formation of a β-arrestin 2 and TRAF6 complex [60–62]. TNF-α is an inflammatory cytokine responsible for the activation of endothelial cells and leukocytes and induction of acute-phase response, among others roles. Also, μ-OR and δ-OR activation in MΦs inhibits TNF-α and IL-1β, IL-6, and IL-12 production after TLR4 triggering via NF-κB inhibition. In addition, opioids reduce the levels of nitric oxide (NO), iNOS, and COX-2 promoted by LPS [63–66], bioactive molecules that, as mentioned, play an important role in vascular function and host defense. Furthermore, activation of ORs in MΦs also impacts the phagocytic activity of those cells [67–69]. For example, morphine enhances MΦs' phagocytic capacity induced by TLR4 activation [70, 71]. However, bacteria clearance is unsuccessful, because morphine suppresses phagosome maturation, which is critical for destroying pathogens [71]. Figure 12.3 summarizes the intracellular mechanisms involved in the impaired response of innate cells against injury stimulus due to OR activation.

Additionally, morphine, fentanyl, and remifentanil decrease neutrophil and monocyte transmigration across endothelial cells via μ-ORs, increasing susceptibility to infections [69, 72–74]. Moreover, opioid agonists acting at μ-ORs in endothelial cells attenuate cell activation during inflammation [72] and suppress the expression of intercellular adhesion molecule (ICAM)-1 and vascular cell adhesion molecule (VCAM)-1 required for cell migration, through a mechanism dependent on NF-κB [74–77].

After a noxious stimulus, μ-ORs also modulate innate immune cells activation in the central nervous system (CNS), including glial cells, microglia, and astrocytes. In most cases, opioid effects are pro-inflammatory, which are opposite to the effects observed on peripheral immune cells. For example, in microglia, morphine enhances the production of TNF-α, IL-1β, IL-6, and NO induced by LPS, by increasing TRAF6 [78], MAPKs [78, 79], and IKK [80] activation (see Box 12.1) promoted by

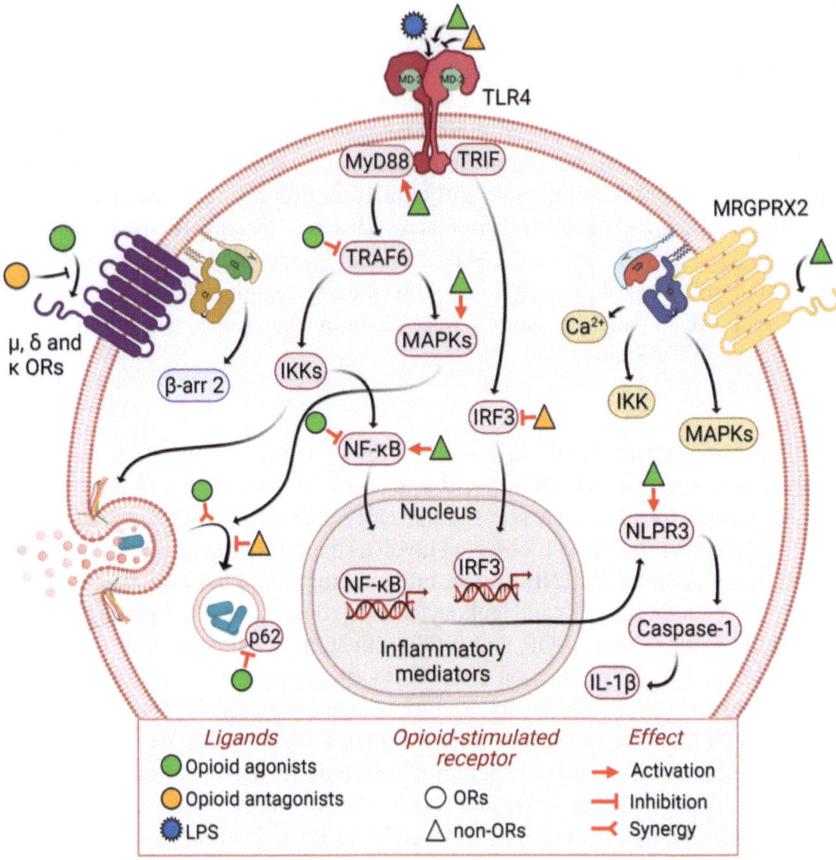

Fig. 12.3 Molecular opioid-controlled check points on TLR4 receptor signaling cascade. Opioids modulate innate immune response to pathogens through the interaction with ORs, such as μ-, δ-, and κ, and non-ORs, including MRGPRX2 and TLR4. Activation of ORs by their ligands (e.g., morphine and fentanyl) activates Gαi/o protein-dependent pathways, but also modifies the activity of several molecules involved in the signaling pathways triggered by TLR4, affecting the effector function of immune cells involved in innate and adaptive immunity. TLR4 signal cascade includes the activation of TRAF6, MAPKs, IKK, and the translocation of NF-κB (see Box 12.1 for more details). Main molecular changes induced by the interaction of ORs and TLR4 signaling cascades are the following: (1) β-arrestin 2 forms a complex with TRAF6, inhibiting the secretion of TNF-α; (2) NF-κB activation is blocked, affecting the production of pro-inflammatory mediators; (3) NLPR3 inflammasome formation and oligomerization is enhanced, inducing the maturation and secretion of IL-1β; (4) although opioids improve phagocytic activity, they limit deubiquitination of p62 protein, leading to a deficient pathogen clearance. Mentioned effects are prevented by (−) opioid antagonist, e.g., naloxone and naltrexone. Opioid ligands also bind non-ORs, activating signaling cascades downstream of those receptors. MRGPRX2 can be activated by morphine and codeine, supporting the influx of Ca^{2+} and the activation of p38 MAPK and IKK and favoring the production of inflammatory mediators. TLR4 is activated by morphine and its metabolite through the interaction with MD-2, promoting a pro-inflammatory signaling pathway that requires the activation of MAPK and NF-κB and leads to the production of pro-inflammatory mediators and the

TLR4. These effects are mediated by classical ORs because they can be blocked by naloxone [79]. Also, in combination with the HIV protein Tat or the bacteria *Streptococcus pneumoniae*, morphine improves the expression of cytokines and chemokines, such as TNF-α, IL-6, CCL2, and RANTES favoring the trafficking of T cells and monocytes into the CNS [81, 82].

Morphine and fentanyl activate the NOD-like receptor protein 3 (NLRP3) inflammasome in microglia, favoring caspase 1 activation and IL-1β production [83, 84]. Furthermore, activation of μ-ORs induces DAMP HSP70 (heat shock protein 70) [85] and HMGB1 (high mobility group box 1) production [86], which, in turn, activates TLR4, causing TNF-α and IL-1β production through NF-κB and the formation of NLRP3 inflammasome [86]. Taken together, these data indicate that μ-ORs mediate a sensitization stage of microglia that enhances its activation by harmful stimuli, favoring the development of neuroinflammation that could result in neuronal damage and even cell death.

Paradoxical effects of opioids seem to be related to their binding to non-classical ORs and non-ORs. For example, morphine activates both MRGPRX2 and TLR4. Similarly, codeine, a low affinity μ-OR agonist, activates MRGPRX2 expressed in MCs, favoring their degranulation and the production of histamine, TNF-α, CCL2, RANTES, and IL-8, by a Ca^{2+}-dependent mechanism and activation of MAPKs and IkB kinase (IKK) [87–89].

Morphine, like LPS, binds to MD-2 protein, promotes TLR4 oligomerization, and activates TLR4's signaling pathway (see Box 12.1) [90, 91]. Moreover, morphine increases TLR4, TAK1, and NLRP3 expression, and these effects do not occur in cells lacking TLR4 [92].

Opioids interact differently with classical ORs and TLR4. Some main differences are:

- Both OR agonists and OR antagonists bind to and activate TLR4.
- Morphine-3-glucuronide (M3G), morphine's inactive metabolite (as analgesic), triggers TLR4 signaling and induces IL-1β production.
- Both dextro- (+) and levo (−) isomers of naloxone and naltrexone block TLR4 [93], but only (−)-naloxone and (−)-naltrexone block ORs.

Both naloxone's and naltrexone's isomers inhibit the production of NO, ROS, TNF-α, and phagocytosis induced by TLR4 signaling in microglia and macrophage cell cultures [90, 94]. Naloxone and naltrexone do not inhibit MAPKs and NF-κB activation but do suppress IFN regulatory factor 3 (IRF3) activation promoted by TLR4 stimulation [94]. These studies open new avenues for studying the cellular and molecular changes that opioid ligands can induce in cells of the IS.

Fig. 12.3 (continued) formation of the NLRP3 inflammasome. Opioid antagonists, both (+ and −) isomers equally, binds to MD-2 as opioid agonists do and inhibit the signaling pathway that depend on TRIF/IRF3, suppressing the production of IFNs and other pro-inflammatory mediators. Ligands are indicated by colors, e.g., green = opioid agonists; opioid-stimulated receptor are illustrated by figures, e.g., circles = ORs; and effect are shown by different line head, e.g., → = activation. Therefore, opioid agonists activate ORs promoting activation. Created with BioRender.com

12.4.2 Direct Actions of Opioids on Lymphocytes

Cells of the lymphocyte lineage express ORs [95]. Morphine and endogenous opioid peptides, including β endorphin, modulate B and T lymphocytes' function. Several researchers have proposed that opioids act like cytokines, orchestrating complex responses where distinct cell types participate. For example, the mixed agonist-antagonist buprenorphine (Chap. 8) suppresses splenic NK cell activity, lymphocyte proliferation, and IFN-γ production in rats [96]. Opioids also suppress the movement and the number of circulating white blood cells [97, 98] and play a role in suppressing a variety of immunological endpoints such as cell proliferation and cytokine synthesis [99].

Studies have evolved from anecdotic reports of immunosuppressive actions of opioids to well-designed and controlled studies in humans. For example, a randomized pilot study in gynecological laparotomy patients evaluated the effect of morphine and oxycodone on immune responses [100]. Patients were randomized to receive morphine, oxycodone, or nonopioid analgesia during and after surgery. Using different molecular techniques, the researchers analyzed gene expression, NK cell activation, and serum cytokine concentration at different times after opioid treatment. The results showed that morphine, but not oxycodone or epidural analgesia, produced immunosuppression 2 hours post incision [100].

Similarly, morphine showed a higher immunosuppressive effect than oxycodone in patients who suffered a radical resection of rectal cancer [101]. Recent studies have shown that morphine antagonizes the chemotaxis induced by TNF-α and IL-1β in human leucocytes, decreases IL-2 and IFN-γ levels, and increases IL-4 and IL-5 plasma concentration, suggesting that morphine can block an effective immune response against pathogens or tissue damage. In addition, long-term use of opiates produces atrophy of lymphoid organs, decreases lymphoid content, alters antigen-specific antibody production, causes loss of T helper (Th) cells, and decreases T cell reactivity [102, 103].

12.5 Indirect Mechanisms of Action of Opioids on the IS

12.5.1 Opioid Actions Through HPA Axis Activation

The hypothalamus-pituitary-adrenal axis (HPA) is the neuroendocrine system involved in mediating the stress response. It starts with the release of corticotropin-releasing hormone (CRH), which is produced in neurons in the hypothalamic paraventricular nucleus (PVN) and travels through the portal vasculature to stimulate the release of adrenocorticotropic hormone (ACTH) from corticotropes of the anterior lobe of the pituitary gland. ACTH acts in the adrenal cortex, promoting glucocorticoid biosynthesis and release, mediating stress responses, and regulating the axis through negative feedback [104]. Acute stress benefits survival of the individual, but chronic stress induces HPA axis dysregulation, leading to the development of various pathologies, including general immunosuppression. The neuroendocrine response of HPA axis activation mainly inhibits pro-inflammatory cytokine production induced by the innate immune response [105]. The main mechanism involved is mediated by glucocorticoids which, by activating their receptors in the cytoplasm of immune cells, act directly in inhibiting the activation of NF-κB or indirectly increasing the transcription and translation of the NF-κB inhibitory protein (IκB) (Fig. 12.4).

The acute immunosuppressive effects of μ-OR agonists (e.g., morphine, fentanyl) mediated by the HPA axis have been studied using animal models or in vitro cell systems [99, 106, 107]. These studies indicate that opioid effects depend on the type of opioid administered, treatment time, and experimental conditions.

In clinical trials, patients treated for at least 6 months with oxycodone, morphine, fentanyl, or buprenorphine had a suboptimal initial cortisol response after stimulating the HPA axis with exogenous ACTH [108]. However, heroin-dependent individuals had elevated ACTH and cortisol levels that returned to basal levels after a single dose of opioids [109]. Few clinical trials have addressed the relationship of the immunosuppressive effects of opioids with cortisol levels in prescription opioid users [110]. Opioid anesthetic and analgesic drugs acutely stimulate the HPA axis during the perioperative period, producing immunosuppression, but not all OR agonists have the same effects. For example, tramadol induces very little immunosuppression compared to remifentanil, fentanyl, or morphine in both animal models and clinical trials [111].

Notably, the immunosuppressive effects of chronic opioids can be dissociated from their analgesic effects even when acting through the same μ-ORs [60]. Thus, it is likely that opioids that produce more significant immunosuppression show a distinctive pattern of HPA axis activation than those with mild effects and that the final outcome depends both on the pharmacological properties of each opioid and the treatment time.

The contribution of the HPA axis to the immunosuppressive effects of prescription opioids remains unclear because studies addressing this mechanism are scarce, and there are confounding factors, such as the stress prior to surgery and other clinical conditions that produce stress mediators in patients. For this reason, it is important to continue the study of opioid effects on the immune system in animal models and clinical trials, especially in conditions where continuous monitoring and pain control is needed.

12.5.2 Actions of Opioids on the IS Via the Sympathetic Nervous System (SNS) and Vagus Nerve

Opioids can also modulate immune responses via the sympathetic nervous system (SNS) (Fig. 12.4). Systemic morphine administration increases acetylcholine (ACh) release in the rat spinal dorsal horn, suggesting that opioids activate the SNS [112]. Most of the evidence that suggests SNS participation in immunosuppressive opioid

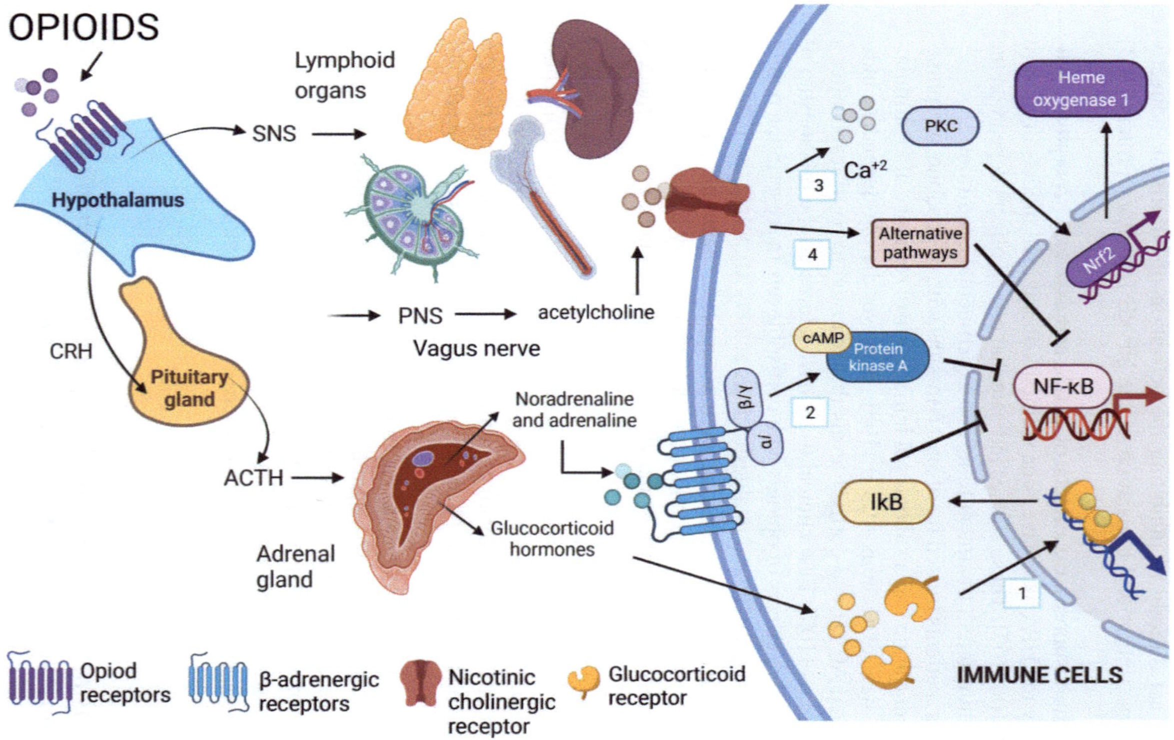

Fig. 12.4 Indirect actions of opioids on the immune system. Opioid receptors lead to activation of the hypothalamus-pituitary-adrenal (HPA) axis, the sympathetic (SNS), and the parasympathetic nervous system (PNS). Vagus nerve innervates lymphoid organs modulating immune cells' differentiation. Activation of adrenergic, acetylcholine, and glucocorticoid receptors in immune cells inhibits pro-inflammatory cytokine synthesis by different intracellular pathways. Glucocorticoid receptors promote the synthesis of NF-κB inhibitor (IkB) protein [1]. β-Adrenergic receptor activation increases cAMP levels and PKA activation [2]. Nicotinic acetylcholine receptor rises intracellular Ca^{2+} concentration, leading to protein kinase C (PKC) activation, and produces Hem Oxygenase 1 [3]. Alternate pathways depending on nicotinic receptors interfere with NFκB signaling [4]. CRH: corticotropin-releasing hormone; ACTH, adrenocorticotropic hormone, nuclear factor erythroid 2-related factor 2 (Nrf2). Created with BioRender.com

effects comes from in vitro and animal studies [113, 114]. ACh binds to nicotinic receptors in postganglionic neurons, promoting the release of norepinephrine and adrenaline from the adrenal medulla. Norepinephrine binds to α- and β-adrenergic receptors located in immune cells. β-Adrenergic agonists suppress lymphocyte and MΦs functions by elevating intracellular cAMP and, consequently, inhibiting pro-inflammatory cytokine synthesis [105, 113].

Another mechanism involved in opioids' immunosuppressive actions is parasympathetic nervous system (PNS) activation, which, through the stimulation of the vagus nerve, initiates a response known as the cholinergic anti-inflammatory reflex. Diverse in vitro studies show that inhibition of TNF-α secretion occurs through nicotinic cholinergic receptors located in murine and human MΦs [115–117]. ACh inhibits immune cell function by binding to ionotropic α_7 nicotinic receptors that promote Ca^{2+} influx, activate several signaling pathways, and cause the synthesis of anti-inflammatory mediators [118]. In MCs, nicotinic receptors block the LPS-induced production of TNF by inhibiting the phosphorylation of the MAPK ERK1/2, which phosphorylates and activates the metalloproteinase responsible for TNF-α processing in the plasma membrane [119]. Other inhibitory mechanisms of ACh involve a direct interaction of ACh receptors with G protein activation of Janus kinase 2 (JAK2)-dependent signaling, which elicits downstream cascades dependent or independent of Ca^{2+} influx, leading to NFκB inhibition [118] (Fig. 12.4).

12.6 Conditions that Modulate Opioid Actions on the Immune System: Genetic Polymorphisms and Aging

12.6.1 Genetic Polymorphisms

Some genetic variants alter HPA axis reactivity, resulting in variations in opioid actions on immune cells. These gene variants exist for $GABA_A$ receptors, μ-ORs, serotonin transporters, the monoamine oxidase enzyme, MAO-A, adrenergic receptors, brain-derived neurotrophic factor (BDNF), and others [120]. In particular, the μ-OR variant containing a single-nucleotide polymorphism (SNP; see Chap. 9) in the extracellular domain (A118G) leads to the change of an amino acid from asparagine to an aspartate during protein translation (Asn40Asp). In vitro studies have found that this variant increases threefold the affinity of μ-OR for endorphins [121]. Thus, carrier individuals of this polymorphism experience less analgesic opioid effects in both postoperative pain and cancer-associated pain and have altered HPA axis activation.

Studies carried out in healthy people with different genotypes (AA, AG, and GG) in the same position of the gene have shown that the produced variants are not associated with changes on baseline cortisol levels but respond to a naloxone challenge with a higher concentration of cortisol [122, 123]. It is possible that individuals carrying this polymorphism, mainly found in European American individuals [123], present a higher inhibitory tone on CRH neurons expressing μ-opioid receptors, and their response to different psychological and physical stressors could be lower when compared to that of individuals with other genetic sequences [124, 125].

Because corticosterone exerts a significant inhibition on the immune system and μ-OR Asp40 polymorphism results in higher inhibition of the HPA axis, it is likely that opioid users with such genetic variation would have low cortisol levels. Hence, it would be interesting to study key immune parameters in patients with this polymorphism and analyze if opioids produce immunosuppression.

12.6.2 Influence of Aging

Cellular stress related to age and senescence induce an immune condition called "inflammaging." This condition refers to a low-grade inflammatory process accompanied by aging, mainly characterized by an increase in pro-inflammatory cytokine levels [126]. During this immune process, cells such as myeloid-derived suppressor cells and other immunosuppressive cell populations become active to counteract the inflammatory response and promote immune system remodeling, in a phenomenon called "immunosenescence" [127].

At this point, we do not know whether immunosenescence is a mechanism that counteracts chronic inflammation or the consequence of the low-grade inflammatory process associated with aging. Several causes and risk factors that lead to immunosenescence include inadequate micronutrient consumption, the decrease in the length of cell telomeres, reactivity to self-antigens, reactive oxygen species accumulation, lifelong stress, and the increase in HPA function along aging [128, 129].

Studies about the effects of aging on the opioid system and opioid actions on immune responses are scarce. What is known is that aging affects the response of the IS, decreasing defenses against pathogens [130–132], lowering vaccination efficacy [133], and impairing anticancer immunity [134]. There is evidence that both endogenous opioid peptides and ORs levels in the brain of rodents decrease with aging [135]. In the elderly, an increased pro-inflammatory general state seems to be related to exacerbated actions of β-endorphin on immune cells, which is evidenced by augmented NK activity, neutrophil adherence, and histamine adherence after opioid administration [135, 136] (Fig. 12.5).

A comparative study of the effects of various prescription opioids in young versus old rats found that systemic fentanyl injection interfered with cellular immunity, by decreasing the lytic activity of NK cells before and after surgery in old rats. In addition, fentanyl administration after surgery did not change the number of lung metastases of chemically induced mammary adenocarcinoma in young animals, but reduced their occurrence in old animals [137]. These results indicate a dual effect of fentanyl on older individuals since, on the one hand, it impairs the function of the cellular immune response, but on the other hand, it protects from cancer metastasis development.

Endogenous opioid peptides appear to enhance cellular immunity in humans (Fig. 12.5). However, endorphin levels in humans and μ-ORs density and affinity increase, not decrease, with age [138, 139]. Moreover, the NK activity and lymphocyte numbers induced by β-endorphins are higher in healthy elderly volunteers (65–89 years) than in young subjects [140].

Oral administration of a sustained-release morphine formulation decreases antibody production but not peripheral mononuclear proliferation in patients (mean age of 50 years old) with chronic pain as compared to healthy controls [141]. However, prolonged treatment (lasting more than 6 months) with morphine, oxycodone, methadone, or buprenorphine does not affect NK cells [142]. In older patients (median age of 77 years old), the susceptibility to respiratory tract infections and complications such as pneumonia occurs mainly during the first 2 weeks of opioid treatment [32]. These results indicate that further studies are needed to address opioid effects on the IS in older individuals, attending variables that frequently occur during this stage of life, such as the presence of comorbidities and polymedication.

12.7 Selected Examples of Pathology-Related Actions of Opioids on the IS

12.7.1 Opioids in the Relationship Between IS and Tumor Growth

Opioids have been used for a long time to manage the perioperative pain associated with tumor ablation or control cancer-induced pain (see Chaps. 10 and 11). Nowadays, the opioid approach to cancer therapy focuses not only on controlling pain but also on fighting tumor growth and metastases. Discrepancies exist among the results of studies analyzing opioid effects on malignant tumor growth (Table 12.2). While some studies indicate that opioids increase tumor mass, others suggest that opioids help block cancer development (Fig. 12.6). Results seem to depend on the type of tumor studied, opioid dose, and administration schedule, as well as the OR subtype activated.

As to positive effects on cancer, evidence indicates that opioids restrict tumor growth by the following mechanisms: (1) inhibition of cytokine synthesis and release; (2) blockage of tumor angiogenesis; and (3) promotion of cancer cell death.

The tumor microenvironment comprises different cells with non-malignant phenotypes, such as tumor-associated MΦs, endothelial cells, and fibroblasts. These cells release vascular endothelial growth factor (VEGF)-A, which increases vascular permeability and promotes endothelial cell survival and proliferation, cell migration, and blood vessel formation. Blood supply is essential for tumors to grow. Morphine decreases the release of VEGF by tumor-associated MΦs when they are

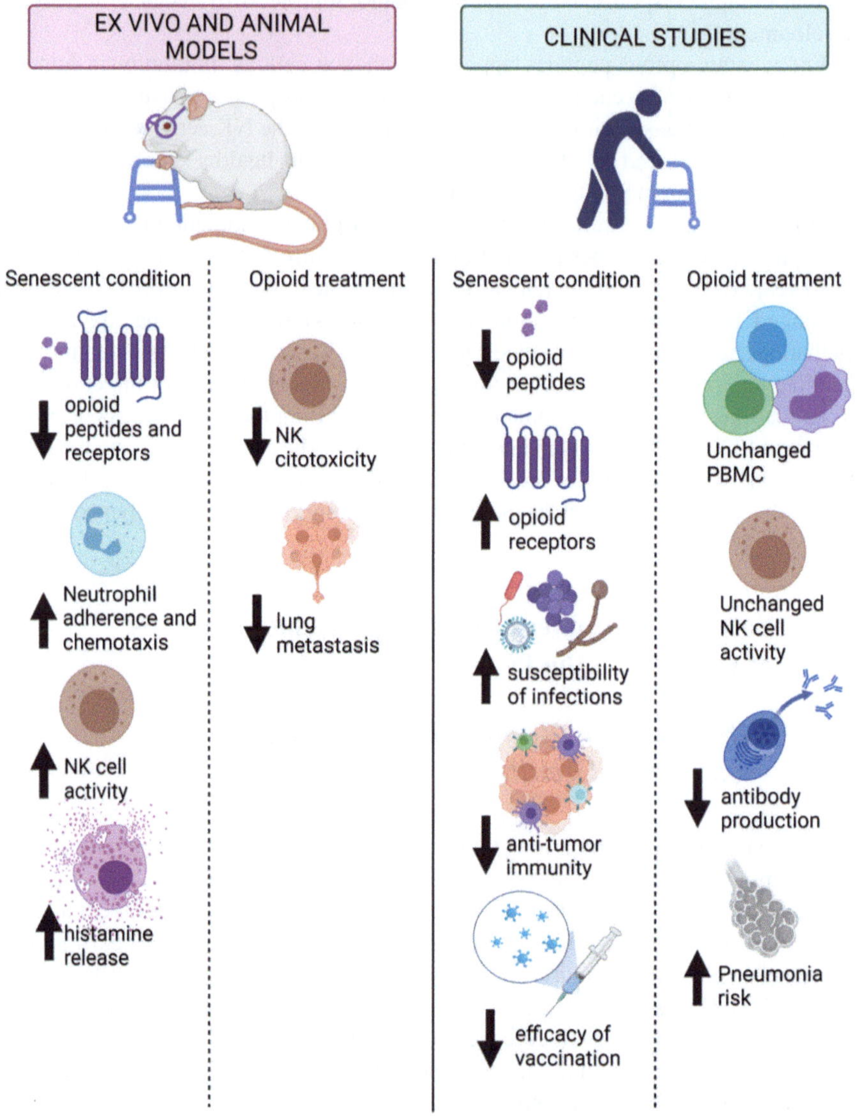

Fig. 12.5 Influence of aging on the opioid system and the effects of opioids on immune system. A comparison of studies carried out in animal models and clinical trials is observed. The senescent condition column shows main results in studies performed in whole healthy elderly individuals or cells isolated from them. The opioid-treated column shows results obtained from studies involving individuals under acute or chronic treatment with exogenous opioids and also studies where ex vivo cells were analyzed (for a more detailed description of these studies, see the text). ↑ = increase, ↓ = decrease. Created with BioRender.com

Table 12.2 Effects of opioids in tumor development

Opioid concentration and treatment	Type of tumor or tumor cells	Type of study	Effect	Reference
Morphine sulfate, 0.714 mg/kg per day for the first 15 days and then 1.43 mg/kg mouse/day to 48 days	MCF-7 cell breast tumor xenograft model	In vivo, mice	Acceleration of tumor growth	[143]
Fentanyl 0.1–0.3 mg/kg	MADB106 mammary adenocarcinoma tumor cell	In vivo, rat	Increases the number of tumor metastases. Suppresses NK cells	[144]
Morphine continuous release pellet implantation	Lewis lung carcinoma cells and human ovarian cancer cells (MA148)	In vitro and in vivo, mice	Reduction tumor cell-induced angiogenesis and tumor growth. It is mediated through the suppression of the hypoxia- induced mitochondrial p38 MAPK pathway	[145]
Morphine 1 nM	Lewis lung carcinoma cells	In vitro and in vivo, mice	Increased proliferation	[146]
Methadone 9–41 μg/mL	Leukemia cells	In vitro	Inhibits the growth and induces apoptosis of leukemia cells through caspases 3, 8, and 9 and mitochondria damage	[147]
Morphine 10 mg/kg I.P. every 12 h for 3 days in mice and 0.1–10 μM in cells for 48 h	Breast tumor	In vitro and in vivo, mice	Inhibits the number of metastatic foci of breast cancer. Decreases release of MMP-9 in MΦs	[148]
Methadone in vitro 10, 3, and 1 μg/mL in combination with doxorubicin. In vivo dosage increased weekly from 60 to 120 to 240 mg/kg/d *bid*	Human glioblastoma cells, glioblastoma stem cells, chemo- and radioresistant glioblastoma cells	In vitro and in vivo, mice	Apoptosis through caspace-3, 9, and -10, strong downregulation of XIAP and Bcl-xL as well a strong upregulation of the pro-apoptotic protein Bcl-xs. In mice reduces tumor size to 49%	[149]
Morphine, slow-release pellet	Lung tumor	In vivo, mice	Inhibit the expression of proteins related to cell adhesion. Decreased tumor growth progression	[150]

(continued)

Table 12.2 (continued)

Opioid concentration and treatment	Type of tumor or tumor cells	Type of study	Effect	Reference
Morphine dose escalation every 2 weeks: 0.75, 1.0, 1.25 mg/kg day and finally to 1.5 mg/kg day for 8 weeks.	Breast cancer	In vivo, rat	Release GM-CSF, RANTES, IL-6, SP	[151]
Morphine 1.5 mg/kg to day for 2 weeks	Breast carcinoma	In vivo, mice	Recruitment, release of cytokines and degranulation of MCs. Increases tumor burden	[151]
DAMGO, morphine, fentanyl	Human non-small cell lung cancer (NSCLC) cells	In vitro	MOR regulates growth factor receptor signaling and epithelial mesenchymal transition (EMT) in human NSCLC cells	[152]
Morphine, 20 μM	Murine mammary breast carcinoma cells (4 T1)	In vitro	Decrease: VEGF-α, TIMP-1, TIMP-2, G-CSF, GM-CSF, IFN-γ, TNF, CCL-2, IL-1β, IL-4, IL-6, IL-13, CXCL4, and THPO. Prevents blood vessels formation	[153, 154]
Methadone hydrochloride 87.8 μm/L to 121.6 μm/mL for 24 and 48 hours	CCRF-CEM (T lymphoblastoid cell line) and HL-60 (human promyelocytic leukemia cells)	In vitro	Apoptosis through induction of Bcl-2, caspase 8 and DNA damage	[155]
Methadone 121.6 μmol/L and 97.18 μmol/L for 24 h and 48 h	Leukemia cell CCRF-CEM and HL-60	In vitro	Dowregulate the expression of Bcl-2, Bid, p21, and survivin. Upregulate the expression of caspase 8. DNA fragmentation and damage	[155]
Morphine-3-glucuronide 10 μM and 20 μM for 24, 48, and 72 h and 10 mg/kg for 14 days	Human lung cancer cells	In vitro and in vivo, mice	Increased PD-L1 in cancer cells and on CD8+ cells increase TIM-3 and reduce IFN-γ	[156]

co-cultured with 4 T1 murine mammary breast carcinoma cells, but not when cultured alone. This effect is not dependent on μ-OR activation, because (−)naloxone pretreatment does not prevent it [157]. Interestingly, morphine treatment inhibits activation of MΦs in co-culture with breast carcinoma, an effect that does depend on μ-OR activation [158]. These results indicate that distinct receptors can mediate opioid effects on tumors (Table 12.2), but the molecular mechanisms involved are still unknown.

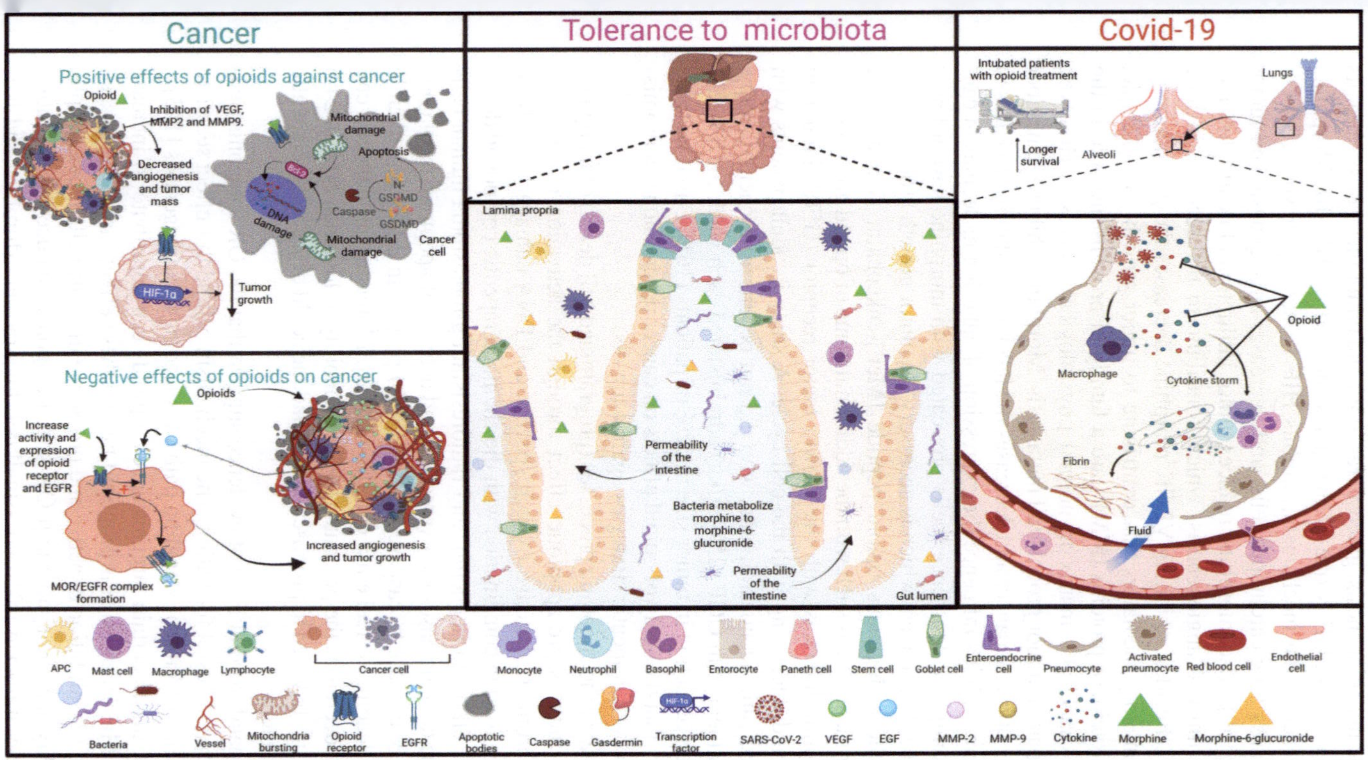

Fig. 12.6 Actions of opioids on specific pathologies. *Left panel*: Opioids inhibit cytokines related to tumor growth that diminish angiogenesis and tumor mass. Activation of ORs generates damage of mitochondria and DNA, to conduce secretion of apoptosis-related molecules (like Bcl-2) and caspase activation. Caspase 1 activates gasdermin D (GSDMD) to form pores in the cell membrane. On the other hand, opioids inhibit transcription factor HIF-1α, which is related to tumor growth. All mentioned effects promote apoptosis of cancer cells. Negative effects of opioids on cancer are related to the induction of pro-inflammatory and pro-angiogenic cytokines that promote tumor growth. Central panel. Bacteria composing gut microbiota metabolize morphine to morphine-6-glucuronide, which alters permeability of the intestinal barrier and promotes systemic inflammation. Opioid induces permeability of the intestine allowing the passage of the bacteria. *Right panel:* In intubated patients, treatment with opioids has been associated with higher probability of survival associated with a blockage of the cytokine storm (see text for details). Created with BioRender.com

The hypoxia-inducible transcription factor 1α (HIF-1α) controls the expression of genes involved in angiogenesis, glucose transport, and tumor metabolism after its nuclear translocation. VEGF is one of the genes induced by HIF-1α. Morphine, through a μ-OR-dependent mechanism, inhibits HIF-1α activity, lowering VEGF production in Lewis lung carcinoma cells, and decreases angiogenesis and tumor growth in mice [159]. Also, morphine reduces tumor infiltration of neutrophils and MΦs due to diminished angiogenesis, which contribute to the decrease in lung tumor size [150]. Besides inhibiting VEGF, morphine decreases the production of other pro-inflammatory mediators that contribute to tumor growth. For example, matrix metalloproteinase proteins (MMP) are a family of proteins involved in the breakdown of extracellular matrix in normal physiological processes. However, they also participate in tumor growth and metastasis. Morphine inhibits the number of metastatic foci of breast cancer and decreases the level of circulating proteases in mice. Furthermore, opioids influence the expression of adhesion molecules in cancer cells. For example, colon cancer cells pretreated with morphine have a lower expression of type IV collagen, which is necessary for tumor growth [160].

In humans, the endogenous opioid peptide [Met^5]-enkephalin is a negative growth regulator identified as such with the additional name of opioid growth factor [OGF]. This peptide interacts with the growth factor receptor (GFR), inhibiting DNA synthesis and interfering with the cell cycle. A study in adult patients with unresectable pancreatic adenocarcinoma, who did not benefit from chemotherapy, showed that treatment with OGF (250 μg/kg) administered each week for 8 weeks led to a significant reduction in tumor size in 62% of patients [161]. Based on these results, it appears that the study of opioid effects on cancer progression is a promising field to find new therapeutic strategies against tumors.

Opioids also favor the apoptotic processes (programmed cell death) of malignant cells. Morphine induces pro-apoptotic effect through the activation of caspases, release of cytochrome C, reactive oxygen species, and DNA damage [155, 160]. Methadone, when used in combination with the chemotherapeutic compound doxorubicin, induces cell death in chemo- and radioresistant glioblastoma human cells [149]. Methadone also increases the expression and activity of apoptotic mediators, such as caspase 3, caspase 8, caspase 9, and HSP70 protein in different leukemic cells, ultimately leading to leukemia cell death [162]. Moreover, in leukemia cell lines, methadone decreases the percentage of cell viability to less than 50% in 24 h [155]. These studies have led to propose that methadone could be a good therapeutic agent against leukemia.

On the other hand, several lines of evidence suggest that the OR antagonist naltrexone inhibits blood vessel formation. The mechanism behind this effect seems to be related to the inactivation of signaling pathways leading to angiogenesis [143].

Paradoxically, under certain circumstances, opioids can contribute to increase the tumor mass. Main mechanisms proposed behind these effects are (1) increased angiogenesis and (2) inhibition of active and protective immune cells. For example, DAMGO, a highly specific synthetic μ-OR-specific agonist, induces proliferation and cell migration in human H358 non-small cell lung cancer. Morphine and fentanyl drive epithelial-mesenchymal transition, a phenomenon associated with increased tumor metastatic capacity [152]. Repeated morphine treatment leads to significant increase in tumor burden, breast carcinoma, and the formation of new lymphatic vessels (lymphangiogenesis) in mice [151]. Also, morphine induces cell recruitment, release of cytokines, and degranulation of MCs inside breast tumors [151]. A proposed mechanism for the increase of malignant tumor size is that morphine inhibits NK cells, which are one of the first lines of defense against transformed cells [163]. To date, evidence indicates that some opioids (like methadone) could be useful for the treatment of certain types of cancers, but methadone effects seem not to be positive in all the types of tumors [144].

Because patients with morphine treatment have elevated levels of morphine's metabolite M3G, some studies have analyzed the effect of this metabolite in cancer. M3G binds to and activates TLR4, which is highly expressed in tumors and is associated with tumor malignancy. In human lung cancer cells, M3G, in a dose-dependent manner and through TLR4 binding, increases the expression of programmed death-ligand 1 (PD-L1), a molecule that transmits an inhibitory signal to the cytotoxic T lymphocytes (CTL) and promotes the inhibition of the protective actions of IS against the tumor. Even opioid metabolites, such as M3G, seem to exert inhibitory actions on immune cells and modulate their interaction with transformed cancer cells, which indicates that opioids alter the participation of the IS on the recognition of transformed cells and the initiation, progress, and success of immune response against tumors, which makes it difficult to predict the final outcome of opioid administration on cancer development.

12.7.2 Opioids, the IS, and Tolerance to Microbiota

Gut microbiota include all microorganisms within the gastrointestinal (GI) tract, including fungi, viruses, bacteria, archaea, and eukaryotes [164]. The microbiota constitutes a complex system that interacts with the host through the secretion of different metabolites, vitamins, and other mediators able to modify metabolic pathways in different organs, such as the brain and the IS [165]. Physiological homeostasis exists when there is an equilibrium between commensal microbiota growth and a low (tolerant) immune response [166]. Intense research in the last years indicates that gut microbiota modulates the development of immune reactions, neural functioning, and the onset of distinct pathologies [167], although the involved mechanisms are not well understood. Alterations in microbiota composition or metabolism, called dysbiosis, are associated with many pathologies, such as autoimmune diseases, neurological disorders, obesity, diabetes, allergies, and other diseases [164].

Opioids modulate GI function. For example, morphine inhibits the intestinal epithelium's protective mucus and bicarbonate secretion [168]. Morphine also attenuates epithelial immune function, inhibiting cytokine secretion in an in vitro model of inflammation [169]. In general, opioid administration causes constipation, leaky intestinal barrier function, nausea, and vomiting [170]. In addition, morphine increases bacterial overgrowth and the PAMP N-formyl-methionyl-leucyl-phenylalanine (FMLP) production, which induces mucosal permeability on the intestine of rats [170]. MCs appear to be the cellular target of FMLP and morphine effects on the rat ileum, since changes in permeability caused by FMLP and their blockage with morphine do not occur in ilea from mice treated with doxantrazole (a stabilizer of MCs) or in ilea from MC-deficient mice [170].

In general, evidence in humans and mouse models indicates that opioid administration leads to increased intestinal barrier permeability, bacterial translocation, increased risk of enteric infection, and life-threatening conditions, such as gut-derived sepsis [171, 172]. The mechanisms through which opioids cause GI dysfunction are not clear. However, some studies have demonstrated that morphine disrupts intestinal barrier function and damages the organization of the tight junction proteins. These events activate TLR, whose signaling cascade leads to the phosphorylation of the myosin light-chain kinase [173] and alters the cell cytoskeleton. Research on the effects of opioids on GI tract has led to propose the use of naltrexone as a therapeutic strategy for active Crohn's disease [174].

Opioids induce gut microbial dysbiosis and lead to sustained systemic inflammation by disrupting the pathway of bile acid metabolism [175]. Opioid metabolism affects microbiota and, in consequence, modifies distinct immune responses. For example, morphine metabolites M6G and M3G are hydrolyzed by beta-glucuronidase in intestinal mucosal cells and gut bacteria, [176, 177]. Anaerobic bacteria such as *Bacteroides* and *Bifidobacteria* are major sources of beta-glucuronidase [177]. In consequence, a gut microbioma enriched with these bacteria could promote an increase degradation of morphine metabolites and limit opioid actions on IS.

Also, morphine can cause lethal gut-derived sepsis in mice [172, 178], presumably by disrupting the gut-barrier function, increasing bacteria translocation and bacterial virulence expression [179].

Finally, in humans, it has been observed that opioid-dependent alteration of GI tract modifies the time course of bacterial and viral infections, such as HIV [173]. An analysis of a cohort of 2933 patients with functional GI disorders found that opioid use was associated with increased vomiting, constipation, and the severity of GI disease [180].

12.8 Opioids and Immune Responses Against COVID-19

The recent global health burden due to SARS-CoV-2 virus infection has led to a worldwide effort toward understanding COVID-19 and the search for therapeutic strategies targeting key aspects of this potentially lethal syndrome. From the beginning of the pandemics, it has been clear that hyper-inflammation relates to symptom severity and mortality. Severely affected COVID-19 patients have GI, respiratory, neuronal, renal, and cardiac symptoms. In addition, several critical immunological parameters are affected, such as the number of leukocytes (especially NK cells) and markers of inflammation [181, 182]. Furthermore, these patients suffer from a significant systemic inflammation that leads to increased blood levels of cytokines and chemokines, something referred to as a "cytokine storm." Elevated cytokines comprise IL1-β, IL1RA, IL7, IL8, IL9, IL-10, basic FGF2, GCSF, GM-CSF, IFNγ, IP10, CCL-2, MIP1α, MIP1β, PDGF-B, TNF-α, and VEGF [183]. Specific therapeutic approaches include immunoglobulins, recombinant human IL-6 receptor monoclonal antibody (tocilizumab), chloroquine, hydroxychloroquine, JAK inhibitors (Ruxolitinib®, Jakotinib®), convalescent plasma, and glucocorticoids, among others [184]. However, when severe inflammation exists, the nonspecific, anti-inflammatory properties of opioids have been used to control damage induced by the over-activation of IS [185].

Different calculations have reported that around 40% of patients with severe COVID-19 require mechanical ventilation, 15% develop acute respiratory distress syndrome (ARDS), and 6% develop septic shock [186]. Sedation is necessary to manage acute agitation, pain treatment, and facilitation of mechanical ventilation. Normally, morphine and fentanyl help patients reach light sedation levels needed for intensive care unit manipulations. In those conditions, the immunosuppressive actions of opioids can help to diminish the cytokine storm [185]. The proposed mechanism for opioid actions includes HPA axis activating because morphine significantly increases plasma corticosterone levels, contributing to damp hyper-inflammation. From this point of view, immunosuppressive actions of opioids make them a valuable tool as a multitarget therapeutic strategy for acute hyper-inflammatory conditions, such as COVID-19.

Like other treatments, the use of opioids in the management of COVID-19 is not free of controversy. For example, opioid-induced inhibition of an initially protective anti-inflammatory response in the lungs can be a risk for COVID-19 complications, and prolonged opioid use has been related to pneumonia development [35]. Due to these effects, caution is recommended for the use of opiates in distinct phases of the COVID-19 disease [187].

12.9 Perspectives

Due to the diverse effects of opioids in IS function, broad terms such as "immunosuppression"or "immunoactivation" referring to the actions of any particular opioid should be used with care, considering the context of the specific immune reaction studied before reaching conclusions on long-term responses. To date, evidence indicates that the opioid system connects and affects immune responses modifying the intensity of critical processes involved in several of the design principles of the IS and, in consequence, opioids alter the direction of a given immune response. This idea is supported by basic studies on the presence of opioid ligands and receptors in distinct groups of organisms, which strongly suggest that opioids constitute a unique evolution-conserved mechanism of cross-regulation between the immune and the nervous systems in metazoans [188]. Thus, alterations on any branch of this communication network, which also participate in the activity of the HPA axis, modify the final immune response against pathogens, tumor recognition, initiation, and resolution of inflammation and affect immunosurveillance or antibody production, besides pain control.

Knowledge of the main effects of opioids on the IS has started to be used to control not only pain but also several pathologic conditions. The potential role of the opioid system on the resolution of inflammatory responses is a promising area for the therapeutics of distinct acute and chronic diseases.

References

1. Orosz CG. An introduction to immuno-ecology and immuno-informatics. In: Segel LA, Cohen IR, editors. Design principles for the immune system and other distributed autonomous systems; 2001. p. 125–49.
2. Lazzarin A, Mella L, Trombini M, Uberti-Foppa C, Franzetti F, Mazzoni G, et al. Immunological status in heroin addicts: effects of methadone maintenance treatment. Drug Alcohol Depend. 1984;13(2):117–23.
3. Mclachlan C, Crofts N, Wodak A, Crowe S. The effects of methadone on immune function among injecting drug users: a review. Addiction. 1993;88(2):257–63.
4. Tubaro E, Avico U, Zuccaro P, Cavallo G, Pac R, Croce C, et al. Morphine and methadone impact on human phagocytic physiology. Int J Immunopharmacol. 1985;7(6):865–74.
5. Delafuente JC, Lindsay DC. Immunologic effects of cocaine and related alkaloids. Immunopharmacol Immunotoxicol. 1991;13(1–2):11–23.
6. Peterson PK, Gekker G, Brummitt C, Pentel P, Bullock M, Simpson M, et al. Suppression of human peripheral blood mononuclear cell function by methadone and morphine. J Infect Dis. 1989;159(3):480–7.
7. Zhang R, Meng J, Lian Q, Chen X, Bauman B, Chu H, et al. Prescription opioids are associated with higher mortality in patients diagnosed with sepsis: a retrospective cohort study using electronic health records. PLoS One. 2018;(1):1–8.
8. Bauchat JR. Neuraxial morphine and oral herpes reactivation in the obstetric population, Anesthesia and analgesia, vol. 111. Lippincott Williams and Wilkins; 2010. p. 1238–41.
9. Østerdal OB, Salminen PR, Jordal S, Sjursen H, Wendelbo O, Haaverstad R. Cardiac surgery for infective endocarditis in patients with intravenous drug use. Interact Cardiovasc Thorac Surg. 2016;22(5):633–40.
10. Barocas JA, Morgan JR, Wang J, McLoone D, Wurcel A, Stein MD. Outcomes associated with medications for opioid use disorder among persons hospitalized for infective endocarditis. Clin Infect Dis. 2021;72(3):472–8.
11. Zhu JW, Liu FL, Mu D, Deng DY, Zheng YT. Heroin use is associated with lower levels of restriction factors and type I interferon expression and facilitates HIV-1 replication. Microbes Infect. 2017;19(4–5):288–94.
12. Do Q-CN, Wallace MS, Ashar N, Mathews C. Long-term methadone treatment: effect on CD4+ lymphocyte counts and HIV-1 plasma RNA level in patients with HIV infection. Eur J Pain. 2001;5(4):415–20.
13. Wang X, Ye L, Zhou Y, Liu MQ, Zhou DJ, Ho WZ. Inhibition of anti-HIV microRNA expression: a mechanism for opioid-mediated enhancement of HIV infection of monocytes. Am J Pathol. 2011;178(1):41–7.
14. Ovalle F, Dembinski D, Yalamanchili S, Minkara A, Stern PJ. Hand and upper extremity infections in intravenous drug users: epidemiology and predictors of outcomes. J Hand Surg Am. 2020;45(6):503–11.

15. Lewer D, Harris M, Hope V. Opiate injection–associated skin, soft tissue, and vascular infections, England, UK, 1997–2016. Emerg Infect Dis. 2017;23(8):1400–3.
16. Suzuki J, Park EM. Buprenorphine/naloxone and dental caries: a case report. Am J Addict. 2012;21:494–5.
17. Farr A, Kiss H, Hagmann M, Holzer I, Kueronya V, Husslein PW, et al. Evaluation of the vaginal flora in pregnant women receiving opioid maintenance therapy: a matched case-control study. BMC Pregnancy Childbirth [Internet]. 2016;16(1):1–8. Available from: https://doi.org/10.1186/s12884-016-1003-z
18. Wiese AD, Griffin MR, Schaffner W, Stein CM, Greevy RA, Mitchel EF, et al. Long-acting opioid use and the risk of serious infections: a retrospective cohort study. Clin Infect Dis. 2019;68(11):1862–9.
19. Liao Y, Jiang J, Liang B, Wei F, Huang J, Pan P, et al. Opiate use inhibits TLR9 signaling pathway in vivo: possible role in pathogenesis of HIV-1 infection. Sci Rep. 2017;7(1)
20. Masvekar RR, El-Hage N, Hauser KF, Knapp PE. GSK3β-activation is a point of convergence for HIV-1 and opiate-mediated interactive neurotoxicity. Mol Cell Neurosci [Internet]. 2015;65:11–20 Available from: https://doi.org/10.1016/j.mcn.2015.01.001
21. Zou S, Fitting S, Hahn YK, Welch SP, El-Hage N, Hauser KF, et al. Morphine potentiates neurodegenerative effects of HIV-1 Tat through actions at μ-opioid receptor-expressing glia. Brain. 2011;134(12):3613–28.
22. Edelman EJ, Gordon KS, Crothers K, Akgün K, Bryant KJ, Becker WC, et al. Association of prescribed opioids with increased risk of community-acquired pneumonia among patients with and without HIV. JAMA Intern Med. 2019;179(3):297–304.
23. Oppeltz RF, Holloway TL, Covington CJ, Schwacha MG. The contribution of opiate analgesics to the development of infectious complications in trauma patients [Internet]. Int J Burn Trauma. 2015;5(2):56–65. Available from: www.IJBT.org.
24. Schwacha MG, McGwin G, Hutchinson CB, Cross JM, MacLennan PA, Rue LW. The contribution of opiate analgesics to the development of infectious complications in burn patients. Am J Surg. 2006;192(1):82–6.
25. Bonilla-García JSL, Cortiñas-Sáenz M, Del Pozo-Gavilán E. Opioids and immunosupression in oncological postoperative patients. Rev Assoc Med Bras. 2017;63(9):753–63.
26. Shao Y-J, Liu W-S, Guan B-Q, Hao J-L, Ji K, Cheng X-J, et al. Contribution of opiate analgesics to the development of infections in advanced cancer patients. Clin J Pain. 2017;33(4):295–9.
27. Acharya C, Betrapally NS, Gillevet PM, Sterling RK, Akbarali H, White MB, et al. Chronic opioid use is associated with altered gut microbiota and predicts readmissions in patients with cirrhosis. Aliment Pharmacol Ther. 2017;45(2):319–31.
28. Lee AA, Baker JR, Wamsteker EJ, Saad R, DiMagno MJ. Small intestinal bacterial overgrowth is common in chronic pancreatitis and associates with diabetes, chronic pancreatitis severity, low zinc levels, and opiate use. Am J Gastroenterol. 2019;114(7):1163–71.
29. Wiese AD, Griffin MR, Stein CM, Mitchel EF, Grijalva CG. Opioid analgesics and the risk of serious infections among patients with rheumatoid arthritis: a self-controlled case series study. Arthritis Rheumatol. 2016;68(2):323–31.
30. Wiese AD, Griffin MR, Schaffner W, Michael Stein C, Greevy RA, Mitchel EF, et al. Opioid analgesic use and risk for invasive pneumococcal diseases a nested case-control study. Ann Intern Med. 2018;168(6):396–404.
31. Hamina A, Taipale H, Karttunen N, Tanskanen A, Tiihonen J, Tolppanen AM, et al. Hospital-treated pneumonia associated with opioid use among community dwellers with Alzheimer's disease. J Alzheimers Dis. 2019;69(3):807–16.
32. Dublin S, Walker RL, Jackson ML, Nelson JC, Weiss NS, Von Korff M, et al. Use of opioids or benzodiazepines and risk of pneumonia in older adults: a population-based case-control study. J Am Geriatr Soc. 2011;59(10):1899–907.
33. Bigliardi-Qi M, Bigliardi P. The roles of opioid receptors in cutaneous wound healing. In: Handbook of experimental pharmacology. Springer New York LLC; 2018. p. 335–45.

34. Franchi S, Moschetti G, Amodeo G, Sacerdote P. Do all opioid drugs share the same immunomodulatory properties? A review from animal and human studies. Front Immunol. 2019;10:1–11.
35. Sacerdote P, Franchi S, E. Panerai A. Non-analgesic effects of opioids: mechanisms and potential clinical relevance of opioid-induced Immunodepression. Curr Pharm Des. 2012;18(37)
36. Cata JP, Bauer M, Sokari T, Ramirez MF, Mason D, Plautz G, et al. Effects of surgery, general anesthesia, and perioperative epidural analgesia on the immune function of patients with non-small cell lung cancer. J Clin Anesth. 2013;25(4)
37. Sacerdote P, Bianchi M, Gaspani L, Manfredi B, Maucione A, Terno G, et al. The effects of tramadol and morphine on immune responses and pain after surgery in cancer patients. Anesth Analg. 2000;90(6):1411–4.
38. Beilin B, Shavit Y, Hart J, Mordashov B, Cohn S, Notti I, et al. Effects of anesthesia based on large versus small doses of fentanyl on natural killer cell cytotoxicity in the perioperative period. Anesth Analg. 1996;82(3):492–7.
39. Inagi T, Suzuki M, Osumi M, Bito H. Remifentanil-based anaesthesia increases the incidence of postoperative surgical site infection. J Hosp Infect. 2015;89(1):61–8.
40. Fukada T, Kato H, Ozaki M, Yagi J. Impact of the timing of morphine administration on lipopolysaccharide-mediated lethal endotoxic shock in mice. Shock. 2016;45(5):564–9.
41. Khosrow-Khavar F, Kurteva S, Cui Y, Filion KB, Douros A. Opioids and the risk of infection: a critical appraisal of the pharmacologic and clinical evidence, Expert opinion on drug metabolism and toxicology, vol. 15. Taylor and Francis Ltd; 2019. p. 565–75.
42. Abbas A, Lichtman A, Pilla S. Cellular and molecular immunology. 9th ed. Elsevier; 2018.
43. Shang Y, Filizola M. Opioid receptors: Structural and mechanistic insights into pharmacology and signaling. Eur J Pharmacol. 2015;763:206–13.
44. Lansu K, Karpiak J, Liu J, Huang XP, McCorvy JD, Kroeze WK, et al. In silico design of novel probes for the atypical opioid receptor MRGPRX2. Nat Chem Biol. 2017;13(5):529–36.
45. Akuzawa N, Obinata H, Izumi T, Takeda S. Morphine is an exogenous ligand for MrgX2, a G protein-coupled receptor for cortistatin. J Cell Anim Biol. 2007;2(1):004–9.
46. Stein C. Opioid receptors. Annu Rev Med. 2016;67:433–51.
47. Darcq E, Kieffer BL. Opioid receptors: drivers to addiction? Nat Rev Neurosci. 2018;19:499–514.
48. Wu H, Zeng M, Cho EYP, Jiang W, Sha O. The origin, expression, function and future research focus of a G protein-coupled receptor, Mas-related Gene X2 (MrgX2). Prog Histochem Cytochem. 2015;50:11–7.
49. Medzhitov R. Origin and physiological roles of inflammation. Nature. 2008;454:428–35.
50. Sugimoto MA, Sousa LP, Pinho V, Perretti M, Teixeira MM. Resolution of inflammation: what controls its onset? Front Immunol [Internet]. 2016;7:160. Available from: www.frontiersin.org.
51. Rittner HL, Hackel D, Voigt P, Mousa S, Stolz A. Mycobacteria attenuate nociceptive responses by formyl peptide receptor triggered opioid peptide release from neutrophils. PLoS Pathog [Internet]. 2009;5(4):1000362. Available from: www.dfg.de.
52. Rittner HL, Labuz D, Richter JF, Brack A, Schäfer M, Stein C, et al. CXCR1/2 ligands induce p38 MAPK-dependent translocation and release of opioid peptides from primary granules in vitro and in vivo. Brain Behav Immun. 2007;21(8):1021–32.
53. Malafoglia V, Ilari S, Vitiello L, Tenti M, Balzani E, Muscoli C, et al. The interplay between chronic pain, opioids, and the immune system. Neurosci [Internet]. 2021;107385842110304. Available from: http://journals.sagepub.com/doi/10.1177/10738584211030493.
54. Pannell M, Labuz D, Celik M, Keye J, Batra A, Siegmund B, et al. Adoptive transfer of M2 macrophages reduces neuropathic pain via opioid peptides. J Neuroinflammation. 2016;13(1):262.
55. Labuz D, Celik M, Seitz V, Machelska H. Interleukin-4 induces the release of opioid peptides from M1 macrophages in pathological pain. J Neurosci. 2021;41(13):2870–82.

56. Brack A, Rittner HL, Machelska H, Beschmann K, Sitte N, Schäfer M, et al. Mobilization of opioid-containing polymorphonuclear cells by hematopoietic growth factors and influence on inflammatoty pain. Anesthesiology. 2004;100(1):149–57.
57. Machelska H, Mousa SA, Brack A, Schopohl JK, Rittner HL, Schäfer M, et al. Opioid control of inflammatory pain regulated by intercellular adhesion molecule-1. J Neurosci. 2002;22(13)
58. Fitzgerald KA, Kagan JC. Toll-like receptors and the control of immunity. Cell. 2020;180(6):1044–66.
59. Akira S, Uematsu S, Takeuchi O. Pathogen recognition and innate immunity. Cell. 2006;124, 4:783–801.
60. Molina-Martínez LM, González-Espinosa C, Cruz SL. Dissociation of immunosuppressive and nociceptive effects of fentanyl, but not morphine, after repeated administration in mice: fentanyl-induced sensitization to LPS. Brain Behav Immun. 2014;1(42):60–4.
61. Madera-Salcedo IK, Cruz SL, Gonzalez-Espinosa C. Morphine decreases early peritoneal innate immunity responses in Swiss-Webster and C57BL6/J mice through the inhibition of mast cell TNF-α release. J Neuroimmunol. 2011;232(1–2):101–7.
62. Madera-Salcedo IK, Cruz SL, Gonzalez-Espinosa C. Morphine prevents lipopolysaccharide-induced TNF secretion in mast cells blocking IκB kinase activation and SNAP-23 phosphorylation: correlation with the formation of a β-Arrestin/TRAF6 complex. J Immunol. 2013;191(6):3400–9.
63. Martucci C, Franchi S, Lattuada D, Panerai AE, Sacerdote P. Differential involvement of RelB in morphine-induced modulation of chemotaxis, NO, and cytokine production in murine macrophages and lymphocytes. J Leukoc Biol. 2007;81(1):344–54.
64. Roy S, Cain KJ, Chapin RB, Charboneau RG, Barke RA. Morphine modulates NFκB activation in macrophages. Biochem Biophys Res Commun. 1998;245(2):392–6.
65. Zeng S, Zhong Y, Xiao J, Ji J, Xi J, Wei X, et al. Kappa opioid receptor on pulmonary macrophages and immune function. Transl Perioper Pain Med. 2020;7(3):225–33.
66. Luan G, Pan F, Bu L, Wu K, Wang A, Xu X. Butorphanol promotes macrophage phenotypic transition to inhibit inflammatory lung injury via κ receptors. Front Immunol. 2021;12:692286.
67. Tubaro E, Avico U, Santiangeli C, Zuccaro P, Cavallo G, Pacifici R, et al. Morphine and methadone impact on human phagocytic physiology. Int J Immunopharmacol. 1985;7(6):865–74.
68. Rojavin M, Szabo I, Bussiere JL, Rogers TJ, Adler MW, Eisenstein TK. Morphine treatment in vitro or in vivo decreases phagocytic functions of murine macrophages. Life Sci. 1993;53(12):997–1006.
69. Wang J, Barke RA, Charboneau R, Roy S. Morphine impairs host innate immune response and increases susceptibility to Streptococcus pneumoniae lung infection. J Immunol. 2005;174(1):426–34.
70. Jana N, Vidhu A, Raini D, Zhang L, Saluja A, Meng J, et al. Differential effects of gram-positive and gram-negative bacterial products on morphine induced inhibition of phagocytosis. Sci Rep. 2016;6:21094.
71. Wan J, Ma J, Anand V, Ramakrishnan S, Roy S. Morphine potentiates LPS-induced autophagy initiation but inhibits autophagosomal maturation through distinct TLR4 dependent and independent pathways HHS Public Access. Acta Physiol. 2015;214(2):189–99.
72. Wang TL, Chang H, Hung CR, Tseng YZ. Attenuation of neutrophil and endothelial activation by intravenous morphine in patients with acute myocardial infarction. Am J Cardiol. 1997;80(12):1532–5.
73. Koodie L, Yuan H, Pumper JA, Yu H, Charboneau R, Ramkrishnan S, et al. Morphine inhibits migration of tumor-infiltrating leukocytes and suppresses angiogenesis associated with tumor growth in mice. Am J Pathol [Internet]. 2014;184(4):1073–84. Available from: https://doi.org/10.1016/j.ajpath.2013.12.019.
74. Hofbauer R, Frass M, Gmeiner B, Sandor N, Schumann R, Wagner O, et al. Effects of remifentanil on neutrophil adhesion, transmigration, and intercellular adhesion molecule expression. Acta Anaesthesiol Scand. 2000;44(10):1232–7.

75. Min TJ, Park S-H, Ji Y-H, Lee Y-S, Kim TW, Kim JH, et al. Morphine attenuates endothelial cell adhesion molecules induced by the supernatant of LPS-stimulated colon cancer cells. J Korean Med Sci [Internet]. 2011;26:747–52. Available from: http://jkms.org
76. Zhang J-N, Ma Y, Wei X-Y, Liu K-Y, Wang H, Han H, et al. Remifentanil protects against lipopolysaccharide-induced inflammation through PARP-1/NF-κB signaling pathway. Mediators Inflamm. 2019;2019:3013716. Available from: https://doi.org/10.1155/2019/3013716.
77. Andrews EJ, Wang JH, Winter DC, Laug WE, Redmond HP. Tumor cell adhesion to endothelial cells is increased by endotoxin via an upregulation of β-1 integrin expression. J Surg Res. 2001;97(1):14–9.
78. Wang L, Yin C, Xu X, Liu T, Wang B, Abdul M, et al. Pellino1 contributes to morphine tolerance by microglia activation via MAPK signaling in the spinal cord of mice. Cell Mol Neurobiol. 2020;40(7)
79. Merighi S, Gessi S, Varani K, Fazzi D, Stefanelli A, Borea PA. Morphine mediates a proinflammatory phenotype via m-opioid receptor-PKCε-Akt-ERK1/2 signaling pathway in activated microglial cells. Biochem Pharmacol. 2013;86(4):487–96.
80. Gessi S, Borea PA, Bencivenni S, Fazzi D, Varani K, Merighi S. The activation of μ-opioid receptor potentiates LPS-induced NF-kB promoting an inflammatory phenotype in microglia. FEBS Lett. 2016;590(17):2813–26.
81. Dutta R, Krishnan A, Meng J, Das S, Ma J, Banerjee S, et al. Morphine modulation of toll-like receptors in microglial cells potentiates neuropathogenesis in a HIV-1 model of coinfection with pneumococcal pneumoniae. J Neurosci. 2012;32(29):9917–30.
82. Dutta R, Roy S. Chronic morphine and HIV-1 Tat promote differential central nervous system trafficking of CD3+ and Ly6C+ immune cells in a murine Streptococcus pneumoniae infection model. J Neuroinflammation. 2015;12:120.
83. Cai Y, Kong H, Pan YB, Jiang L, Pan XX, Hu L, et al. Procyanidins alleviates morphine tolerance by inhibiting activation of NLRP3 inflammasome in microglia. J Neuroinflammation. 2016;13(1):53.
84. Carranza-Aguilar CJ, Hernández-Mendoza A, Mejias-Aponte C, Rice KC, Morales M, González-Espinosa C, et al. Morphine and fentanyl repeated administration induces different levels of NLRP3-dependent pyroptosis in the dorsal raphe nucleus of male rats via cell-specific activation of TLR4 and opioid receptors. Cell Mol Neurobiol. 2020;42(3):677–94.
85. Qian J, Zhu Y, Bai L, Gao Y, Jiang M, Xing F, et al. Chronic morphine-mediated upregulation of high mobility group box 1 in the spinal cord contributes to analgesic tolerance and hyperalgesia in rats. Neurotherapeutics. 2020;17(2):722–42. Available from: https://doi.org/10.1007/s13311-019-00800-w.
86. Qu J, Tao X-Y, Teng P, Zhang Y, Guo C-L, Hu L, et al. Blocking ATP-sensitive potassium channel alleviates morphine tolerance by inhibiting HSP70-TLR4-NLRP3-mediated neuroinflammation. Journal of Neuroinflammation. 2017;14:228.
87. Babina M, Wang Z, Roy S, Guhl S, Franke K, Artuc M, et al. MRGPRX2 is the codeine receptor of human skin mast cells: desensitization through β-Arrestin and lack of correlation with the FcεRI pathway. J Invest Dermatol. 2021;141(5):1286–96.
88. Lazki-Hagenbach P, Ali H, Sagi-Eisenberg R. Authentic and ectopically expressed MRGPRX2 elicit similar mechanisms to stimulate degranulation of mast cells. Cells. 2021;10(2):376. Available from: https://doi.org/10.3390/cells10020376.
89. Sheen CH, Schleimer RP, Kulka M. Codeine induces human mast cell chemokine and cytokine production: involvement of G-protein activation. Allergy. 2007;62(5):532–8.
90. Hutchinson MR, Zhang Y, Shridhar M, Evans JH, Buchanan MM, Zhao TX, et al. Evidence that opioids may have toll-like receptor 4 and MD-2 effects. Brain Behav Immun. 2010;24(1):83–95.
91. Wang X, Loram LC, Ramos K, De Jesus AJ, Thomas J, Cheng K, et al. Morphine activates neuroinflammation in a manner parallel to endotoxin. Proc Natl Acad Sci U S A. 2012;109(16):6325–30.

92. Wang H, Huang M, Wang W, Zhang Y, Ma X, Luo L, et al. Microglial TLR4-induced TAK1 phosphorylation and NLRP3 activation mediates neuroinflammation and contributes to chronic morphine-induced antinociceptive tolerance. Pharmacol Res. 2021;165:6325–30.
93. Lewis SS, Hutchinson MR, Rezvani N, Loram LC, Zhang Y, Maier SF, et al. Evidence that intrathecal morphine-3-glucuronide may cause pain enhancement via toll-like receptor 4/ MD-2 and interleukin-1β. Neuroscience. 2010;165(2)
94. Wang X, Zhang Y, Peng Y, Hutchinson MR, Rice KC, Yin H, et al. Pharmacological characterization of the opioid inactive isomers (+)-naltrexone and (+)-naloxone as antagonists of toll-like receptor 4. Br J Pharmacol [Internet]. 2016;173:856–69. Available from: www.brjpharmacol.org.
95. Sibinga NES, Goldstein A. Opioid peptides and opioid receptors in cells of the immune system. Annu Rev Immunol. 1988;6:219–49.
96. Bidlack JM, Khimich M, Parkhill AL, Sumagin S, Sun B, Tipton CM. Opioid receptors and signaling on cells from the immune system. J Neuroimmune Pharmacol. 2006;
97. Miyagi T, Chuang LF, Lam KM, Kung HF, Wang JM, Osburn BI, et al. Opioids suppress chemokine-mediated migration of monkey neutrophils and monocytes – an instant response. Immunopharmacology. 2000;47(1):53–62.
98. Pérez-Castrillón JL, Pérez-Arellano JL, García-Palomo JD, Jiménez-López A, De CS. Opioids depress in vitro human monocyte chemotaxis. Immunopharmacology. 1992;23(1)
99. Vallejo R, de Leon-Casasola O, Benyamin R. Opioid therapy and immunosuppression: a review. Am J Ther. 2004;11(5):354–65.
100. Wodehouse T, Demopoulos M, Petty R, Miraki-Moud F, Belhaj A, Husband M, et al. A randomized pilot study to investigate the effect of opioids on immunomarkers using gene expression profiling during surgery. Pain. 2019;160(12):2691–8.
101. Cui JH, Jiang WW, Liao YJ, Wang QH, Xu M, Li Y. Effects of oxycodone on immune function in patients undergoing radical resection of rectal cancer under general anesthesia. Medicine (Baltimore). 2017;96(31):e7519.
102. McDonough RJ, Madden JJ, Falek A, Shafer DA, Pline M, Gordon D, et al. Alteration of T and null lymphocyte frequencies in the peripheral blood of human opiate addicts: in vivo evidence for opiate receptor sites on T lymphocytes. J Immunol. 1980;125(6):2539–43.
103. Donahoe RM, Bueso-Ramos C, Donahoe F, Madden JJ, Falek A, Nicholson JKA, et al. Mechanistic implications of the findings that opiates and other drugs of abuse moderate T-cell surface receptors and antigenic markers. Ann N Y Acad Sci. 1987;496(1):711–21.
104. Herman JP, Figueiredo H, Mueller NK, Ulrich-Lai Y, Ostrander MM, Choi DC, et al. Central mechanisms of stress integration: hierarchical circuitry controlling hypothalamo-pituitary-adrenocortical responsiveness. Front Neuroendocrinol. 2003;24(3):151–80.
105. Sternberg EM. Neural regulation of innate immunity: a coordinated nonspecific host response to pathogens. Nat Rev Immunol. 2006;6:318–28.
106. Mellon RD, Bayer BM. Evidence for central opioid receptors in the immunomodulatory effects of morphine: review of potential mechanism(s) of action. J Neuroimmunol. 1998;83:19–28.
107. Zubelewicz B, Muc-Wierzgoń M, Harbuz MS, Brodziak A. Central single and chronic administration of morphine stimulates corticosterone and interleukin (IL)-6 in adjuvant-induced arthritis. J Physiol Pharmacol. 2000;51(4):897–906.
108. Gibb FW, Stewart A, Walker BR, Strachan MWJ. Adrenal insufficiency in patients on long-term opioid analgesia. Clin Endocrinol. 2016;85(6):831–5.
109. Walter M, Gerber H, Kuhl HC, Schmid O, Joechle W, Lanz C, et al. Acute effects of intravenous heroin on the hypothalamic-pituitary-adrenal axis response: a controlled trial. J Clin Psychopharmacol. 2013;33(2):193–8.
110. Bakr MAEM, Amr SAER, Mohamed SA, Hamed HB, EL-Rahman AMA, Mostafa MAM, et al. Comparison between the effects of intravenous morphine, tramadol, and ketorolac on stress and immune responses in patients undergoing modified radical mastectomy. Clin J Pain. 2016;32(10):889–97.

111. Moyano J, Aguirre L. Opioids in the immune system: from experimental studies to clinical practice. Rev Assoc Med Bras. 2019;65:262–9.
112. Nallu R, Radhakrishnan R. Spinal release of acetylcholine in response to morphine. J Pain [Internet]. 2007;8(4):S19. Available from: https://linkinghub.elsevier.com/retrieve/pii/S1526590007004543.
113. Hall DM, Suo J-L, Weber RJ. Opioid mediated effects on the immune system: sympathetic nervous system involvement. J Neuroimmunol. 1998;83:29–35.
114. Fecho K, Dykstra LA, Lysle DT. Evidence for beta adrenergic receptor involvement in the immunomodulatory effects of morphine. J Pharmacol Exp Ther. 1993;265:1079–87.
115. Wang H, Yu M, Ochani M, Amelia CA, Tanovic M, Susarla S, et al. Nicotinic acetylcholine receptor α7 subunit is an essential regulator of inflammation. Nature. 2003;421(6921):384–8.
116. Bernik TR, Friedman SG, Ochani M, Diraimo R, Ulloa L, Yang H, et al. Pharmacological stimulation of the cholinergic antiinflammatory pathway [Internet]. J Exp Med. The Rockefeller University Press; 2002 195:781–8. Available from: http://www.jem.org/cgi/content/full/195/6/781.
117. Parrish WR, Rosas-Ballina M, Gallowitsch-Puerta M, Ochani M, Ochani K, Yang LH, et al. Modulation of TNF release by choline requires α7 subunit nicotinic acetylcholine receptor-mediated signaling. Mol Med. 2008;14(9–10):567–74.
118. Fujii T, Mashimo M, Moriwaki Y, Misawa H, Ono S, Horiguchi K, et al. Expression and function of the cholinergic system in immune cells. Front Immunol. Frontiers Media S.A.;. 2017;8:1085.
119. Guzmán-Mejía F, López-Rubalcava C, González-Espinosa C. Stimulation of nAchRα7 receptor inhibits TNF synthesis and secretion in response to LPS treatment of mast cells by targeting ERK1/2 and TACE activation. J Neuroimmune Pharmacol. 2018;13(1):39–52.
120. Derijk RH. Single nucleotide polymorphisms related to HPA axis reactivity. Neuroimmunomodulation. 2009;16:340–52.
121. Bond C, Laforge KS, Tian M, Melia D, Zhang S, Borg L, et al. Single-nucleotide polymorphism in the human mu opioid receptor gene alters-endorphin binding and activity: possible implications for opiate addiction [Internet]. Proc Natl Acad Sci U S A. 1998;95(16):9608–13. Available from: www.pnas.org.
122. Wand GS, Mccaul M, Yang X, Reynolds J, Gotjen D, Lee S, et al. The Mu-opioid receptor gene polymorphism (A118G) alters HPA Axis activation induced by opioid receptor blockade. Neuropsychopharmacology. 2002;26(1):106–14.
123. Hernandez-Avila CA, Covault J, Wand G, Zhang H, Gelernter J, Kranzler HR. Population-specific effects of the Asn40Asp polymorphism at the l-opioid receptor gene (OPRM1) on HPA-axis activation [Internet]. Vol. 17, Pharmacogenetics and Genomics. Wolters Kluwer Health | Lippincott Williams & Wilkins; 2007. Available from: http://www.ncbi.nlm.nih.gov/SNP/.
124. Chong RY, Oswald L, Yang X, Uhart M, Lin PI, Wand GS. The mu-opioid receptor polymorphism A118G predicts cortisol responses to naloxone and stress. Neuropsychopharmacology. 2006;31(1):204–11.
125. Ducat E, Ray B, Bart G, Umemura Y, Varon J, Ho A, et al. Mu-opioid receptor A118G polymorphism in healthy volunteers affects hypothalamic-pituitary-adrenal axis adrenocorticotropic hormone stress response to metyrapone. Addict Biol. 2013;18(2):325–31.
126. Franceschi C, Campisi J. Chronic inflammation (Inflammaging) and its potential contribution to age-associated diseases. J Gerontol Ser A Biol Sci Med Sci. Oxford University Press;. 2014;69:S4–9.
127. Salminen A, Kaarniranta K, Kauppinen A. Immunosenescence: the potential role of myeloid-derived suppressor cells (MDSC) in age-related immune deficiency. Cell Mol Life Sci. Birkhauser Verlag AG;. 2019;76:1901–18.
128. Bauer ME. Stress, glucocorticoids and ageing of the immune system. Stress. 2005;8:69–83.
129. Castelo-Branco C, Soveral I. The immune system and aging: a review. Gynecol Endocrinol. 2014;30:16–22.

130. Murciano C, Villamón E, Yáñez A, O'Connor JE, Gozalbo D, Gil ML. Impaired immune response to Candida albicans in aged mice. J Med Microbiol. 2006;55(12):1649–56.
131. Goldmann O, Lehne S, Medina E. Age-related susceptibility to Streptococcus pyogenes infection in mice: underlying immune dysfunction and strategy to enhance immunity. J Pathol. 2010;220(5):521–9.
132. Inomata M, Xu S, Chandra P, Meydani SN, Takemura G, Philips JA, et al. Macrophage LC3-associated phagocytosis is an immune defense against Streptococcus pneumoniae that diminishes with host aging. Proc Natl Acad Sci U S A. 2020;117(52):33561–9. Available from: www.pnas.org/cgi/doi/10.1073/pnas.2015368117.
133. Pinti M, Appay V, Campisi J, Frasca D, Fülöp T, Sauce D, et al. Aging of the immune system: focus on inflammation and vaccination. Eur J Immunol. Wiley-VCH Verlag;. 2016;46:2286–301.
134. Zhang X, Meng X, Chen Y, Leng SX, Zhang H. The biology of aging and cancer frailty, inflammation, and immunity the biology of aging [internet]. Cancer J. 2017;23(4):201–5. Available from: www.mmaapf.org.
135. Morley JE, Flood JF, Silver AJ. Opioid peptides and aging. Ann N Y Acad Sci. 1990;579(1):123–32.
136. SG B, Kimura T, Stefano JM III, Patrick Finn J, Leung MK, Smith E, et al. Autoimmunomodulation: age-related opioid differences in vertebrate and invertebrate immune systems. Ann N Y Acad Sci. 1992;663(1):396–402.
137. Forget P, Collet V, Lavand'homme P, De Kock M. Does analgesia and condition influence immunity after surgery? Effects of fentanyl, ketamine and clonidine on natural killer activity at different ages. Eur J Anaesthesiol. 2010 Mar;27(3):233–40.
138. Gross-Isseroff R, Dillon KA, Israeli M, Biegon A. Regionally selective increases in opioid receptor density in the brains of suicide victims. Brain Res. 1990;530(2):312–6.
139. González-Maeso J, Torre I, Rodríguez-Puertas R, García-Sevilla JA, Guimón J, Meana JJ. Effects of age, postmortem delay and storage time on receptor-mediated activation of G-proteins in human brain [Internet]. Neuropsychopharmacology. 2002;26:468–78. Available from: www.acnp.org/citations/Npp.
140. Solomon GF, Fiatarone MA, Benton D, Morley JE, Bloom E, Makinodan T. Psychoimmunologic and endorphin function in the aged. Ann N Y Acad Sci. 1988;521(1):43–58.
141. Palm S, Lehzen S, Mignat C, Steinmann J, Leimenstoll G, Maier C. Does prolonged oral treatment with sustained-release Morphine tablets influence immune function? Anesth Analg. 1998;86(1):166–72.
142. Tabellini G, Borsani E, Benassi M, Patrizi O, Ricotta D, Caimi L, et al. Effects of opioid therapy on human natural killer cells. Int Immunopharmacol. 2014;18(1):169–74.
143. Gupta K, Kshirsagar S, Chang L, Schwartz R, Law PY, Yee D, et al. Morphine stimulates angiogenesis by activating proangiogenic and survival-promoting signaling and promotes breast tumor growth. Cancer Res. 2002;62(15):4491–8.
144. Shavit Y, Ben-Eliyahu S, Zeidel A, Beilin B. Effects of fentanyl on natural killer cell activity and on resistance to tumor metastasis in rats: dose and timing study. Neuroimmunomodulation. 2004;11(4):255–60.
145. Koodie L, Ramakrishnan S, Roy S. Morphine suppresses tumor angiogenesis through a HIF-1α/p38MAPK pathway. Am J Pathol [Internet]. 2010;177(2):984–97. Available from: https://doi.org/10.2353/ajpath.2010.090621.
146. Mathew B, Lennon FE, Siegler J, Mirzapoiazova T, Mambetsariev N, Sammani S, et al. The novel role of the mu opioid receptor in lung cancer progression: a laboratory investigation. Anesth Analg. 2011;112(3):558–67.
147. Singh A, Jayanthan A, Farran A, Elwi AN, Kim SW, Farran P, et al. Induction of apoptosis in pediatric acute lymphoblastic leukemia (ALL) cells by the therapeutic opioid methadone and effective synergy with Bcl-2 inhibition. Leuk Res [Internet]. 2011;35(12):1649–57. Available from: https://doi.org/10.1016/j.leukres.2011.06.035.

148. Afsharimani B, Baran J, Watanabe S, Lindner D, Cabot PJ, Parat MO. Morphine and breast tumor metastasis: the role of matrix-degrading enzymes. Clin Exp Metastasis. 2014;31(2):149–58.
149. Friesen C, Hormann I, Roscher M, Fichtner I, Alt A, Hilger R, et al. Opioid receptor activation triggering downregulation of cAMP improves effectiveness of anti-cancer drugs in treatment of glioblastoma. Cell Cycle. 2014;13(10):1560–70.
150. Koodie L, Yuan H, Pumper JA, Yu H, Charboneau R, Ramkrishnan S, et al. Morphine inhibits migration of tumor-infiltrating leukocytes and suppresses angiogenesis associated with tumor growth in mice. Am J Pathol. 2014;184(4):1073–84.
151. Nguyen J, Luk K, Vang D, Soto W, Vincent L, Robiner S, et al. Morphine stimulates cancer progression and mast cell activation and impairs survival in transgenic mice with breast cancer. Br J Anaesth. 2014;113(1):4–13.
152. Lennon FE, Mirzapoiazova T, Mambetsariev B, Poroyko VA, Salgia R, Moss J, et al. The Mu opioid receptor promotes opioid and growth factor-induced proliferation, migration and Epithelial Mesenchymal Transition (EMT) in human lung cancer. PLoS One. 2014;9(3):1–13.
153. Khabbazi S, Goumon Y, Parat MO. Morphine modulates interleukin-4- or breast cancer cell-induced pro-metastatic activation of macrophages. Sci Rep [Internet]. 2015;5:1–12. Available from: https://doi.org/10.1038/srep11389.
154. Khabbazi S, Nassar ZD, Goumon Y, Parat MO. Morphine decreases the pro-angiogenic interaction between breast cancer cells and macrophages in vitro. Sci Rep [Internet]. 2016;6:1–10. Available from: https://doi.org/10.1038/srep31572.
155. Kua VMD, Rasul A, Sreenivasan S, Rasool B, Younis T, Lai NS. Methadone hydrochloride and leukemia cells: effects on cell viability, DNA fragmentation and apoptotic proteins expression level. Pak J Pharm Sci. 2019;32(4):1797–803.
156. Wang K, Wang J, Liu T, Yu W, Dong N, Zhang C, et al. Morphine-3-glucuronide upregulates PD-L1 expression via TLR4 and promotes the immune escape of non-small cell lung cancer. Cancer Biol Med. 2021;18(1):155–71.
157. Khabbazi S, Nassar ZD, Goumon Y, Parat MO. Morphine decreases the pro-angiogenic interaction between breast cancer cells and macrophages in vitro. Sci Rep. 2016;6:1–10.
158. Khabbazi S, Goumon Y, Parat MO. Morphine modulates interleukin-4- or breast cancer cell-induced pro-metastatic activation of macrophages. Sci Rep. 2015;5(June):1–12.
159. Koodie L, Ramakrishnan S, Roy S. Morphine suppresses tumor angiogenesis through a HIF-1α/p38MAPK pathway. Am J Pathol. 2010;177(2):984–97.
160. Afsharimani B, Cabot P, Parat MO. Morphine and tumor growth and metastasis. Cancer Metastasis Rev. 2011;30(2):225–38.
161. Smith JP, Bingaman SI, Mauger DT, Harvey HH, Laurence M, Zagon IS. Patients with advanced pancreatic. Cancer. 2010;2010(2):37–48.
162. Singh A, Jayanthan A, Farran A, Elwi AN, Kim SW, Farran P, et al. Induction of apoptosis in pediatric acute lymphoblastic leukemia (ALL) cells by the therapeutic opioid methadone and effective synergy with Bcl-2 inhibition. Leuk Res. 2011;35(12):1649–57.
163. Riou B, Lennon FE, Moss J, Singleton PA. Clinical concepts and commentary the-opioid receptor in cancer progression is there a direct effect? Anesthesiology. 2012;116:940–5.
164. Gordon JI. Honor thy gut symbionts redux. Science. 2012;336:1251–3.
165. Kau AL, Ahern PP, Griffin NW, Goodman AL, Gordon JI. Human nutrition, the gut microbiome and the immune system. Nature. 2011;474(7351):327–36.
166. Brown EM, Sadarangani M, Finlay BB. The role of the immune system in governing host-microbe interactions in the intestine. Nat Immunol. 2013;14:660–7.
167. Sommer F, Bäckhed F. The gut microbiota-masters of host development and physiology. Nat Rev Microbiol. 2013;11:227–38.
168. Rogers DF, Barnes PJ. Opioid inhibition of neurally mediated mucus secretion in human bronchi. Lancet. 1989;333(8644):930–2.

169. Brosnahan AJ, Jones BJ, Dvorak CM, Brown DR. Morphine attenuates apically-directed cytokine secretion from intestinal epithelial cells in response to enteric pathogens. Pathogens. 2014;3(2)
170. Harari Y, Weisbrodt NW, Moody FG. The effect of morphine on mast cell-mediated mucosal permeability. Surgery. 2006;139(1)
171. Mora AL, Salazar M, Pablo-Caeiro J, Frost CP, Yadav Y, Dupont HL, et al. Moderate to high use of opioid analgesics are associated with an increased risk of Clostridium difficile infection. Am J Med Sci. 2012;343(4)
172. Babrowski T, Holbrook C, Moss J, Gottlieb L, Valuckaite V, Zaborin A, et al. Pseudomonas aeruginosa virulence expression is directly activated by morphine and is capable of causing lethal gut-derived sepsis in mice during chronic morphine administration. Ann Surg. 2012;255(2)
173. Meng J, Yu H, Ma J, Wang J, Banerjee S, Charboneau R, et al. Morphine induces bacterial translocation in mice by compromising intestinal barrier function in a TLR-dependent manner. PLoS One. 2013;8(1)
174. Smith JP, Bingaman SI, Ruggiero F, Mauger DT, Mukherjee A, McGovern CO, et al. Therapy with the opioid antagonist naltrexone promotes mucosal healing in active Crohn's disease: a randomized placebo-controlled trial. Dig Dis Sci. 2011;56(7)
175. Banerjee S, Sindberg G, Wang F, Meng J, Sharma U, Zhang L, et al. Opioid-induced gut microbial disruption and bile dysregulation leads to gut barrier compromise and sustained systemic inflammation. Mucosal Immunol. 2016;9(6)
176. Hawksworth G, Drasar BS, Hill MJ. Intestinal bacteria and the hydrolysis of glycosidic bonds. J Med Microbiol. 1971;4(4)
177. Walsh CT, Levine RR. Studies of the enterohepatic circulation of morphine in the rat. J Pharmacol Exp Ther. 1975;195(2)
178. Hilburger ME, Adler MW, Truant AL, Meissler JJ, Satishchandran V, Rogers TJ, et al. Morphine induces sepsis in mice. J Infect Dis. 1997;176(1)
179. Banerjee S, Meng J, Das S, Krishnan A, Haworth J, Charboneau R, et al. Morphine induced exacerbation of sepsis is mediated by tempering endotoxin tolerance through modulation of miR-146a. Sci Rep. 2013;3:1977. Available from: www.nature.com/scientificreports,
180. Melchior C, Desprez C, Wuestenberghs F, Leroi AM, Lemaire A, Goucerol G. Impact of opioid consumption in patients with functional gastrointestinal disorders. Front Pharmacol. 2020;11:596467.
181. Del Rio C, Malani PN. Novel coronavirus – important information for clinicians. JAMA. 2019, 2020;323(11):1039–40.
182. Henry BM, Vikse J. Clinical characteristics of Covid-19 in China. N Engl J Med. 2020;382(19)
183. Ye Q, Wang B, Mao J. The pathogenesis and treatment of the Cytokine Storm in COVID-19. J Infect. 2020;80(6):607–13.
184. Sanders JM, Monogue ML, Jodlowski TZ, Cutrell JB. Pharmacologic treatments for coronavirus disease 2019 (COVID-19): a review. JAMA. 2020;323:1039–40.
185. Hudzik B, Nowak J, Zubelewicz-Szkodzinska B. Consideration of immunomodulatory actions of morphine in COVID-19 – short report. Eur Rev Med Pharmacol Sci. 2021;25(24):13062–4.
186. Poston JT, Patel BK, Davis AM. Management of critically ill adults with COVID-19. JAMA. 2020;323:1839–41.
187. Ataei M, Shirazi FM, Lamarine RJ, Nakhaee S, Mehrpour O. A double-edged sword of using opioids and COVID-19: a toxicological view. Subst Abuse Treat Prev Policy. 2020;15:1–4.
188. Esch T, Kream RM, Stefano GB. Emerging regulatory roles of opioid peptides, endogenous morphine, and opioid receptor subtypes in immunomodulatory processes: metabolic, behavioral, and evolutionary perspectives. Immunol Lett. Elsevier B.V;. 2020;227:28–33.

Chapter 13
Opioid Dependence, Tolerance, and Withdrawal

César J. Carranza-Aguilar, María Teresa Rivera-García, and Silvia L. Cruz

Abstract This chapter describes the main adaptive changes occurring after repeated opioid administration at different levels: behavioral, pharmacological, cellular, and molecular. The first part provides an overview of opioid dependence characteristics and tools for its diagnosis in humans. It also defines the concepts of tolerance, physical dependence, and withdrawal and explains some animal models to study opioid dependence. The second section includes updated information about brain regions involved in addiction, the development of tolerance and withdrawal, and the associated adaptive changes at the intracellular level. The chapter also reviews other mechanisms involved in the long-term effects of opioids, such as neuroplasticity, changes in neurotransmitter release, activation of anti-opioid systems, and neuroinflammation. Finally, it presents clinical and preclinical strategies to reduce the undesirable side effects of opioids.

Keywords Opioid dependence · Tolerance · Withdrawal · COWS · Neuroadaptations

C. J. Carranza-Aguilar · M. T. Rivera-García
Department of Pharmacobiology, Center for Research and Advanced Studies (Cinvestav), IPN, Mexico City, Mexico

S. L. Cruz (✉)
Department of Pharmacobiology, Center for Research and Advanced Studies (Cinvestav), IPN, Mexico City, Mexico

Opioids Working Group, Global Studies Seminar, Faculty of Medicine, National Autonomous University of Mexico, Mexico City, Mexico
e-mail: slcruzm@cinvestav.mx

S. L. Cruz (ed.), *Opioids*, https://doi.org/10.1007/978-3-031-09936-6_13

13.1 General Concepts

13.1.1 Opioid Dependence (Addiction)

Opioids are potent psychoactive drugs and effective pharmacological tools for treating acute and cancer pain (see Chap. 10); however, poor efficacy and the risk of adverse effects limit their use in chronic non-cancer pain management (see Chap. 11). In addition, opioids can produce dependence (addiction) and tolerance when administered chronically or without medical supervision. In the last decades, opioid use disorders (OUDs) have increased until developing into an epidemic, causing a severe economic impact and decreasing the population's life expectancy in the United States (see Chap. 5).

OUDs are more prevalent in men than in women. However, some reports suggest that women are more vulnerable to opioid effects since they develop addiction faster, experience worse withdrawal symptoms, and have a higher rate of comorbidity with other psychiatric illnesses than men. Furthermore, women have a greater risk of developing chronic pain conditions and are twice as likely to take prescription opioids than men [1].

The International Classification of Diseases (ICD-11) defines opioid dependence as: "a disorder of regulation of opioid use arising from repeated or continuous use of opioids. The characteristic feature is a strong internal drive to use opioids, which is manifested by impaired ability to control use, increasing priority given to use over other activities, and persistence of use despite harm or negative consequences. These experiences are often accompanied by a subjective sensation of urge or craving to use opioids. Physiological features of dependence may also be present, including tolerance to the effects of opioids, withdrawal symptoms following cessation or reduction in the use of opioids, or repeated use of opioids or pharmacologically similar substances to prevent or alleviate withdrawal symptoms. The features of dependence are usually evident over a period of at least 12 months, but the diagnosis may be made if opioid use is continuous (daily or almost daily) for at least 3 months."

The most recent edition of the *Diagnostic Statistical Manual of Psychiatric Disorders (DSM 5.0)* includes ten criteria for diagnosing OUDs:

- Taking opioids in larger amounts or over more extended periods than intended.
- Persistent desire or unsuccessful efforts to cut down or control opioid use.
- Spending a great deal of time in activities necessary to obtain or use the opioid.
- Craving or a strong desire to use opioids.
- Failure to fulfill significant role obligations at work, school, or home due to recurrent opioid use.
- Giving up or reducing important social, occupational, or recreational activities due to opioid use.
- Recurrent opioid use in physically hazardous situations.

- Continued opioid use despite knowing or having persistent or recurrent physical or psychological problems associated with opioids.
- Tolerance.
- Withdrawal.

The disorder is mild with 2–3 symptoms, moderate with 4–5, and severe when the patient fulfills six or more criteria [2].

The vulnerability to develop opioid dependence is related to several factors, including genetics, age of initiation, adverse social environments, and psychiatric comorbidities. Also, the use of specific routes of administration contributes to the faster development of OUD. For example, injecting opioids allows a quicker uptake into the brain, enhancing their dependence potential [3, 4].

Addiction is a chronic, relapsing disorder characterized by an uncontrolled craving and compulsive drug-seeking behavior.

13.1.2 Tolerance and Opioid-Induced Hyperalgesia

Tolerance is a decrease in the effects of a particular drug after repeated or prolonged administration [5]. The development of tolerance to opioid analgesic effects limits their therapeutic use. For example, patients with chronic pain can need up to tenfold the initial opioid dose to obtain the same effect, increasing the risk of addiction, respiratory depression, and overdose [6]. Similarly, heroin-dependent subjects can take amounts of heroin that would be fatal for non-opioid users.

Due to tolerance development, opioid users increase the drug dose to reach the desirable effects, initiating an escalating path that does not always provide relief but increases the probability of experiencing adverse effects [6]. Tolerance for other opioid-induced actions occurs with different time courses. Tolerance for euphoria and analgesia develops first, followed by tolerance to nausea and vomiting. In contrast, tolerance for constipation is slow and partial; therefore, decreased gastrointestinal motility persists among patients with chronic pain conditions. On the other hand, miosis (pinpoint pupil), a typical sign of heavy opiate use, never disappears [7].

Although opioids are effective analgesics, under certain circumstances, they can produce hyperalgesia, i.e., increased sensitivity to painful stimuli as a result of prolonged administration. Although the outcome (an increase in pain sensitivity) is technically the same when considering the typical tolerance to opioids' analgesic effects, there are essential differences between the two processes [5]. For example, opioid-induced hyperalgesia (OIH), also known as acute tolerance, can occur within a few hours after starting opioid administration. In contrast, opioid tolerance develops over months or years. Besides, OIH does not improve with a higher opioid dose, which can be a problem during perioperative procedures. The exact mechanism of this paradoxical effect is unclear but likely multifactorial, with several molecular targets and neurobiological circuits proposed to be included in this phenomenon [5].

13.1.3 *Physical Dependence and Withdrawal*

Physical dependence is not synonymous with addiction (or dependence). Instead, this condition refers to the adaptative physiological changes occurring due to continuous exposure to drugs as an attempt to restore homeostasis. These changes are usually opposite to those produced by the acute drug administration and become evident when the drug is not present. Because of this, excitatory withdrawal signs appear in chronic opioid users after abrupt opioid cessation. The withdrawal syndrome causes significant discomfort, including myalgia, joint pain, lacrimation, rhinorrhea, sneezing, fatigue, dysphoria, fever, and piloerection (gooseflesh). In addition, stomach cramps, diarrhea, sweating, elevated heart rate, increased blood pressure, irritability, insomnia, and persistent yawning are common [8].

Some reports indicate that after abrupt heroin or morphine cessation, the withdrawal syndrome develops within a day and lasts over a week, declining its severity after 10 days. However, dysphoria and anhedonia can persist for longer [7].

The severity of withdrawal syndrome depends on the chronicity of opioid use, the amount consumed, individual variability, and drug characteristics; usually, more potent and short-acting drugs (e.g., fentanyl and heroin) induce more severe withdrawal symptoms. In this sense, opioid users continue taking the drug to avoid withdrawal symptoms, perpetuating the addiction cycle [9].

13.1.3.1 The Clinical Opiate Withdrawal Scale (COWS)

COWS helps clinicians rate the severity of opioid withdrawal by assigning a number from zero to five depending on the magnitude of each of the following signs or symptoms: resting pulse rate, sweating, restlessness, pupil size, bone or joint aches, runny nose or tearing, gastrointestinal upset over the 30 minutes before applying the questionnaire, tremor when outstretching hands, yawning, anxiety or irritability, and gooseflesh skin. For example, for pulse rate 0 = 80 beats per minute (bpm) or below; 1; 2 = 101–120; and 4 = greater than 120. The COWS questionnaire has a similar rating system for the other ten common withdrawal signs to get a total score that is the sum of the 11 items. Withdrawal is mild with a score = 5–12; moderate = 13–24; moderately severe = 25–36; and severe with a score higher than 36 [10].

Withdrawal symptoms are not life-threatening except for a fetus of a mother with OUD. Different medication-assisted treatment programs with either buprenorphine or methadone are recommended as a first-line option to improve maternal and neonatal outcomes [11] (Chap. 14).

Usually, acute withdrawal begins within hours or days after opioid discontinuation and resolves in 4–10 days for most opioids. However, methadone withdrawal may last up to 3 weeks. Protracted (extended, chronic, or late) withdrawal refers to signs that drug users experience after the acute withdrawal stage and may persist for many weeks or months. Some symptoms experienced during protracted abstinence from opioids are mood disturbances (anxiety, depression), sleep disturbances, irritability, persistent fatigue, and deficits in some executive functions [12].

13.1.3.2 The Neonatal Abstinence Syndrome (NAS) and Finnegan Scoring System

NAS consists of signs and symptoms involving disturbances at several physiological systems in infants from mothers with substance use disorders. Opioid drugs are small and lipophilic and, therefore, cross the placenta and the blood-brain barrier of the fetus quickly, causing adaptations at several neurotransmitter systems that produce hyperexcitability when the baby ceases to be in contact with the drug.

The *Finnegan neonatal abstinence scoring system* is the most commonly used withdrawal scale for neonates exposed to drugs in utero. It helps clinicians to make decisions about babies' management after birth [13]. The signs of neonatal abstinence syndrome include irritability, tremors, high-pitched crying, and feeding problems. In addition, newborns can present hyperactive reflexes, diarrhea, excessive sucking, increased muscle tone, sneezing, yawning, vomiting, fever, restlessness, short periods of sleep, and slow body weight gain. A number from 1 to 5 is assigned to each withdrawal sign depending on its intensity or frequency. It is recommended to record the first abstinence score 2 hours after birth and then every 4 or 2 hours, depending on each case [14]. Between 50% and 70% of affected neonates need pharmacological treatment; however, specific treatments depend on the drug (or drugs) used during the prenatal period and the overall newborn health [11, 15].

13.2 Animal Models to Study Opioid Dependence Liability

Drug dependence relies on a strong association between the effects of the drug itself, different stimuli from the environment, and the behavior [4]. These factors can be studied in well-characterized animal models, some of which are briefly described here.

Drug self-administration in animals allows the identification of drugs with abuse liability in humans [3]. The key feature in this model is that the animal controls drug intake. In these studies, animals are trained to perform a specific behavior, such as pressing a lever or doing a nose poke, to receive a drug infusion through an intravenous catheter [16]. Continuous reinforcement is the most straightforward and most commonly used paradigm. In this model, the experimenter establishes the ratio, meaning the amount of work needed to obtain a reinforcer. For example, under a fixed ratio 1 schedule, animals receive a single drug infusion after every active behavior (e.g., a lever press or nose poke) [3], and the number of responses provides an estimate of the drug-rewarding effects [3, 16].

All opioid agonists produce self-administration, reflecting their high addiction liability, but fentanyl is more potent to induce this behavior than oxycodone, heroin, and morphine.

The progressive ratio (PR) task is a different but complementary approach to assess both the effectiveness of a reinforcer, and the animals' motivation to take the drug is the progressive-ratio (PR) task [17]. Under this schedule, the number of active responses needed to have a drug infusion increases exponentially until the animal stops performing the behavior. Thus, the maximum number of responses is the breakpoint and directly reflects the animal's motivational state and the magnitude of the drug rewarding effects [3, 17].

As described elsewhere, the PR approach is commonly used to study the rewarding mechanisms of new opioid-like drugs or opioids combined with other drugs [3, 17]. Interestingly, despite their high addictive potential, some studies suggest that opioids have a lower breakpoint value than psychostimulants [17].

The self-administration model also allows testing for relapse. After self-administration, it is possible to extinguish drug-seeking behavior after several sessions during which active responses do not result in drug administration. Once the extinction is stable, drug-seeking behavior can be reinstated either by administering the drug itself (drug-induced reinstatement) or by presenting different cues previously paired with the infusion of the drug (cue-induced reinstatement). Opioids induce drug-seeking during abstinence and both cue and drug-induced relapse [3].

The drug discrimination (DD) test is a paradigm used to determine how similar a new compound is compared to others already known. During this assay, food-deprived animals are placed into two-lever chambers and trained to press a lever to receive food. Thus, hunger is the motivational state in this test, and food is the reward for performing the behavior. The reinforcement to press a lever is a food pellet. The response is correct if the animal presses the lever associated with the drug or the one associated with the vehicle (usually saline solution) after receiving the drug or the vehicle, respectively. After several training sessions, rodents associate the correct lever side administration because they can accurately identify the interoceptive cues induced by the drug.

During the discriminative test, rats perform active lever presses after the novel drug administration. If the new drug has a similar pharmacologic profile to the training drug, the animals will press the drug-associated lever instead of the one associated with the vehicle [18]. The DD approach is a powerful tool to predict the potentially harmful effects of the new compounds in the illegal market.

The conditioned place preference (CPP) test is a commonly used paradigm to study drug dependence liability [19]. In contrast to self-administration, this test relies on a passive association between the drug effects and the experimental stimuli. A typical CPP paradigm consists of three phases: preconditioning, conditioning, and testing sessions.

The preconditioning phase evaluates the animal's preference for one of two environments with different visual and tactile contextual cues. The conditioning part involves pairing the drug effects with a specific environment and the drug vehicle with a different one. After several conditioning pairings, the experimenter evaluates the rewarding drug properties during a test session by allowing the animal to freely move between the two environments in the absence of acute drug effects. The amount of time spent in the drug-paired environment directly reflects the drug rewarding properties [19, 20]. Although a single pairing with morphine induces conditioned place preference, the response increases with the number of conditioning trials [21].

Since a particular environment could be rewarding, some researchers use the called biased approach of conditioning. In this variant, the animal is placed in the less preferred environment after drug administration. Regardless of using a biased or unbiased conditioning approach, opioids reliably induce CPP [20, 21]. In the CPP paradigm is also possible to extinguish the association between the context and drug effects. Extinction occurs when animals receive saline in every trial, decreasing the time spent in the drug-paired side. After extinction, it is also possible to reinstate the rewarding memory of the drug, which usually occurs, giving a single challenge drug dose, lower than the one used for conditioning [20].

Animal models have helped understand the neurobiological aspects of opioid dependence and withdrawal. The abstinence response in animals occurs after the abrupt cessation of opioid administration (as it usually occurs in humans) or after administering an opioid receptor antagonist to counteract the opioid agonist's effects. Antagonist-precipitated abstinence has a shorter time to peak, lasts less, and is more intense than spontaneous withdrawal. Precipitated withdrawal depends not only on the agonist used and on the exposure time but also on the antagonist dose, with more intense responses elicited with higher opioid antagonist doses.

Withdrawal signs include defecation/diarrhea, weight loss, salivation, lacrimation, rhinorrhea, jumping, burrowing, wet-dog shakes, hyperactivity, vocalization, teeth chatter, piloerection, decreased rectal temperature, pain threshold, vocalizing on touch, and ptosis. These signs can be grouped as behavioral (e.g., wet-dog shakes and jumping), physiological (e.g., hypothermia and weight loss), somatic (e.g., salivation and rhinorrhea), aversive responses (e.g., diminished pain threshold, vocalizing on touch), or total signs. Withdrawal scales are methods for quantifying opioid withdrawal intensity. In these scales, individual signs, groups of signs, and total signs receive a score number over an observation period that varies from minutes to hours, depending on the type of withdrawal studied (spontaneous or precipitated). Thus, the number assigned depends on the occurrence of some signs and the mean number of events observed during a given time.

13.3 Adaptations Resulting from Repeated Opioid Administration

Prolonged or repeated exposure to exogenous opioids causes allostatic neuroadaptations that intend to maintain the cells in suitable conditions for their survival. Allostasis refers to "a superordinate system by which stability is achieved through change" [22]. Under physiological conditions, opioid receptors located at neurons bind endogenous opioid peptides, responding briefly to a specific need. Prolonged exposure to exogenous opioids attenuates the pharmacological response to subsequent opioid administration resulting in an apparent "normal" functionality and is responsible for the hyperexcitability observed during withdrawal. The following sections describe the main changes produced by repeated opioid exposure at the cellular level and in specific brain pathways.

13.3.1 Changes at the Cellular Level

Compensatory mechanisms at the cellular level include (a) adaptations in μ-opioid receptor signaling and (b) opioid receptor desensitization and internalization. The following sections explain each of these mechanisms (Fig. 13.1).

13.3.1.1 Receptor Signaling

AC, cAMP, PKA, and CREB pathway As described in Chap. 9, opioids bind to specific receptors and inhibit the AC enzyme, reducing cAMP and protein kinase A (PKA) levels and cAMP response element-binding (CREB) activity. Chronic opioid administration in vivo or persistent opioid exposure in vitro (ranging from hours to days) hypertrophies this signaling pathway by synthesizing new AC isoforms and increasing cAMP levels (cAMP overshoot) and PKA activity. These events promote the transcription of new proteins by increasing CREB activation contributing to opioid tolerance. The opioid-induced cAMP increase can also occur by switching the coupling of inhibitory (Gi/o) proteins to excitatory (Gs) proteins after chronic opioid treatment [23]. Furthermore, the pharmacological inhibition of AC, PKA, or CREB reduces antinociceptive tolerance to morphine in preclinical models.

Calcium and potassium channels Chronic morphine treatment reduces GIRK channels expression and increases P/Q- and L-type calcium channel currents. Accordingly, in animal models, pharmacological inhibition of calcium channels reduces morphine tolerance and dependence. Also, activation of potassium channels by selective agonists has been suggested as a promising strategy against opioid tolerance [24].

GRKs and β-arrestin G protein-coupled receptor kinases (GRKs) phosphorylate opioid receptors at the serine/threonine residues and prepare them for β-arrestin binding, which ends G protein signaling and promotes opioid receptors endocytosis. Once internalized, opioid receptors are transported to lysosomes for degradation or recycled to the cell membrane (see Chap. 9). The fate of opioid receptors depends on the receptor subtype and the conformational changes produced by specific ago-

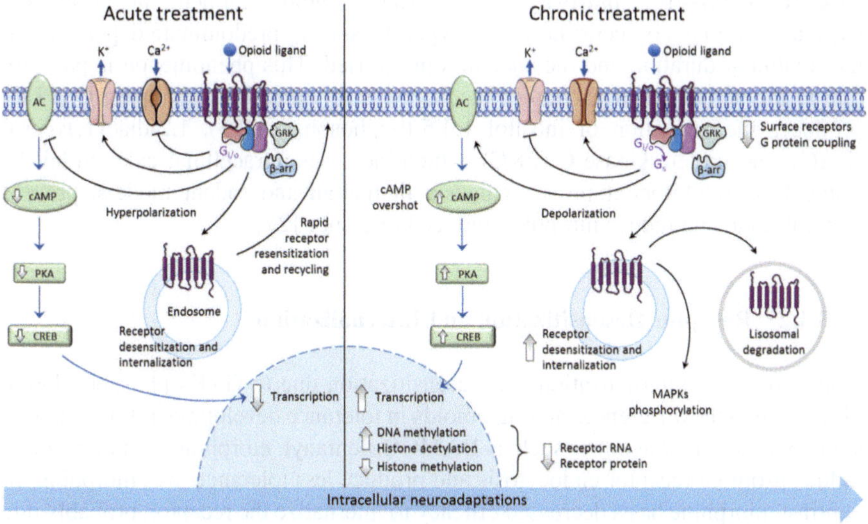

Fig. 13.1 Cellular neuroadaptations produced by opioids. The left side shows the main signals evoked after acute opioid exposure. Blockade of Ca^{2+} channels and increased K^+ currents produce neuronal hyperpolarization. Inhibition of the AC-cAMP-PKA pathway reduces the transcriptional activity of CREB. In addition, the GRK and β-arrestin proteins induce desensitization, internalization, and a rapid resensitization and recycling of opioid receptors. The right side shows main neuroadaptations produced after chronic opioid treatments such as increased Ca^{2+} currents and blockade of K^+ channels. Furthermore, AC enzyme activation produces an excess of cAMP levels and an increase in CREB-dependent transcriptional activity. Persistent desensitization and internalization of the opioid receptors can result in lysosomal degradation. Once internalized, opioid receptors induce endocytic signals such as the activation of MAPKs. Additionally, prolonged exposure to opioids produces epigenetic changes such as reduced histone methylation and increased DNA methylation and histone acetylation that result in reduced opioid receptor expression (RNA and protein). Ca^{2+}, calcium; K^+, potassium; Gi/o, inhibitory G protein; Gs, excitatory G protein; GRK, G protein-coupled receptor kinases; β-arr, β-arrestin; AC, adenylyl cyclase; cAMP, cyclic adenosine monophosphate; PKA, protein kinase A; CREB, cAMP response element-binding; DNA, deoxyribonucleic acid; RNA, ribonucleic acid; MAPK, mitogen-activated protein kinase

nists after binding [25]. Interestingly, mice lacking β-arrestin2 have enhanced analgesic function and decreased morphine tolerance.

Mitogen-activated protein kinases (MAPKs) The extracellular signal-regulated kinase (ERK), Jun N-terminal kinase (JNK), and p38 kinase activate transcription factors that control gene expression. Increased phosphorylation of p38, ERK, and JNK contributes to tolerance by inducing the expression of calcitonin gene-related peptide (CGRP), substance P (SP), and CREB. Moreover, the administration of specific inhibitors of p38, ERK, and JNK attenuates the tolerance to morphine antinociceptive effects in rats [26].

PLC-IP3-DAG-PKC pathway After chronic opioid treatment, analgesia and hyperalgesia may co-occur; however, hyperalgesia can predominate depending on the treatment duration and the specific opioid used. This phenomenon is partially due to the activation of excitatory mechanisms in which phospholipase C (PLC) produces the activation of inositol 1,4,5-trisphosphate (IP3), 1,2-diacylglycerol (DAG), and protein kinase C (PKC), which increases intracellular calcium levels. Some PLC inhibitors improve analgesia in human and rodent models, prevent hyperalgesia, and reduce morphine-induced tolerance [27].

13.3.1.2 Receptor Desensitization and Internalization

Opioid receptor internalization and desensitization due to GRKs phosphorylation play a role in the differences among opioids in tolerance development. For example, some biased opioid agonists such as DAMGO, fentanyl, etorphine, and methadone induce μ-opioid receptor endocytosis and produce less tolerance than morphine. In contrast, morphine has a decreased efficacy to internalize the receptor, probably due to its inability to induce receptor conformational changes that serve as recognition sites for GRKs [7, 24, 28].

Some authors have suggested that opioid receptor endocytosis diminishes tolerance by inducing rapid receptor resensitization and recycling. In this sense, the relative activity versus endocytosis (RAVE) index has been proposed to determine tolerance liability. For example, morphine has a high RAVE index value due to its ability to induce intracellular modifications but low ability to promote opioid receptors' internalization. In contrast, DAMGO, fentanyl, and etorphine promote rapid opioid receptor internalization and have a low RAVE value and low potential for developing tolerance. Although the RAVE index has been accepted for many years and is a theoretical framework that can explain various experimental findings, several studies show that this index differs not only depending on the type of opioid used but also on the duration of treatment and the effect evaluated. For example, a study suggested that the RAVE values predict the liability of some opioids to develop addiction but not physical dependence or cAMP overshoot [29].

13.3.1.3 Structural and Functional Changes in Opioid Receptors

R* opioid receptors overexpression Similar to other GPCRs, free opioid receptors exist in two conditions, in an inactive state (R) and in an active state (R*), already signaling (i.e., with constitutive activity). Under physiological conditions, R predominates. However, repeated opioid exposure promotes R* overexpression. The high concentration of constitutively active μ- and δ-opioid receptors plays a role in tolerance development and withdrawal because R* signal in the absence of an agonist, promoting regulatory neuroadaptations [30]. Furthermore, the constitutive activity of opioid receptors may contribute to tolerance by facilitating receptor desensitization and internalization.

Splice variants Alternative splicing produces structural variations at the C-terminal domain of opioid receptors that can alter intracellular mechanisms related to opioid tolerance (see Chap. 9). Preclinical studies have demonstrated that structural variations in the C-terminal tail of the μ-opioid receptor change the pharmacological profile of morphine and alter its ability to develop tolerance and physical dependence. Furthermore, truncated 6-transmembrane and single-transmembrane variants of the μ-opioid receptor are involved in their expression and traffic. These mechanisms could explain the high individual variability among users to develop side effects after prolonged opioid administration [28].

13.3.2 Changes in Brain Circuitries

13.3.2.1 Opioid Dependence and the Mesolimbic System

All drugs produce rewarding effects by activating the mesocorticolimbic dopaminergic system, promoting behavior repetition. This system includes the ventral tegmental area (VTA), from which myelinated dopaminergic fibers innervate brain regions involved in executive, affective, and motivational functions, such as the prefrontal cortex (PFC), amygdala, and nucleus accumbens (NAc). Unlike psychostimulant drugs (e.g., cocaine, methamphetamine), which act directly at dopaminergic neurons, opioids act on opioid receptors located at GABAergic neurons that exert a tonic inhibition on dopaminergic neurons in the VTA. The activation of these opioid receptors inhibits GABA release preventing the tonic inhibitory actions on dopaminergic neurons, which results in DA release in the NAc, dorsal striatum, and prefrontal cortex.

Persistent activation of the reward system causes neuronal adaptations that ultimately diminish opioids rewarding effects and their actions at other neurochemical systems. For example, chronic opioid administration dysregulates the hypothalamic-pituitary-adrenal (HPA) axis involved in the stress response. Specifically, while acute morphine administration stimulates the HPA axis and releases corticosterone in rodents, high doses or chronic opioid administration inhibits glucocorticoid release. During acute withdrawal, there is an increase in the release of the corticotropin-releasing factor, adrenocorticotropic hormone, and cortisol, which are the typical hallmarks of the stress response [9].

13.3.2.2 Opioid Withdrawal, the Locus Coeruleus, and Periaqueductal Gray (PAG)

The abstinence syndrome unmasks the adaptive changes occurring at different neurotransmitter systems due to repeated or prolonged opioid administration. Abstinence manifestations are directly related to changes in the locus coeruleus (LC). The LC (blue spot in the brainstem) contains noradrenergic neurons with projections to other brain areas. This nucleus regulates wakefulness, breathing, blood pressure, and general alertness, among other functions. When opioids bind to receptors located in the LC, they hyperpolarize neurons (see Chap. 9), reducing noradrenaline (NA) release. This inhibition produces drowsiness, slowed respiration, and orthostatic hypotension (a decrease in blood pressure when standing up), characteristic of opioid intoxication (see Chap. 8). Repeated opioid exposure produces cellular changes that turn adrenergic neurons hyperexcitable [7]. As a result, opioid discontinuation results in sweating, increased blood pressure, and tachycardia. Pharmacological inhibition of AC activity with the α_2-adrenergic receptor agonists clonidine or lofexidine diminishes most of these signs (see Chap. 15).

13.3.2.3 Tolerance, the Spinal Cord, Rostral Ventromedial Medulla, and PAG

The brain regions involved in the descending pain modulatory pathway are the PAG, the rostral ventromedial medulla (RVM), and the dorsal horn of the spinal cord (see Chap. 10). Activation of this pathway inhibits nociception and plays a role in tolerance development. Under basal conditions, the GABAergic neurons within the PAG tonically release GABA into the RVM. Acute opioid administration activates postsynaptic opioid receptors and G-protein inwardly rectifying potassium (GIRK) channels via Gα proteins resulting in potassium release and neuronal hyperpolarization (see Chap. 9). In addition, opioids bind to presynaptic μ-opioid receptors inhibiting voltage-dependent calcium currents and activating voltage-dependent potassium channels (K_v). These mechanisms result in a GABA release blockage and, therefore, suppression of the inhibition of the PAG neurons projecting to the RVM.

After repeated opioid administration, there are changes in the properties of GABAergic neurons in the PAG, including the uncoupling of the μ-opioid receptor from G protein-mediated effects on GIRK channels and adenylyl cyclase (AC). These changes produce an upregulation of cyclic adenosine monophosphate (cAMP) at the postsynapsis. On the other hand, at the presynaptic level, there is a blockade of calcium channel-mediated inhibition and K_v channel-mediated activation. Therefore, μ-opioid receptor activation no longer suppresses GABA release, contributing to opioid tolerance [31].

In addition to the mechanisms mentioned above, once tolerance has developed, there is a reduction in dopamine release in the NAc, which may contribute to a lack of pleasure in previously rewarding activities in patients with OUDs. Although not completely understood, changes in GABAergic and glutamatergic function at the VTA, as well as molecular adaptations in the LC and PAG, also underlie craving and compulsive drug use [25, 32].

13.3.3 Other Adaptive Changes

In addition to cellular and brain circuitry changes, other processes contribute to opioid tolerance, hyperalgesia, physical dependence, and addiction. These include changes in neuronal plasticity, alterations in neurotransmitter release, activation of anti-opioid systems, and the influence of glial cells.

13.3.3.1 Neural Plasticity

Neuroplasticity is the ability of neurons to modify their structure and functions in response to experience, injury, and intrinsic or extrinsic stimuli. Sustained opioid activity can produce plastic changes such as neurogenesis, neuronal connections generation, the formation or elimination of synapses, and cell death (Fig. 13.2). Neuroplastic changes induced by opioids are brain region-specific and depend on age, duration of drug treatment, and the drug used.

Neurogenesis is a continuous progressive process that involves the proliferation, differentiation, and survival of neurons. Adult neurogenesis in the dentate gyrus of the hippocampus contributes to opioid reward and relapse. In various preclinical paradigms, prolonged morphine, heroin, or buprenorphine administration decreases adult neurons' proliferation and differentiation at the hippocampus. Conversely, multiple fentanyl injections increase proliferation, whereas buprenorphine increases the survival of hippocampal neurons. Interestingly, methadone does not affect neurogenesis in rats. Positive regulators of neurogenesis, such as exercise and an enriched environment, diminish morphine and heroin rewarding effects [33].

Synaptic remodeling The reduction of dendritic density and complexity is a phenomenon related to addiction. In vitro studies have demonstrated that chronic morphine causes the loss of neuronal dendritic spines in the hippocampus, PFC, and NAc [34]. Furthermore, morphine exposure increases autophagy in cultured hippocampal neurons, a necessary process for synaptic pruning [35]. These opioid-induced dendritic changes promote the reorganization of synaptic connections and regulate glutamatergic excitatory transmission.

Cell death Opioids regulate cell death in brain regions that control learning, emotional memory, rewarding, and nociceptive processing. In mice, chronic morphine administration increases programmed cell death (apoptosis) of the hippocampus's neural stem/progenitor cells, influencing addiction liability [33]. Morphine also causes apoptosis in neurons of the cortex, amygdala, spinal cord, and PAG in models of antinociceptive tolerance and dependence in neonatal rats [36]. In adult rats, opioids cause glial and neuronal death after chronic treatment. For example, repeated morphine and fentanyl administration induces different levels of inflammatory cell death (pyroptosis) in the dorsal raphe nucleus, which is related to the ability of both opioids to develop antinociceptive tolerance or hyperalgesia [37].

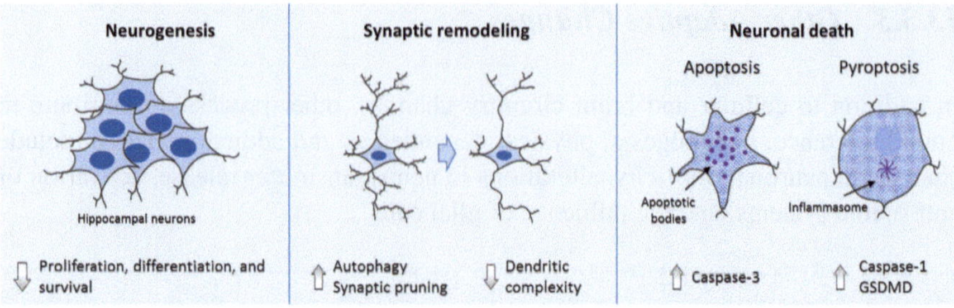

Fig. 13.2 Changes in neuroplasticity produced by long-term opioid administration. Chronic exposure to opioids produces changes in neurogenesis, such as a reduction in the proliferation, differentiation, and survival of neurons in the hippocampus. After repeated chronic exposure, synaptic remodeling, characterized by a reduction in dendritic complexity (fewer neural branches and dendritic spines) and an increase in autophagy-related synaptic pruning, occurs. Finally, the chronic administration of opioids induces the expression of cell death markers, such as caspase-3 and apoptotic bodies (apoptosis) and inflammasomes, caspase-1, and GSDMD (pyroptosis). These changes are brain region-specific and depend on the age, duration of treatments, and specific drug used. GSDMD, Gasdermin D

13.3.3.2 Changes in Neurotransmitters Release

Persistent opioid signaling can produce an excessive release of excitatory neurotransmitters such as CGRP, SP, aspartate, and glutamate. Chronic opioid treatments also modify the release of GABA, the most important inhibitory neurotransmitter. In addition, monoaminergic neurotransmitters such as dopamine and serotonin play crucial roles in opioid tolerance, dependence, hyperalgesia, and withdrawal (Fig. 13.3).

Glutamatergic neurotransmission Repeated opioid administration induces N-Methyl-D-Aspartate (NMDA) receptor activation. Increased glutamatergic activity promotes pain and opioid tolerance via protein kinase C upregulation and intracellular calcium increase. Preclinical studies indicate that competitive (LY235959) and non-competitive (ketamine, memantine, and MK-801) NMDA receptor antagonists delay the development of analgesic tolerance and attenuate hyperalgesia induced by chronic morphine administration. In addition, hyperactivation of NMDA receptors leads to neurotoxicity. The resultant apoptotic cell death in the spinal cord is associated with hyperalgesia [38].

GABA In opposition to acute effects, chronic morphine administration decreases GABA release at the supraspinal level and increases the expression of the reuptake protein GABA transporter-1 (GAT-1) in the spinal cord. These changes in the GABAergic system contribute to the development of tolerance to the analgesic effects of opioids [31].

Dopamine Morphine-induced changes in the mesolimbic dopamine system are associated with rewarding effects during opioid addiction. The participation of dopamine in morphine tolerance has been demonstrated by pharmacologically blocking neuronal dopamine D_2 receptors in mice. Furthermore, increased dopamine levels in the reward system during morphine injection and their decrease after cessating administration explain the dysphoric effects experienced during withdrawal [38].

Serotonin 5-Hydroxy tryptamine (5-HT) or serotonin is a neurotransmitter involved in the descending inhibitory pain pathways. Some studies indicate that 5-HT reuptake inhibitors attenuate the development of morphine-induced antinociceptive tolerance by increasing spinal levels of serotonin [39]. Serotonin neurons from the dorsal raphe nucleus (DRN) and the median raphe nucleus (MRN) are required for morphine analgesia. Recent evidence indicates that the death of the DRN serotonergic neurons contributes to morphine- or fentanyl-induced antinociceptive tolerance and hyperalgesia in rats [37].

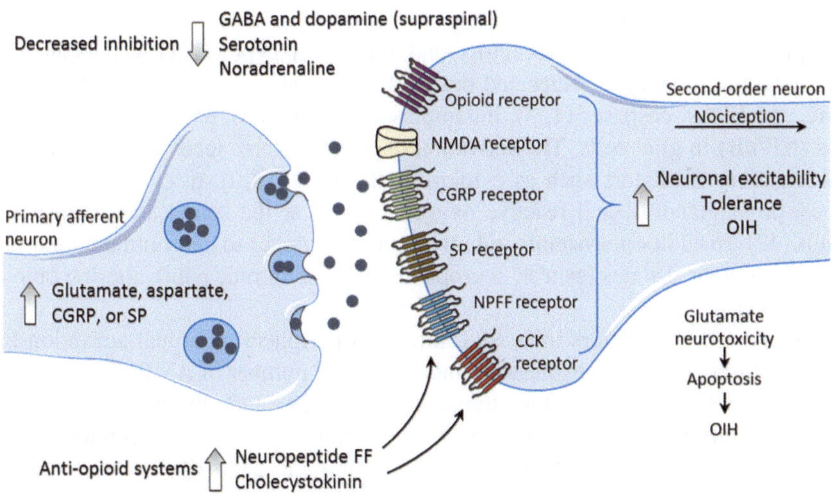

Fig. 13.3 Changes in neurotransmitters produced by chronic opioid exposure. The increased release of excitatory neurotransmitters such as CGRP, SP, aspartate, and glutamate from the presynapsis and the activation of anti-opioid systems (cholecystokinin and FF neuropeptide) are significant neuroadaptations produced by long-term opioids. The activation of NMDA, CGRP, SP, NPFF, and CCK receptors at the postsynapsis terminal increases neuronal excitability and facilitates nociceptive transmission. Also, chronic opioid treatment decreases the release of inhibitory neurotransmitters such as GABA, serotonin, and noradrenaline. All these changes contribute to the development of tolerance and opioid-induced hyperalgesia (OIH). Additionally, the increased release of glutamate can produce toxicity and apoptotic cell death of neurons in various regions of the descending pain pathway, which contribute to OIH

13.3.3.3 Anti-opioid Systems Activation

Chronic opioid administration upregulates pronociceptive systems. In particular, repeated administration of morphine induces the synthesis and release of peptides such as cholecystokinin (CCK) or neuropeptide FF (NPFF) in the spinal cord, promoting a pronociceptive or "anti-opioid" effect (Fig. 13.3).

Different mechanisms explain CKK's contribution to opioid tolerance. On the one hand, forming a heterodimer between the CCK receptor and the μ-opioid receptor reduces the binding of opioid ligands after chronic opioid administration. On the other hand, CCK receptor activation increases intracellular calcium, an effect opposite to that induced by exogenous opioids [40].

In addition, NPFF release is associated with morphine-induced hyperalgesia because blockade of NPFF receptors with the RF9 antagonist increases acute analgesia and prevents the development of analgesic tolerance in mice [41].

13.3.3.4 Neuronal-Glia Cross Talk

Microglia and astrocytes are the immunocompetent glial cells in the CNS, and increasing evidence indicates that neuroinflammation plays a significant role in OIH and opioid tolerance (Fig. 13.4).

Opioids bind to opioid receptors and toll-like receptors (TLRs) located in microglia and astrocytes. Repeated opioid administration activates MAPKs (p38, JNK, and ERK), caspases (1, 3), inflammasomes (NLRP3), and transcription factors (NF-κB) in glial cells. The activation of these proteins leads to the release of inflammatory mediators such as cytokines (IL-1β, IL,18, IL-6, or TNF), chemokines, prostaglandins, and reactive oxygen species at the spinal and supraspinal levels [42]. In addition, systemic inflammation contributes to neuroinflammation in some pathological states (cancer, neuropathic pain, or chronic pain), altering opioid effects [43].

Inflammatory mediators increase central and peripheral neuronal activation by reducing GABA receptor expression, increasing the number of AMPA and NMDA receptors, decreasing glutamate transporter proteins, and diminishing outward potassium currents [42]. Preclinical reports indicate that astrocyte inhibitors (fluorocitrate) and NF-κB inhibitors (e.g., minocycline and pentoxifylline) effectively delay tolerance development and opioid-induced hyperalgesia [37, 44, 45].

In addition to neuroinflammation, chronic morphine regulates the expression and activation of diverse proteins and channels in glial cells. In particular, the morphine-induced activation of the glial purinergic P2X4 receptor induces BDNF release. This neurotrophic factor interacts with the tropomyosin receptor kinase B (TrkB) in spinal neurons and downregulates the potassium-chloride co-transporter 2 (KCC2), producing hyperalgesia. Conversely, inhibition of P2X4 receptor expression prevents tolerance to continuous morphine infusion and attenuates the expression of the microglia and astrocytes markers in rats [46]. In addition, chronic morphine potentiates microglial P2X7 receptor-mediated calcium responses, and this effect can contribute to the development of morphine analgesic tolerance [47]. Together, these data highlight the relevance of purinergic receptors in morphine-induced neuroinflammation and the development of tolerance.

13.3.3.5 Epigenetic Modifications

Epigenetic mechanisms such as deoxyribonucleic acid (DNA) methylation, histone acetylation, or methylation (see Chap. 9) have been implicated in opioid-induced hyperalgesia, dependence, and tolerance.

DNA methylation generally leads to gene silencing, which could be related to the decreased opioid-receptor expression produced by several opioids. For example, heroin-dependent persons have high methylation levels in the coding regions of the μ-opioid receptor gene in neurons [48]. In contrast, DNA demethylation increases brain-derived neurotrophic factor (BDNF) in dorsal root ganglion neurons in rats, which correlates with OIH development [49].

Acetylation of lysine residues on histone tails causes chromatin remodeling to allow access to transcriptional activators and promote gene expression. Several studies have found that brains from animals that self-administered heroin and opioid users have high histone acetylation levels. Enhanced opioid-induced histone acetylation may be due to the downregulation of the histone deacetylase (HDAC) enzyme or activation of the histone acetyltransferase (HAT) enzyme. In mice, an inhibitor of the HAT enzyme attenuated morphine-induced hyperalgesia, tolerance, and physical dependence [50].

Contrary to histone acetylation, histone methylation represses gene expression. Interestingly, brain areas of rodents show low levels of histone methylation after repeated opioid treatment. Specifically, morphine downregulates G9a histone methyltransferase, which demethylates histone 3 at lysine 9 (H3K9) [51, 52].

13.4 Clinical and Preclinical Approaches to Prevent the Development of Opioid Dependence and Tolerance

The great diversity of patients requiring opioid treatment hinders the establishment of universal opioid therapies. In particular, different schedules of opioid administration for pain management are necessary for patients who use opioids for the first time and people with SUD.

Tolerance, dependence, and hyperalgesia produced by chronic opioid administration represent significant challenges for service providers. For example, differentiating opioid-induced hyperalgesia, tolerance, and preexisting pain can determine the selection and continuity of specific opioid treatments. Furthermore, some opioid treatments are effective in delaying analgesic tolerance but not physical dependence. Table 13.1 shows novel therapeutic approaches used in the clinic to avoid some opioid long-term effects, and the following sections address these topics more extensively.

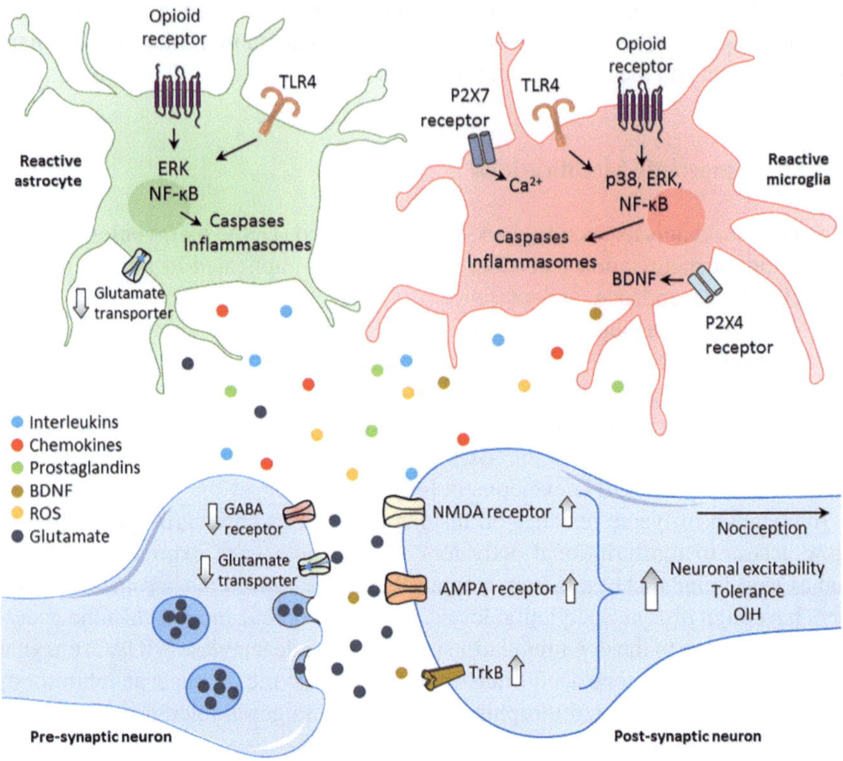

Fig. 13.4 The role of neuroinflammation in the long-term effects of opioids. Sustained activation of opioid receptors or TLR4 in glial cells activates several neuroinflammatory mechanisms. In astrocytes, activation of ERK and NF-κB leads to inflammasome oligomerization and caspase maturation. Also, opioids induce a decrease in the glutamate transporter in astrocytes. In microglia, opioids promote the activation of p38, ERK, and NF-κB activating inflammasomes and caspases. Furthermore, in these cells, opioids induce the activation of purinergic receptors (P2X4 and P2X7), which increases intracellular calcium and BDNF production. The release of inflammatory mediators (cytokines, chemokines, and prostaglandins) and BDNF and ROS induces a decrease in the expression of the GABA receptor at the presynaptic terminal. Also, these inflammatory mediators increase the expression of NMDA, AMPA, and TrkB receptors in the postsynaptic neuron. The decrease in glutamate transporters in neurons and astrocytes induces an increase in extracellular glutamate levels. These changes increase the transmission of pain and contribute to the development of tolerance and opioid-induced hyperalgesia (OIH). TLR4, Toll-like receptor 4; ERK, extracellular signal-regulated kinase; JNK, Jun N-terminal kinase; p38, p38 kinase; NF-κB, nuclear transcription factor κB; P2X4, P2X purinoceptor 4; P2X7, P2X purinoceptor 7; BDNF, brain-derived neurotrophic factor; ROS, reactive oxygen species; GABA, γ-aminobutyric acid; NMDA, N-methyl-D-aspartate; AMPA, α-amino-3-hydroxy-5-methyl-4-isoxazolepropionic acid; TrkB, tropomyosin receptor kinase B

13.4.1 Administration of Drugs that Enhance Opioid Analgesic Effects

Antidepressants Reinforcing descending systems that control nociceptive transmission can help diminish pain perception. Antidepressants produce analgesia by increasing the amount of norepinephrine and serotonin, which inhibit pain transmission at spinal levels (see Chap. 10). There is evidence that antidepressants used as opioid adjuvants prolong analgesia in patients with cancer-related pain. However, this effect is short-lasting and depends on the dosage, specific opioid, and antidepressant used [53]. Moreover, some opioid-antidepressant combinations cause potentially serious adverse effects, including neuromuscular and autonomic hyperactivity and altered mental states due to serotonin toxicity. Tramadol, pethidine, dextromethorphan, tapentadol, fentanyl, and methadone can induce serotonin syndrome when combined with antidepressants [54].

Adrenergic agonists Combining opioids and adrenergic agonists produces supra-additive analgesic effects [55]. Specifically, clonidine and lofexidine ($\alpha 2$-receptor adrenergic agonists) have been used to control chronic pain and block acute withdrawal symptoms in chronic opioid users (see Chap. 15). In addition, the infusion of dexmedetomidine, a potent $\alpha 2$ adrenergic agonist, effectively reduces OIH and contributes to the recovery of normal nociceptive and antinociceptive responses in opioid-tolerant patients [56].

Cannabinoids Recent evidence has emerged on the use of cannabis and cannabinoids to reduce the harms associated with long-term opioid use because some opioid-cannabinoid combinations produce analgesic synergy. Reducing opioid doses when combined with cannabis can delay the development of opioid-induced analgesic tolerance. Also, several preclinical studies have demonstrated that some cannabinoids relieve acute opioid withdrawal symptoms [57, 58]. However, further studies are needed to establish optimal doses for individual cannabinoids and opioids to improve opioid analgesic effects.

13.4.2 Inhibition of Neurotransmitter Systems that Counteract Opioid Analgesic Effects

Anticonvulsants As mentioned above, chronic opioid exposure causes hyperactivation of neurons that transmit nociceptive signals. Anticonvulsant agents such as gabapentin and pregabalin inhibit the release of excitatory neurotransmitters by activating the $\alpha 2\delta$ subunit of voltage-gated calcium channels located in the spinal cord. There is evidence that gabapentin reduces opioid intake in chronic pain patients and block OIH [59]. As to pregabalin, only preclinical studies have demonstrated its efficacy in preventing morphine tolerance and dependence. However,

Table 13.1 Clinical approaches to avoid long-term opioid effects

	Mechanisms of action	Therapeutic strategy	Clinical model	Effects and side effects
Analgesic enhancers				
Tricyclic antidepressants	SERT and NET inhibition	Coadministration with opioids	Patients with cancer-related pain	Short-term increased analgesia Serotonin toxicity [53, 54]
Clonidine	α2-Adrenoceptor activation	Adjuvant for morphine	Patients with chronic pain	Improved analgesia [55]
Lofexidine or clonidine	α2-Adrenoceptor activation	Opioid withdrawal treatment	Chronic opioid users	Blocking of acute withdrawal symptoms [55]
Dexmedeto-midine	α2-Adrenoceptor activation	Administration after chronic opioid treatments	Patients with OIH	Recovery of normal nociceptive responses [55, 56]
Cannabis and cannabinoids	CB_1 and CB_2 receptors activation	Combination with opioids	Chronic opioid users	Improved analgesia Tolerance prevention [57, 58]
Neuroexcitability inhibitors				
Gabapentin	Voltage-gated calcium channels activation	Combination with opioids	Patients with chronic pain	Opioid consumption reduction [59]
			Patients with OIH	Hyperalgesia blockade [59]
			Poly-drug users	Gabapentin addiction Increased opioid withdrawal [60]
Ziconotide	N-type Ca_v blockade	Combination with morphine	Patients with cancer-related pain	Improved analgesia [61]
Verapamil or diltiazem	L-type Ca_v blockade	Opioid withdrawal treatment	Opioid-dependent patients with or without methadone maintenance treatment	Less withdrawal symptoms and craving [62]
Ketamine	NMDA antagonism	Pre- and post-surgery administration	Pre- and postoperative patients	Less respiratory depression and sedation [63]
		Drug rotation	Children with opioid tolerance	Reduction in the infusion of fentanyl [64]

(continued)

Table 13.1 (continued)

	Mechanisms of action	Therapeutic strategy	Clinical model	Effects and side effects
Proglumide or devazepide	CCK_A and CCK_B receptors inhibition	Combination with morphine	Patients with neuropathic pain	Improved analgesia Tolerance prevention Reversal of established tolerance [65, 66]
Opioid combinations				
Morphine and oxycodone	μ- and κ-opioid receptor activation	Combination	Patients with cancer-related pain	Improved analgesia Reduction in the consumption of morphine [68]
Morphine and buprenorphine	κ- and δ-opioid receptor blockade and μ-opioid receptor activation	Systemic buprenorphine and spinal morphine	Patients with chronic pain	Improved analgesia [68]
			Heroin users	Buprenorphine-induced hyperalgesia [69]
Fentanyl and tramadol	μ-Opioid receptor activation and SERT and NET inhibition	Transdermal fentanyl and oral tramadol	Pain management in patients with advanced cancer	Fentanyl dose escalation reduction [68, 70]
Fentanyl and morphine or oxycodone	μ-Opioid receptor activation	Transdermal fentanyl and oral morphine or oxycodone	Patients with cancer-related pain	Improved analgesia Less hyperalgesia [68, 70]
Neuroinflammation inhibitors				
Pentoxifylline	Glial inhibition	Pre-surgery administration	Postoperative-care patients	Reduction in postoperative morphine consumption [73]
Minocycline	Glial inhibition NF-κB inhibition	Acute administration in combination with oxycodone	Non-tolerant, recreational opioid users	Attenuation of positive subjective responses [74]
		Two-week minocycline therapy	Opioid-dependents with methadone or buprenorphine/methadone maintenance treatment	Cognitive enhancer No changes in OIH, craving or withdrawal [75]

SERT serotonin transporter, *NET* noradrenaline transporter, *OIH* opioid-induced hyperalgesia, *CB* cannabinoid, Ca_v voltage-gated calcium channels, *NMDA* N-methyl-D-aspartic acid, *CCK* cholecystokinin, *NF-κB* nuclear transcription factor κB some reports indicate that subjects with a history of heroin use can also misuse gabapentin and pregabalin to achieve relaxation, euphoria, and sedation [60]. In such cases, the use of pregabalin may increase withdrawal symptoms.

Calcium channels blockers Preclinical studies have shown that selective L- and N-type calcium channel blockers enhance morphine antinociception, reverse tolerance, and prevent hyperalgesia. In clinical settings, the combination of ziconotide (an N-type calcium channels blocker) and morphine has been used to produce analgesia when opioid monotherapy is ineffective [61]. Furthermore, the L-type calcium channel blockers verapamil and diltiazem effectively alleviate withdrawal symptoms and craving in opioid-dependent patients [62].

NMDA receptor antagonists NMDA receptor antagonism is a proposed method to neutralize the neuronal hyperexcitability that exists after chronic opioid treatments. Preclinical studies suggest that NMDA antagonists reduce tolerance development and OIH. For example, the non-competitive NMDA receptor antagonist ketamine effectively reduced some side effects produced by morphine in postoperative patients, such as respiratory depression, nausea, and sedation [63]. Moreover, drug rotation with ketamine effectively reduced the rate of fentanyl infusion in children tolerant to opioid analgesic effects [64].

CCK receptor antagonists The release of anti-opioid peptides such as CCK can counteract opioid analgesic effects. Several clinical reports indicate that the administration of CCK antagonists (proglumide, devazepide) with morphine improves analgesia. Also, these drugs prevent the development of tolerance and reverse the established tolerance in patients with neuropathic pain [65, 66]. However, targeting only one anti-opioid system has been ineffective in preventing analgesic tolerance in humans; thus, multi-target therapies should be investigated.

13.4.3 Coadministration of Opioids with Different Pharmacological Profiles

Preclinical evidence indicates that combining opioids with different pharmacological profiles can produce analgesia with fewer side effects. For example, concurrent administration of the μ-opioid agonist morphine with the δ-opioid antagonist naltrindole increases antinociception while preventing tolerance and dependence. Also, combining morphine with fentanyl significantly increases analgesia and delays tolerance development [67].

Clinical evidence indicates that coadministration of morphine (μ-, δ-, and κ-opioid receptor agonist) with oxycodone (μ- and κ-opioid receptor agonist) enhances the analgesic effect and reduces morphine use in patients with cancer-related pain [68].

It is necessary to take into account the route of administration because different opioids interact at different levels. Morphine activates μ-opioid receptors mainly at the spinal level. Thus, the systemic administration of buprenorphine (κ- and δ-opioid receptor antagonist and partial μ-opioid receptor agonists) and spinal morphine administration provide supra-additive analgesia in patients with chronic pain [68]. However, medication-assisted therapy in heroin users with buprenorphine can produce hyperalgesia [69]. These conflicting results indicate that particular opioid combinations require careful evaluation considering the history of opioid use, drug doses, treatment duration, and administration route.

Fentanyl is commonly used in anesthesia and as transdermal patches for chronic and cancer-related pain treatment. For example, coadministration of transdermal fentanyl with the weak opioid receptor agonist tramadol (μ-opioid receptor agonist and serotonin/norepinephrine transporter inhibitor) effectively reduces conventional fentanyl escalation in patients with advanced cancer. Furthermore, the combination of transdermal fentanyl with oral morphine or oxycodone in cancer patients improves pain relief with fewer adverse effects [68, 70].

13.4.4 Coadministration of Opioids and Anti-inflammatory Drugs

Chronic opioid exposure induces central and peripheral immune changes that contribute to central sensitization, tolerance, and dependence. Furthermore, inflammatory processes play a significant role in pathological conditions such as cancer, neuropathic, and dental pain. Some nonsteroidal anti-inflammatory drugs (NSAIDs) reduce peripheral pro-inflammatory mediators that increase pain, cytokine release, and prostaglandin production. Therefore, using NSAIDs as opioid adjuvants may provide better analgesia or prevent dose escalation, but opioids can still produce tolerance, hyperalgesia, and addiction [24, 71]. This is not surprising considering that some NSAIDs also produce endogenous opioid peptides release [72].

Targeting neuroinflammation with glial inhibitors has been proposed to reduce opioid-induced hyperalgesia and tolerance. As mentioned before, minocycline and pentoxifylline are drugs that reduce neuroinflammatory responses by inhibiting NF-κB. Also, the inhibition of the NLRP3 inflammasome, a protein complex required for IL-1β release and pyroptotic cell death, represents an emerging therapeutic target for treating chronic pain and reducing opioid-induced hyperalgesia and tolerance [37]. In preclinical models, minocycline and pentoxifylline attenuate microglia and astrocytes' activation and the development of antinociceptive tolerance induced by chronic opioid administration. In addition, administering pentoxifylline before surgery to patients reduces morphine consumption for postoperative pain relief [73]. As to minocycline, this NF-κB inhibitor attenuates the positive subjective responses to oxycodone in non-tolerant patients [74]. Although minocycline may attenuate the abuse liability of opioids, additional studies are needed to evaluate its role in other long-term effects of opioids.

References

1. Serdarevic M, Striley CW, Cottler LB. Sex differences in prescription opioid use. Curr Opin Psychiatry. 2017;30:238–46. https://doi.org/10.1097/YCO.0000000000000337.
2. American Psychiatric Association. Diagnostic and statistical manual of mental disorders. American Psychiatric Association; 2013. https://doi.org/10.1176/appi.books.9780890425596.
3. Panlilio LV, Goldberg SR. Self-administration of drugs in animals and humans as a model and an investigative tool. Addiction. 2007;102:1863–70. https://doi.org/10.1111/j.1360-0443.2007.02011.x.
4. Koob GF, Volkow ND. Neurobiology of addiction: a neurocircuitry analysis. Lancet Psychiatry. 2016;3:760–73. https://doi.org/10.1016/S2215-0366(16)00104-8.
5. Hayhurst CJ, Durieux ME. Differential opioid tolerance and opioid-induced hyperalgesia: a clinical reality. Anesthesiology. 2016;124:483–8. https://doi.org/10.1097/ALN.0000000000000963.
6. Voon P, Karamouzian M, Kerr T. Chronic pain and opioid misuse: a review of reviews. Subst Abuse Treat Prev Policy. 2017;12:36. https://doi.org/10.1186/s13011-017-0120-7.
7. Cruz SL, Granados-Soto V. Opioids and opiates: pharmacology, abuse, and addiction. In: Pfaff DW, Volkow ND, editors. Neuroscience in the 21st century. New York: Springer; 2016. p. 3625–57. https://doi.org/10.1007/978-1-4939-3474-4_156.
8. Chevlen E. Opioids: a review. Curr Pain Headache Rep. 2003;7:15–23. https://doi.org/10.1007/s11916-003-0005-5.
9. Koob GF. A role for brain stress systems in addiction. Neuron. 2008;59:11–34. https://doi.org/10.1016/j.neuron.2008.06.012.
10. Wesson DR, Ling W. The clinical opiate withdrawal scale (COWS). J Psychoactive Drugs. 2003;35:253–9. https://doi.org/10.1080/02791072.2003.10400007.
11. Harter K. Opioid use disorder in pregnancy. Ment Health Clin. 2019;9:359–72. https://doi.org/10.9740/mhc.2019.11.359.
12. Center for Substance Abuse Treatment. Protracted withdrawal. Subst Abus Treat. 2010;9:1–8.
13. Finnegan LP, Connaughton JF, Kron RE, Emich JP. Neonatal abstinence syndrome: assessment and management. Addict Dis. 1975;2:141–58.
14. Anbalagan S, Mendez MD. Neonatal abstinence syndrome. In: StatPearls [Internet]. StatPearls Publishing; 2021. https://www.ncbi.nlm.nih.gov/books/NBK551498/.
15. Jones HE, Dengler E, Garrison A, O'Grady KE, Seashore C, Horton E, et al. Neonatal outcomes and their relationship to maternal buprenorphine dose during pregnancy. Drug Alcohol Depend. 2014;134:414–7. https://doi.org/10.1016/j.drugalcdep.2013.11.006.
16. Kuhn BN, Kalivas PW, Bobadilla A-C. Understanding addiction using animal models. Front Behav Neurosci. 2019;13:262. https://doi.org/10.3389/fnbeh.2019.00262.
17. Stoops WW. Reinforcing effects of stimulants in humans: sensitivity of progressive-ratio schedules. Exp Clin Psychopharmacol. 2008;16:503–12. https://doi.org/10.1037/a0013657.
18. Young R. Drug discrimination. In: Buccafusco JJ, editor. Methods of behavior analysis in neuroscience. 2nd ed. Boca Raton: CRC Press/Taylor & Francis; 2009.
19. Sun Y, Chen G, Zhou K, Zhu Y. A conditioned place preference protocol for measuring incubation of craving in rats. J Vis Exp. 2018;6(141):e58384. https://doi.org/10.3791/58384.
20. McKendrick G, Graziane NM. Drug-induced conditioned place preference and its practical use in substance use disorder research. Front Behav Neurosci. 2020;14:582147. https://doi.org/10.3389/fnbeh.2020.582147.
21. Sun H, Maze I, Dietz DM, Scobie KN, Kennedy PJ, Damez-Werno D, et al. Morphine epigenomically regulates behavior through alterations in histone H3 lysine 9 dimethylation in the nucleus accumbens. J Neurosci. 2012;32:17454–64. https://doi.org/10.1523/JNEUROSCI.1357-12.2012.
22. McEwen BS. Stressed or stressed out: what is the difference? J Psychiatry Neurosci. 2005;30:315–8.

23. Gintzler AR, Chakrabarti S. Chronic morphine-induced plasticity among signalling molecules. Novartis Found Symp. 2004;261:167–76; discussion 176–80, 191–3
24. Mercadante S, Arcuri E, Santoni A. Opioid-induced tolerance and hyperalgesia. CNS Drugs. 2019;33:943–55. https://doi.org/10.1007/s40263-019-00660-0.
25. Zhou J, Ma R, Jin Y, Fang J, Du J, Shao X, et al. Molecular mechanisms of opioid tolerance: from opioid receptors to inflammatory mediators (review). Exp Ther Med. 2021;22:1004. https://doi.org/10.3892/etm.2021.10437.
26. King T, Ossipov MH, Vanderah TW, Porreca F, Lai J. Is paradoxical pain induced by sustained opioid exposure an underlying mechanism of opioid antinociceptive tolerance? Neurosignals. 2005;14:194–205. https://doi.org/10.1159/000087658.
27. Listos J, Łupina M, Talarek S, Mazur A, Orzelska-Górka J, Kotlińska J. The mechanisms involved in morphine addiction: an overview. Int J Mol Sci. 2019;20:4302. https://doi.org/10.3390/ijms20174302.
28. Gledhill LJ, Babey AM. Synthesis of the mechanisms of opioid tolerance: do we still say no? Cell Mol Neurobiol. 2021;41:927–48. https://doi.org/10.1007/s10571-021-01065-8.
29. Wang Y, Sun J, Tao Y, Xu X, Chi Z, Liu J. Paradoxical relationship between RAVE (relative activity versus endocytosis) values of several opioid receptor agonists and their liability to cause dependence. Acta Pharmacol Sin. 2010;31:393–8. https://doi.org/10.1038/aps.2010.19.
30. Cruz SL, Villarreal JE, Volkow ND. Further evidence that naloxone acts as an inverse opiate agonist: implications for drug dependence and withdrawal. Life Sci. 1996;58:PL381–9. https://doi.org/10.1016/0024-3205(96)00250-0.
31. Lueptow LM, Fakira AK, Bobeck EN. The contribution of the descending pain modulatory pathway in opioid tolerance. Front Neurosci. 2018;12:886. https://doi.org/10.3389/fnins.2018.00886.
32. Serafini RA, Pryce KD, Zachariou V. The mesolimbic dopamine system in chronic pain and associated affective comorbidities. Biol Psychiatry. 2020;87:64–73. https://doi.org/10.1016/j.biopsych.2019.10.018.
33. Zhang Y, Loh HH, Law P-Y. Effect of opioid on adult hippocampal neurogenesis. Sci World J. 2016;2016:1–7. https://doi.org/10.1155/2016/2601264.
34. Beltrán-Campos V, Silva-Vera M, García-Campos ML, Díaz-Cintra S. Effects of morphine on brain plasticity. Neurologia (Barcelona, Spain). 2015;30:176–80. https://doi.org/10.1016/j.nrl.2014.08.004.
35. Cai Y, Yang L, Hu G, Chen X, Niu F, Yuan L, et al. Regulation of morphine-induced synaptic alterations: role of oxidative stress, ER stress, and autophagy. J Cell Biol. 2016;215:245–58. https://doi.org/10.1083/jcb.201605065.
36. Bajic D, Commons KG, Soriano SG. Morphine-enhanced apoptosis in selective brain regions of neonatal rats. Int J Dev Neurosci. 2013;31:258–66. https://doi.org/10.1016/j.ijdevneu.2013.02.009.
37. Carranza-Aguilar CJ, Hernández-Mendoza A, Mejias-Aponte C, Rice KC, Morales M, González-Espinosa C, et al. Morphine and fentanyl repeated administration induces different levels of NLRP3-dependent pyroptosis in the dorsal raphe nucleus of male rats via cell-specific activation of TLR4 and opioid receptors. Cell Mol Neurobiol. 2020;42(3):677–94. https://doi.org/10.1007/s10571-020-00957-5.
38. Mehrabadi S, Karimiyan SM. Morphine tolerance effects on neurotransmitters and related receptors: definition, overview and update. Int J Pharm Res. 2018;23:1–11. https://doi.org/10.9734/JPRI/2018/41936.
39. Ozdemir E, Gursoy S, Bagcivan I. The effects of serotonin/norepinephrine reuptake inhibitors and serotonin receptor agonist on morphine analgesia and tolerance in rats. J Physiol Sci. 2012;62:317–23. https://doi.org/10.1007/s12576-012-0207-x.
40. Zhang L, Zhang J, Hang L, Liu T. Mu opioid receptor heterodimers emerge as novel therapeutic targets: recent progress and future perspective. Front Pharmacol. 2020;11:1–11. https://doi.org/10.3389/fphar.2020.01078.

41. Elhabazi K, Trigo J, Mollereau C, Moulédous L, Zajac J-M, Bihel F, et al. Involvement of neuropeptide FF receptors in neuroadaptive responses to acute and chronic opiate treatments. Br J Pharmacol. 2012;165:424–35. https://doi.org/10.1111/j.1476-5381.2011.01563.x.
42. Eidson LN, Murphy AZ. Inflammatory mediators of opioid tolerance: implications for dependency and addiction. Peptides. 2019;115:51–8. https://doi.org/10.1016/j.peptides.2019.01.003.
43. Varrassi G, Fusco M, Skaper SD, Battelli D, Zis P, Coaccioli S, et al. A pharmacological rationale to reduce the incidence of opioid induced tolerance and hyperalgesia: a review. Pain Ther. 2018;7:59–75. https://doi.org/10.1007/s40122-018-0094-9.
44. Harada S, Nakamoto K, Tokuyama S. The involvement of midbrain astrocyte in the development of morphine tolerance. Life Sci. 2013;93:573–8. https://doi.org/10.1016/j.lfs.2013.08.009.
45. Fukagawa H, Koyama T, Kakuyama M, Fukuda K. Microglial activation involved in morphine tolerance is not mediated by toll-like receptor 4. J Anesth. 2013;27:93–7. https://doi.org/10.1007/s00540-012-1469-4.
46. Horvath RJ, Romero-Sandoval AE, De Leo JA. Inhibition of microglial P2X4 receptors attenuates morphine tolerance, Iba1, GFAP and μ opioid receptor protein expression while enhancing perivascular microglial ED2. Pain. 2010;150:401–13. https://doi.org/10.1016/j.pain.2010.02.042.
47. Leduc-Pessah H, Weilinger NL, Fan CY, Burma NE, Thompson RJ, Trang T. Site-specific regulation of P2X7 receptor function in microglia gates morphine analgesic tolerance. J Neurosci. 2017;37:10154–72. https://doi.org/10.1523/JNEUROSCI.0852-17.2017.
48. Browne CJ, Godino A, Salery M, Nestler EJ. Epigenetic mechanisms of opioid addiction. Biol Psychiatry. 2020;87:22–33. https://doi.org/10.1016/j.biopsych.2019.06.027.
49. Chao Y-C, Xie F, Li X, Guo R, Yang N, Zhang C, et al. Demethylation regulation of BDNF gene expression in dorsal root ganglion neurons is implicated in opioid-induced pain hypersensitivity in rats. Neurochem Int. 2016;97:91–8. https://doi.org/10.1016/j.neuint.2016.03.007.
50. Liang D-Y, Li X, Clark JD. Epigenetic regulation of opioid-induced hyperalgesia, dependence, and tolerance in mice. J Pain. 2013;14:36–47. https://doi.org/10.1016/j.jpain.2012.10.005.
51. Zhang Z, Tao W, Hou Y-Y, Wang W, Kenny PJ, Pan ZZ. MeCP2 repression of G9a in regulation of pain and morphine reward. J Neurosci. 2014;34:9076–87. https://doi.org/10.1523/JNEUROSCI.4194-13.2014.
52. Gerra MC, Dallabona C, Arendt-Nielsen L. Epigenetic alterations in prescription opioid misuse: new strategies for precision pain management. Genes (Basel). 2021;12:1226. https://doi.org/10.3390/genes12081226.
53. Bennett MI. Effectiveness of antiepileptic or antidepressant drugs when added to opioids for cancer pain: systematic review. Palliat Med. 2011;25:553–9. https://doi.org/10.1177/0269216310378546.
54. Perananthan V, Buckley N. Opioids and antidepressants: which combinations to avoid. Aust Prescr. 2021;44:41–4. https://doi.org/10.18773/austprescr.2021.004.
55. Srivastava AB, Mariani JJ, Levin FR. New directions in the treatment of opioid withdrawal. Lancet. 2020;395:1938–48. https://doi.org/10.1016/S0140-6736(20)30852-7.
56. Belgrade M, Hall S. Dexmedetomidine infusion for the management of opioid-induced hyperalgesia. Pain Med. 2010;11:1819–26. https://doi.org/10.1111/j.1526-4637.2010.00973.x.
57. Wiese B, Wilson-Poe AR. Emerging evidence for cannabis' role in opioid use disorder. Cannabis Cannabinoid Res. 2018;3:179–89. https://doi.org/10.1089/can.2018.0022.
58. MacCallum CA, Eadie L, Barr AM, Boivin M, Lu S. Practical strategies using medical cannabis to reduce harms associated with long term opioid use in chronic pain. Front Pharmacol. 2021;12:633168. https://doi.org/10.3389/fphar.2021.633168.
59. Stoicea N, Russell D, Weidner G, Durda M, Joseph NC, Yu J, et al. Opioid-induced hyperalgesia in chronic pain patients and the mitigating effects of gabapentin. Front Pharmacol. 2015;6:104. https://doi.org/10.3389/fphar.2015.00104.
60. Lyndon A, Audrey S, Wells C, Burnell ES, Ingle S, Hill R, et al. Risk to heroin users of polydrug use of pregabalin or gabapentin. Addiction. 2017;112:1580–9. https://doi.org/10.1111/add.13843.

61. Deer TR, Pope JE, Hanes MC, McDowell GC. Intrathecal therapy for chronic pain: a review of morphine and ziconotide as firstline options. Pain Med. 2019;20:784–98. https://doi.org/10.1093/pm/pny132.
62. Oliveto A, Mancino M, Sanders N, Cargile C, Benjamin Guise J, Bickel W, et al. Effects of prototypic calcium channel blockers in methadone-maintained humans responding under a naloxone discrimination procedure. Eur J Pharmacol. 2013;715:424–35. https://doi.org/10.1016/j.ejphar.2013.03.007.
63. Sveticic G, Eichenberger U, Curatolo M. Safety of mixture of morphine with ketamine for postoperative patient-controlled analgesia: an audit with 1026 patients. Acta Anaesthesiol Scand. 2005;49:870–5. https://doi.org/10.1111/j.1399-6576.2005.00740.x.
64. Neunhoeffer F, Hanser A, Esslinger M, Icheva V, Kumpf M, Gerbig I, et al. Ketamine infusion as a counter measure for opioid tolerance in mechanically ventilated children: a pilot study. Pediatr Drugs. 2017;19:259–65. https://doi.org/10.1007/s40272-017-0218-4.
65. McCleane G. Cholecystokinin antagonists a new way to improve the analgesia from old analgesics? Curr Pharm Des. 2004;10:303–14. https://doi.org/10.2174/1381612043386419.
66. Wiesenfeld-Hallin Z, Xu X-J, Hökfelt T. The role of spinal cholecystokinin in chronic pain states. Pharmacol Toxicol. 2002;91:398–403. https://doi.org/10.1034/j.1600-0773.2002.910619.x.
67. Silva-Moreno A, Gonzalez-Espinosa C, León-Olea M, Cruz SL. Synergistic antinociceptive actions and tolerance development produced by morphine-fentanyl coadministration: correlation with μ-opioid receptor internalization. Eur J Pharmacol. 2012;674:239–47. https://doi.org/10.1016/j.ejphar.2011.10.034.
68. Mercadante S. Opioid combination: rationale and possible clinical applications. Ann Palliat Med. 2013;2:189–96. https://doi.org/10.3978/j.issn.2224-5820.2013.09.04.
69. Athanasos P, Ling W, Bochner F, White JM, Somogyi AA. Buprenorphine maintenance subjects are hyperalgesic and have no antinociceptive response to a very high morphine dose. Pain Med. 2019;20:119–28. https://doi.org/10.1093/pm/pny025.
70. Bairaktari A, Al Harbi M, Dimitriou V. Combined use of strong opioids for pain relief in cancer patients-a prospective randomized comparative study. Middle East J Anesthesiol. 2018;25:31–6.
71. Varrassi G, Yeam CT, Rekatsina M, Pergolizzi JV, Zis P, Paladini A. The expanding role of the COX inhibitor/opioid receptor agonist combination in the management of pain. Drugs. 2020;80:1443–53. https://doi.org/10.1007/s40265-020-01369-x.
72. Hernández-Delgadillo GP, Cruz SL. Endogenous opioids are involved in morphine and dipyrone analgesic potentiation in the tail flick test in rats. Eur J Pharmacol. 2006;546:54–9. https://doi.org/10.1016/j.ejphar.2006.07.027.
73. Lu C-H, Chao P-C, Borel CO, Yang C-P, Yeh C-C, Wong C-S, et al. Preincisional intravenous pentoxifylline attenuating perioperative cytokine response, reducing morphine consumption, and improving recovery of bowel function in patients undergoing colorectal cancer surgery. Anesth Analg. 2004;99(5):1465–71. https://doi.org/10.1213/01.ANE.0000132974.32249.C8.
74. Mogali S, Askalsky P, Madera G, Jones JD, Comer SD. Minocycline attenuates oxycodone-induced positive subjective responses in non-dependent, recreational opioid users. Pharmacol Biochem Behav. 2021;209:173241. https://doi.org/10.1016/j.pbb.2021.173241.
75. Arout CA, Waters AJ, MacLean RR, Compton P, Sofuoglu M. Minocycline does not affect experimental pain or addiction-related outcomes in opioid maintained patients. Psychopharmacology. 2019;236:2857–66. https://doi.org/10.1007/s00213-018-5146-7.

Chapter 14
Medication-Assisted Treatment (MAT) 1: Opioid Substitution Therapy

James Tidder, Alexander M. Baldacchino, and Joseph Tay Wee Teck

Abstract This chapter is the first of two parts describing the advantages, standards, and goals of medication-assisted treatment (MAT) for opioid use disorders. MAT is a complex biopsychosocial intervention with the provision of opioid substitution treatment (OST) at its core. These chapters aim to provide the clinician with charts, tables, and clinical guidance around the prescribing of OST within MAT. This first part (this chapter) covers the assessment of patients with opioid use disorders (OUDs), initiation, titration, and maintenance with OST, as well as physical monitoring and the treatment of those with OUDs requiring admission to hospital.

Data presented here is based on clinical guidelines and the experience in treating persons with OUDs in Scotland, United Kingdom (UK), and reflects evolution in practice in response to the high levels of drug deaths and Scottish national MAT standards published in May 2021. Within the UK, the National Health Service (NHS) exists as a single-payer healthcare system, providing universal care to all residents. This of course is not the case worldwide, and there is substantial variation in availability of services between countries. Clinicians should always ensure that they are familiar with what services are available locally as well as local licensing and legal requirements around prescribing OST, as substitute medications are usually controlled drugs.

Keywords MAT · OST · OAT · OUD · Methadone · Buprenorphine · Low threshold

---

J. Tidder (✉)
NHS Borders Addiction Service, Galashiels, Scotland, UK
e-mail: james.tidder@borders.scot.nhs.uk

A. M. Baldacchino
University of St Andrews, St Andrews, Scotland, UK

NHS Fife Addiction Services, Leven, Scotland, UK

J. T. W. Teck
Forward Leeds, Leeds, UK

S. L. Cruz (ed.), *Opioids*, https://doi.org/10.1007/978-3-031-09936-6_14

14.1 Introduction

Drug dependence is a complex multifactorial biological and behavioral disorder. Scientific advances are making it possible to develop treatments that help normalize the brain functioning of affected individuals and support them in changing their behavior. Offering treatments based on scientific evidence is now helping millions of affected individuals to regain control over their lives. Unfortunately, outdated views about drug use disorders (DUD) persist in many parts of the world. Stigma and discrimination commonly applied to opioid-dependent individuals and professionals working with them have significantly compromised the implementation of quality treatment interventions in this area, undermining the development of treatment facilities, health professionals' training, and investment in recovery programs. Like other medical problems such as HIV infection or hypertension, opioid use disorders (OUDs) are best managed within a public health system. Despite this evidence, the inclusion of addiction treatment in the healthcare system is still challenging in many countries where a considerable gap exists between science, policy, and clinical practice.

The goals of treatment in the management of OUDs are:

1. To reduce opioid use and cravings for drug use
2. To improve the health, well-being, and social functioning of the affected individual
3. To prevent future harm by decreasing the risk of complications and relapse

In addition to these criteria that have a clinical effectiveness focus, the treatment of OUDs should meet the common standards of all healthcare:

1. Be consistent with the United Nations (UN) Declaration of Human Rights and existing UN Conventions.
2. Promote personal autonomy.
3. Promote individual and societal safety.

14.1.1 What Is Medication-Assisted Treatment (MAT)?

MAT for OUD is an evidence-based complex intervention that uses substitute medication, alongside behavioral and psychological therapies effective in supporting the cessation, stabilization, and reduction of harms from illicit opioid use [1, 2]. Prescribing within MAT follows a framework including (a) assessment and initiation of opioid substitution therapy (OST), also known as opioid agonist treatment, or OAT, (b) maintenance treatment, and (c) detoxification. The first two stages are reviewed in this chapter. Chapter 15 covers detoxification, aftercare, and the care of special populations requiring MAT.

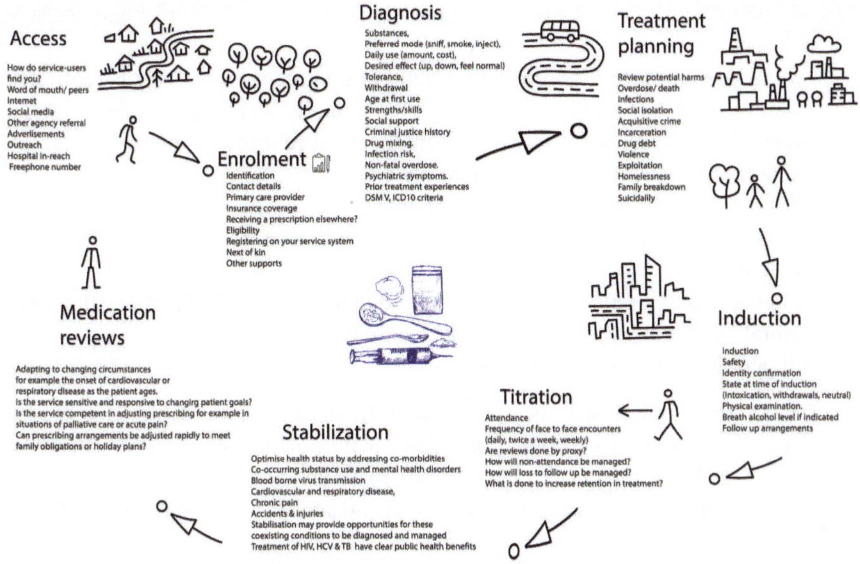

Fig. 14.1 Summary of medication-assisted treatment processes

Methadone is currently the most established pharmacological option, although other licensed drugs exist, such as buprenorphine and naltrexone, and unlicensed use of medications, such as slow-release morphine [3].

Figure 14.1 summarizes the whole MAT process.

14.1.2 MAT Standards

Comprehensive MAT provision can reduce drug-related deaths by 50% [4, 5]. However, there is an increased awareness that desirable outcomes from complex interventions such as MAT are contingent on contexts such as societal perspectives on addiction, institutional attitudes, ease of access into services, and availability and capacity of providers. For this reason, the Scottish Government's Drug Deaths Taskforce introduced a set of MAT standards in response to an ongoing drug deaths public health crisis [6]. The ten MAT standards are:

1. All people accessing services have the option to start MAT from the same day of presentation.
2. All people are supported to make an informed choice on what medication to use for MAT and the appropriate dose.
3. All people at high risk of drug-related harm are proactively identified and offered support to commence, re-commence, or continue MAT.

4. All people are offered evidence-based harm reduction at the point of MAT delivery.
5. All people will receive support to remain in treatment for as long as requested.
6. The system that provides MAT is psychologically and trauma-informed, routinely delivers evidence-based low-intensity psychosocial interventions, and supports the development of social networks.
7. All people have the option of MAT shared with primary care.
8. All people have access to advocacy and support for housing, welfare, and income needs.
9. All people with co-occurring drug use and mental health difficulties can receive mental healthcare at the point of MAT delivery.
10. All people receive trauma-informed care.

MAT is an intervention that goes far beyond the prescribing of OST. This chapter and Chap. 15 aim to consider the prescribing within MAT and focus on the above standards 1–5; however services that provide MAT should ensure that they consider the whole range of prescribing, psychological and social interventions they are able to offer.

14.2 Opioid Substitution Therapy (OST)

14.2.1 Assessment and Initiation of OST

Treatment thresholds can be defined as barriers a patient may face to and during treatment [7]. The traditional practice, or high threshold assessment, requires a complete full assessment before initiating OST. This approach often involves attending multiple appointments and the requirement for the patients to give detailed information around drug use, mental and physical health, social circumstances, and personal background. In addition, to prove OUD in this model often required multiple drug tests and attending in a clear withdrawal state at a prescribing appointment.

There is increasing recognition that patients often present to services seeking treatment within a limited window of motivation or at a time of crisis. Exhaustive interviews and in-depth assessment processes may result in losing the opportunity to engage patients in treatment and build a therapeutic relationship. Assessment must, therefore, always seek to balance the need to obtain adequate information with the risk of introducing unnecessary delay [8, 9].

14.2.1.1 Low Threshold Assessment for OST

The low threshold assessment model prioritizes gathering information to diagnose OUD and make a prescribing decision on the same day of the initial assessment.

The *Diagnostic and Statistical Manual of Mental Disorders*, 5th Edition [10], describes OUD as:

> …A problematic pattern of opioid use leading to problems or distress, with at least two of the following occurring within a 12-month period [10] (see Chap. 13):

1. **Taking larger amounts or taking drugs over a longer period than intended.**
2. **Persistent desire or unsuccessful efforts to cut down or control opioid use.**
3. **Spending a great deal of time obtaining or using the opioid or recovering from its effects.**
4. **Craving, or a strong desire or urge to use opioids.**
5. Problems fulfilling obligations at work, school, or home.
6. Continued opioid use despite having recurring social or interpersonal problems.
7. **Giving up or reducing activities because of opioid use.**
8. Using opioids in physically hazardous situations.
9. **Continued opioid use despite ongoing physical or psychological problems likely to have been caused or worsened by opioids.**
10. **Tolerance (i.e., need for increased amounts or diminished effect with continued use of the same amount).**
11. **Experiencing withdrawal (opioid withdrawal syndrome) or taking opioids (or a closely related substance) to relieve or avoid withdrawal symptoms.**

Therefore, the assessing clinician should focus initially on taking a clear history of the patient's substance use. Features suggestive, especially of opioid dependence (highlighted in bold), should be carefully considered. To be appropriate for same-day prescribing, patients would report using opioids regularly, usually daily.

Initial assessment should include:

- A clear account of the patient's opioid use, including frequency of use, quantity, and route of administration
- The time that an opioid was last used, including details of the substance and amount
- An account of all other substances being used, including alcohol
- Medical history, including details of any conditions that might affect a prescribing decision, e.g., chronic obstructive pulmonary disease (COPD) or cardiac disease
- Details of all currently prescribed medications, including any that may prolong the QTc interval
- Details of any previous overdose
- Details of any previous experiences in treatment with OST

Initial assessment should also include (a) physical examination looking for evidence of any objective withdrawal state or other signs indicating significant hazardous opioid misuse, such as injection sites; (b) drug testing, either oral fluid or urine; and (c) pregnancy test in women of reproductive age (see Fig. 14.2).

Patients testing negative for opioids should not usually be prescribed on the same day and require additional assessment.

While there are strategies to initiate OST when patients do not present in a withdrawal state, evidence of an objective withdrawal state may form a key part of the assessment. When a patient is not previously known to a service and does not

INITIAL ASSESSMENT / SAME DAY PRESCRIBING DECISION

- The goal of the initial assessment is to establish if it is safe and appropriate to prescribe OST on that day.
- Full assessment is an ongoing process and can be completed over time, there is not an expectation that the full psychosocial assessment will be completed at first attendance, or before OST is started.

1. DRUG AND ALCOHOL USE over the last 30 days (initial history)

What is being used?
How much is being used?
How often are opioids (& alcohol / other drugs) being used?
Routes of administration

Confirm use with a point of care drug screen (oral or urine) and Breath Alcohol Concentration test where appropriate.
IF INITIAL SCREEN IS NEGATIVE FOR OPIATES THEN NOT USUALLY SUITABLE FOR SAME DAY PRESCRIBING

2. ESTABLISH DIAGNOSIS / DEPENDENCE

1. Have you ever taken any of the following: Heroin, methadone, buprenorphine, morphine, oxycodone, tramadol, dihydrocodeine, codeine, and any other opioid analgesic?
2. Do you ever need to use more opioids to get the same high as when you first started using opioids?
3. Does the idea of missing a fix (or dose) ever make you feel anxious or worried?
4. In the morning, do you ever use opioids to keep from having "withdrawals" or do you ever have withdrawals?
5. Do you worry about your use of opioids?
6. Do you find it difficult to stop or not use opioids?
7. Do you ever need to spend a lot of time/energy on finding opioids or recovering from feeling high?
8. Do you ever miss important things like doctor's appointments, family/friends' activities, or other things because of opioids?

If answer to Q1 is yes and 3 or more of Q2 - 8 are yes = likely opioid dependent
If answer to Q1 is no or less than 3 answers to Q2 – 8 are yes = likely not opioid dependent.
Rapid Opioid Dependence Screen, (Wickersham et al 2015)(4)

Consider further available information to support same-day prescribing decision and diagnosis
Injection sites
Near fatal overdose presentations to other services
Collateral information from other agencies or healthcare providers
Previously history of treatment with OST
Assessment of physical withdrawal state (can use COWS)

3. ADDITIONAL INITAL INFORMATION

- Identification of physical health problems that may impact immediate prescribing decision e.g., COPD, heart disease
- Other prescribed medications
  - Any that prolong QTc?
- Identification of significant mental health problems
- Safe storage of medications / access to children
- Pregnancy test if needed
- Establish with the patient their expectations and view on starting OST
  - Do they wish to start OST today?

4. DECISION ON SAME DAY PRESCRIBING

1. Can a diagnosis of opioid use disorder, in particular opioid dependence be made today?
2. Would the benefits of starting OST immediately likely outweigh the risks?

There must not be any pressure on a clinician to start OST on the same day where it is not felt safe or clinically appropriate to do so.

5. CHOICE OF OPIOID REPLACEMENT THERAPY AND TITRATION

- If a decision is taken to start OST the choice of medication used should be taken jointly with the patient. Further sections of this chapter provide information to support this decision and guidance on titration.
  - If OST is not suitable to be started on the same day ongoing assessment should continue, with OST started as soon as possible, if then determined to be clinically appropriate.
- Full psychosocial assessment should be continued once OST has been started with a comprehensive individual treatment plan being formed as part of this.

Fig. 14.2 Suggested structure for the initial assessment

present in withdrawals, clinicians considering same-day prescribing should carefully satisfy themselves that there is other adequate evidence to support the diagnosis.

The initial assessment aims to allow the clinician to answer two straightforward questions:

1. Can a diagnosis of an opioid use disorder, in particular with opioid dependence, be made today?
2. Would the benefits of starting OST immediately likely outweigh the risks?

When the answer to both questions is "yes," the clinician should consider with the patient starting OST immediately and offer other essential interventions such as the training and supply of naloxone, provision of safe injecting equipment, and blood-borne virus testing, if not on the day of initial assessment, then as soon as possible.

It is worth emphasizing two points where the low threshold assessment model is in use. Firstly, the model aims to give the clinician the option to start OST on the day of the initial assessment. It does not, however, aim to place pressure on the clinician to do so. On the contrary, OST should only be started on the same day if the clinician is satisfied that the patient meets the criteria for OUD and that the benefits of doing so outweigh the risks. Secondly, the model changes the order of the traditional assessment approach, prioritizing the diagnosis of OUD and start of OST; however, the conventional comprehensive physical, psychological, and social assessment around the patient remains essential and should be continued following the start of OST.

Patients already in treatment with OST will likely engage better with further assessment and interventions.

Less commonly, in people who struggle to stabilize on oral medication, injectable options such as diacetylmorphine are available in some countries (Table 14.1). The World Health Organization (WHO), in 2005, added both buprenorphine and methadone to its Model List of Essential Drugs [11]. In addition, in 2009, the WHO emphasized OST as the key form of treatment for people who inject opioid drugs [12].

OST effectively supports the cessation of injecting drug use, significantly reducing the risk of HIV transmission [13]. For example, methadone maintenance therapy can result in a 54% reduction in HIV transmission risk among people who inject drugs [14]. OST is also associated with broader health, economic, and psychosocial benefits, including reducing crime and recidivism [8, 15–17]. In many countries, methadone is the preferred opioid agonist for OST, but buprenorphine is an excellent alternative. The following sections detail some differences between these medications.

14.2.1.2 Clinical Decision-Making: Methadone or Buprenorphine?

- *Methadone*

Methadone is a synthetic full μ-opioid receptor agonist, though it also acts on δ- and, to some degree, on κ-opioid receptors [18]. Methadone maintenance treatment first began in New York City in 1964, and it remains a primary treatment option in opioid dependence [19]. The pharmacokinetics of methadone varies considerably between individuals, but it is effective when orally administered, has a slow onset of action, and has a prolonged half-life (13–55 hours) [20]. Following

Table 14.1 An international perspective on OST

Medication	Formulation	Availability	Reference
Methadone	Oral formulations (liquid, tablets)	Widely available but illegal in some countries, for example, Russia	
	Single isomer levo-methadone	Germany	[41]
Buprenorphine	Sublingual tablets	Widely available but illegal in some countries, for example, Russia	[42]
	Oral lyophilisate freeze-dried wafer	UK, India	[43]
	Buprenorphine with naloxone	Widely available but illegal in some countries, for example, Russia	[44]
	Buprenorphine depot injection (subcutaneous) (given weekly or monthly)	UK, Europe, USA, Australia	[45]
	Buprenorphine implant (6 monthly)	Europe	[31]
Naltrexone	Tablets	Widely available	[46]
	Implant	Russia, US	[47]
Dihydrocodeine	Tablets	UK	[3]
Morphine	Sustained release oral morphine tablets or capsules	A range of European jurisdictions, Canada	[48]
Intravenous OAT	Heroin-assisted treatment (diacetylmorphine or diamorphine)	UK, the Netherlands, and Switzerland	[49]
	Injectable methadone	UK	[14]
	Injectable hydromorphone	Canada	[50]

long-term administration, methadone induces its own metabolism [21]; however, its long half-life allows for once-daily administration for most patients. Liquid preparations also allow for easier supervision and make diversion from supervised self-consumption more difficult.

Methadone effectively suppresses the craving to use opioids and relieves opioid withdrawal symptoms. Adequate dosing creates cross-tolerance, reducing the risk of overdose and the euphoric effects of other opioids, such as heroin. As a full opioid agonist, overdose with methadone can produce respiratory depression, and particular care should be taken around its initiation. Methadone can cause prolongation of the QTc interval, most especially in patients with other cardiac risk factors and on higher doses [22].

- *Buprenorphine*

Buprenorphine is a semisynthetic opioid that primarily acts on μ-opioid receptors as a partial agonist; it is also a κ-antagonist with a ceiling effect [23]. This flattening of the dose/effect curve in causing respiratory depression, sedation, and subjective euphoria distinguishes it from full opioid agonists such as methadone.

It also makes it safer to use in patients with significant physical comorbidities or ongoing comorbid harmful opioid, polydrug, or alcohol use [24].

In addition, buprenorphine has a high receptor affinity; therefore, it can displace some full opioid agonists from opioid receptors, but it is not easily displaced by them. This property often leads to buprenorphine being described as having a "blocking" effect on other opioids, with full agonists unable to produce dose-related opioid effects on receptors already occupied by buprenorphine. However, equally this means that when introduced, buprenorphine may precipitate a withdrawal syndrome in patients dependent upon full opioid agonists, such as morphine, heroin, or fentanyl.

Buprenorphine has proven efficacy; its half-life, high receptor binding affinity, and slow disassociation make it suitable for daily, alternate day, and three times weekly administration [25]. Buprenorphine is available as sublingual tablets, films, and oral lyophilisates (wafers that disperse buprenorphine rapidly on the tongue). Preparations combined with naloxone can be used when there is concern around potential diversion and injection. Several different long-acting preparations are also available, removing the need to attend for supervised self-consumption or for patients to maintain the routine of taking regular oral medication.

Arguments to consider buprenorphine as the first-choice treatment over methadone, especially for patients not previously treated with OST, include:

- The higher intrinsic safety of buprenorphine in overdose and low incidences of fatal overdose [26]
- The lower risk to the community in case of buprenorphine diversion, due to its relative inherent safety
- The opiate receptor "blocking" effect of buprenorphine, reducing the incentive to use other opioids "on top" of treatment, due to decreased euphoric effects
- The greater ease of detoxification from buprenorphine for patients who wish to move toward abstinence
- The "clear headedness" of buprenorphine, which potentially increases the likelihood of normalizing social and occupational functioning for those in treatment

The clinician and patient should jointly choose the preferred agent, strongly considering the patient's preferences. However, there are factors that clinicians should consider when making this decision; among them, it is essential to determine opioid use, level of tolerance, and if the patient is currently presenting in opioid withdrawal.

Table 14.2 compares the advantages and risks of using methadone or buprenorphine in OST.

14.2.1.3 Initiating Treatment with Methadone

Methadone is available as both tablets and oral solutions. Tablet forms of methadone vary considerably between different countries and regions, for example, in the UK only 5 mg tablets are available, while in the USA a 40 mg dispersible tablet is available. If tablet forms are used, clinicians should ensure that they are familiar

Table 14.2 Clinical comparison between methadone and buprenorphine

	Methadone	Buprenorphine
Requirement for natural withdrawal state when started	No risk of precipitated withdrawals.	Risk of precipitated withdrawals
Overdose risk	Full μ-opioid receptor agonist, greater risk of overdose (intentional or accidental) when used with other opioids or in polydrug use	Partial μ-opioid receptor agonist and κ-opioid receptor antagonist, lower risk of overdose (intentional or accidental) when used with other opioids or in polydrug use. Should be considered as preferable where the risk of overdose is particularly high (e.g., previous overdose, chaotic polysubstance use, and high dose, groin injecting)
Comorbid alcohol use disorder	Higher risk (more sedative)	Lower risk (less sedative)
QTc prolongation	More likely to prolong QTc	Less likely to prolong QTc [51]
Interaction with other medications	Inducers/inhibitors of CYP3A4 may alter plasma levels, e.g., some SSRIs and erythromycin	Less likely to be affected by interactions with other medications
Retention in treatment	May be more likely to retain patients in treatment than low dose buprenorphine (<7 mg) [33]	Associated with worse retention in treatment if doses <7 mg are used. No difference to methadone for doses >7 mg [33]
Clear headedness/ level of sedation from treatment	Does not give clear-headedness. Patients with comorbid mental health symptoms (e.g., anxiety or trauma symptoms) may benefit from the greater sedative and anxiolytic effect	Gives clear-headedness and less sedation
Patients desire for a period of stability	Suitable for patients seeking a longer period of stability	Suitable for patients seeking for longer periods of stability. Patients aiming for more rapid detoxification (within 12 months) are likely to find this better tolerated
Withdrawal symptoms when withdrawn	More marked and prolonged compared to buprenorphine	Less marked and prolonged compared to methadone. It may be easier for patients to tolerate detoxification
Extended-acting preparations/ dosing	No long-acting preparation available. Daily dosing required	Long-acting preparations available. Less than daily oral dosing is possible
Pregnancy	If pregnant and already on methadone, this should be maintained. If not already on OST, should be offered methadone	If pregnant and already on buprenorphine, this should be maintained. Should not be started during pregnancy

with what is available locally and the associated licensing and prescribing guidelines. A potential disadvantage of tablets is that they are relatively easy to crush and inject. Generally, oral solutions are preferred as they are easier to monitor in supervised self-consumption, with a lower risk of diversion. The concentration of oral solutions can again vary between different countries and regions, most commonly 1, 5, or 10 mg/ml. Prescribers and dispensers should again be careful to make themselves familiar with the preparation used locally. As a general recommendation, services should use a single strength of preparation wherever possible to reduce the risk of prescribing or dispensing errors.

- *Community titration: Patients presenting in opioid withdrawal*

Where possible, patients should be encouraged to present to begin treatment in an objectively assessable withdrawal state. This is evidence to aid in diagnosing opioid dependence and assists, alongside the history taken, in assessing the likely level of an individual's tolerance. Tools such as the Clinical Opiate Withdrawal Scale (COWS) [27] can aid the clinician in this assessment (see Chap. 13).

When starting methadone, the rule "start low and go slow" should always be followed. Too high of an initial dose or too rapid dose increases can elevate the overdose risk. The initial starting dose should be between 10 and 30 mg daily, based on clinical assessment of likely tolerance, considering the reported frequency of opioid use, route of administration, and use of other drugs. Starting doses of greater than 20 mg daily should not be used if there is uncertainty around tolerance or if this is likely to be low. Deaths have occurred with starting doses of under 30 mg [8]. Dose increases should not be more than 5–10 mg on a single day, and doses should be maintained for at least 2 days before further increases, longer if there is doubt over opioid tolerance. There should be a maximum of 30 mg increase in any 1 week. Table 14.3 shows examples of methadone titration regimes.

Patients should be counseled around the increasing effect of multiple doses while reaching a steady-state, so they do not excessively "top up" with illicit drugs. Clinicians should explain that achieving a steady state of methadone takes around 3–10 days; during this time, methadone plasma levels rise even when someone remains on the same dose. This means that doses tolerated on day one can become toxic several days later.

A therapeutic dose of methadone is generally between 60 and 120 mg daily to give adequate cross-tolerance to other opioids and provide a protective effect against overdose. While the "high-dose methadone start" regime demonstrates titration up to 80 mg daily in 2 weeks, for patients who have not received methadone previously, it is advisable to have an initial titration goal of 60 mg daily. Clinicians can consider

Table 14.3 Examples of methadone titration regimes

Low-dose methadone start. For low opioid use and tolerance uncertain														
Day	1	2	3	4	5	6	7	8	9	10	11	12	13	14
Dose (mg)	10	10	10	15	15	20	20	20	25	25	25	30	30	30
High-dose methadone start. For patients with high opioid use and tolerance likely														
Day	1	2	3	4	5	6	7	8	9	10	11	12	13	14
Dose (mg)	30	30	40	40	50	50	50	60	60	70	70	70	80	80

further titration if patients are unable to stabilize on this dose. Some may require doses over 120 mg daily to stabilize, and this should be done cautiously.

Patients receiving methadone 100 mg daily and above should have ECG monitoring to check for a prolonged QTc; this should also be considered for patients on lower doses if they have existing cardiac disease or other risk factors.

- *Community titration: Patients not presenting in opioid withdrawal*

Some patients may struggle to tolerate a withdrawal state and may not present in evident, objective withdrawal, even when asked to attend on multiple occasions. Where clinicians following assessment are satisfied that there is clear evidence of opioid dependence, they can begin a cautious induction of methadone using the "low-dose methadone start," as detailed above. This induction should be done only where experienced clinicians consider that the risks of a further delay to starting methadone justify this and the patient is not intoxicated and understands and consents to the risks involved.

Clinicians should exercise extreme caution if patients present as intoxicated, either with opioids or other substances, especially depressants. These patients would not be suitable to be started on that day and should be encouraged to return as soon as possible when not in an intoxicated state. In addition, given the nature of methadone as a full opioid receptor agonist and the overdose risks when used alongside other opioids or depressants, "self-titration" of initial doses without supervision should not be considered.

- *Tolerance testing with methadone*

Methadone tolerance testing titrates a methadone dose against the subjective symptoms and objective signs of opioid withdrawal. Tolerance testing aims to determine the dose of methadone, which can fully relieve and prevent withdrawal without producing sedation. It is an alternative to community titration. In this time-intensive procedure, the patient must be willing and able to attend a clinical setting for 2 full days, abstain from opioid use, and be in withdrawal on the first day. Tolerance testing allows for an experienced prescriber to judge an initial starting dose of methadone and then, following a period of observation on site, administer further doses if indicated. With a similar procedure, on the second day of attendance (at least 72 hours, but not more than 7 days from the first), patients can more rapidly achieve an effective treatment dose of methadone.

Methadone tolerance testing has potential advantages to a community titration where:

1. The patient is reporting high levels of opioid use or complex patterns of substance use, making the assessment of target dose especially difficult.
2. The stability of the patient's drug use cannot be confirmed. The procedure, however, should only be undertaken by services with clear protocols and structures in place (including the availability of naloxone).

The importance of close supervision and observation of patients undergoing tolerance testing cannot be understated, given overdose risks. Changes in the patient's conscious level are a cause of concern, and any signs of unexplained intoxication require an urgent response. There are examples of protocols from services experienced in this area [28].

14.2.1.4 Titration with Oral Buprenorphine

As previously mentioned, buprenorphine is available in different oral preparations, including sublingual tablets, buprenorphine/naloxone combination in sublingual and buccal films, and oral lyophilisates. Therefore, although all doses refer to buprenorphine as generic sublingual tablets within this chapter, dose adjustment may be needed with other preparations, which frequently do not have equivalent bioavailability. In addition, the availability of brands and preparations can vary considerably between different countries.

- *Community titration: Patients presenting in opioid withdrawal*

As with methadone, where possible, patients should be encouraged to present to begin treatment in an objectively assessable withdrawal state; this both presents evidence to aid in making the diagnosis of opioid dependence and reduces the chance of patients experiencing a precipitated withdrawal state when starting buprenorphine. Tools such as the Clinical Opiate Withdrawal Scale (COWS) again can be used to aid clinicians in their assessment of the degree of withdrawal [27]. With a COWS score of 12 or above, it is unlikely that initiating buprenorphine will cause precipitated withdrawal. Clinicians should also clarify that the last dose of a short-acting opioid, such as heroin, has been taken at least 6–12 hours before starting buprenorphine. Patients who report using long-acting opioids such as methadone should ensure that at least 24 hours have elapsed before the initial dose of buprenorphine.

The starting dose of buprenorphine should be between 4 and 8 mg, depending on the clinician's assessment of the patient's likely opioid tolerance. While dose increases of 2–4 mg daily are often adequate, dose increases of up to 8 mg daily are considered safe. Patients initiated on 8 mg on day one of treatment can be safely increased to 16 mg on day two.

A therapeutic dose of buprenorphine is generally between 8 and 16 mg daily, though some patients may require up to 32 mg daily. Doses of 8 mg daily and over are likely to have a more significant effect in blocking the effect of additional opioid use through occupying opioid receptors and are more protective against opioid overdose [29]. As with methadone, patients should be encouraged to stabilize on a dose within the therapeutic range aiming to achieve relief from cravings to use opioids, relief from withdrawal symptoms, and at a level where cross-tolerance exists and, in the case of buprenorphine, the "blocking" effects reduce the experience of euphoria when other opioids are used.

- *Patients not presenting in opioid withdrawal*

It must be accepted that in services offering same-day prescribing at a low threshold, patients will present seeking treatment with buprenorphine who are not in an adequate withdrawal state to have this started. Given its better relative safety and the risk of precipitating a withdrawal state if using other full opioid receptor agonists, clinicians can consider dispensing the initial dose of buprenorphine without supervised self-consumption. With this approach, the patients can take buprenorphine home and "self-titrate" when they are satisfied that they have entered an adequate withdrawal state. Then, buprenorphine can continue the following day, and supervised self-consumption can begin, with no dose adjustments to the usual titration outlined above.

Where clinicians consider this as an option, care should be taken to ensure that patients clearly understand why precipitated withdrawals may occur and that while not life-threatening, this is an extremely unpleasant experience. In addition, clinicians must ensure that patients have an understanding of the level of natural withdrawal expected before they take an initial dose of buprenorphine. The initial dose can be dispensed as lower-dose (usually 2 mg) tablets to reduce this risk further, advising the patient to take 2 mg as an initial "test dose" and allow at least an hour to see if they experience any withdrawal features before taking the remainder. Consideration could also be given to using a lower total starting dose, e.g., 4 mg for the first day with subsequent titration.

- *Buprenorphine micro-dosing*

Conventional buprenorphine induction recommendations require a patient to be in moderate opioid withdrawal before initiation. This can take 8–24 hours for short-acting opioids and 48–72 hours for long-acting opioids such as methadone. The opioid withdrawal symptoms experienced during this wait time may be intolerable or impractical for some patients. This situation may lead to lower overall utilization of buprenorphine even though it has a superior safety profile compared with methadone, specifically in those with:

- Respiratory compromise, for example, COPD
- Cardiovascular disease
- Uncertain or low tolerance
- Short treatment histories or intending rapid detoxification
- Polysubstance use with multiple sedative drugs

Buprenorphine micro-dosing, also commonly referred to as the "Bernese method" [30], involves overlapping low doses of buprenorphine with the patient's continued use of a full opioid agonist, which effectively means that the patient does not need to reach moderate withdrawal (see Table 14.4 for examples of published buprenorphine microdosing regimens; Table 14.5 gives a suggested slower schedule for patients on higher doses on methadone).

Buprenorphine micro-dosing is currently an off-label method of OST induction, so there are limited national guidance documents. However, countries such as

Table 14.4 Examples of various buprenorphine micro-dosing schedules

	Day	1	2	3	4	5	6	7	8	9	10	11
Bernese method [20]	Dose (mg)	0.2	0.2	0.8 + 2	2 + 2.5	2.5 + 2.5	2.5 + 4[a]	4 + 4	4 + 4	8 + 4[b]	Titrate PRN	
Terasaki et al. (2019) [25]	Dose (mg)	0.5	0.5 bd	1 bd	4 bd	8	8 + 4	12[c]	Titrate PRN			
VCH [22]	Dose (mg)	0.25	0.25 bd	0.5 bd	1 bd	2 bd	4 bd	12[b]	Titrate PRN			
Lu and Cho (2018) [27]	Dose (mg)	0.5 bd	1 bd	2 bd	3 bd	4 bd	12	16[b]	Titrate PRN			
Tay et al. (2021) [30]	Dose (mg)	0.4	0.4	0.8	1.2	1.6	1.6	2[a]	4	6	8–12	16[b]

VCH Vancouver Coastal Health, *bd* twice a day, *PRN* as required
[a]Tapering of primary or illicit opioid
[b]Primary or illicit opioid stopped here
[c]Methadone dose stopped here

Table 14.5 A suggested slower buprenorphine micro-dosing regime[a] where the patient is on a high methadone dose

Day	1	2	3	4	5	6	7	8	9	10	11	12	13	14
Dose (mg)	0.4	0.4	0.8	1.2	1.6	1.6	2[b]	4	6	6	8	8	16	24[c]

[a]Daily doses should be split to twice or three times a day from day three onwards
[b]Start tapering down methadone (sooner if tolerated); mg = milligram
[c]Cease all other opioids at this point

Canada, the US, and Germany have used this method extensively, and more recently, the UK and Australia have been developing expertise in this method. For this reason, the patient must understand the risks and benefits of micro-dosing and the implications of off-label use of medications, such as product information leaflets or labeling on medication not matching instructions given by their clinicians.

Patients eligible for micro-dosing induction onto buprenorphine include:

- Patients who fear withdrawal or experience severe withdrawal symptoms during conventional induction (moderate withdrawal is required to begin a traditional induction)
- Patients who have failed a conventional induction due to inability to tolerate mild withdrawal
- Patients with significant social instability make attending a scheduled clinic appointment for induction challenging (examples may include no access to reliable transport or a lack of financial resources for transportation, chaotic lifestyle, or lack of social support, among others)

Particularly good candidates for micro-dosing include:

- Patients being switched from methadone, or other high-dose long-acting opioids such as slow release oral morphine or slow release oxycodone, to buprenorphine. Due to these medication switches' complexity and case-by-case variability, specialist guidance must be sought to ensure appropriate customization of the micro-dosing schedule, including appropriate cross-titration of methadone or other high-dose long-acting opioids being discontinued. There are several schedules with evidence listed above.
- Patients who use illicit fentanyl or fentanyl analogs (due to the uncertain risk of precipitated withdrawal).
- Patients who may be unable to tolerate moderate withdrawal due to a medical or mental health condition.
- Pregnant women who are not currently in withdrawal and for whom methadone is contraindicated, refuse treatment with methadone, or do not have access to methadone treatment in their home community. Inpatient admission for induction is preferable because withdrawals can result in miscarriage or premature labor (see Chap. 15).
- Individuals who present to the emergency department or other acute or unscheduled care facilities with severe complications of opioid use disorder such as overdose, infectious complications, or a mental health crisis. These patients may not be in enough withdrawal to facilitate a conventional buprenorphine induction nor be appropriate for a home induction. Providing a limited supply of buprenorphine with precise micro-dosing instructions may effectively engage this high-risk population in care. Appropriate arrangements for follow-up care in the community must be in place.

Clear communication with the dispensing service or community pharmacy is essential to facilitate a smooth micro-dosing induction process. In addition, the community dispensary that the patient attends for continued treatment must be contacted and made aware of the plan since the patient may seek support from them at any time during the micro-dosing induction process. Further, as is their duty, pharmacists or dispensers may well decline to dispense medications without clearly understanding the rationale.

- *Long-acting buprenorphine preparations*

Long-acting buprenorphine preparations are now available and licensed in multiple countries, more commonly as prolonged-release subcutaneous injections and in some areas also as an implant. For some patients with work or study commitments, a need to frequently travel, or a strong preference against attendance for oral dosing, these preparations are effective and share the same advantages and benefits of buprenorphine treatment, such as the blockade effect when other full opioid agonists are used. Some patients have also reported that the more steady plasma levels achieved with these products reduce the experience of not feeling "held" and entirely comfortable between oral doses, as well as initial data suggesting generally high rates of satisfaction with treatment and positivity around these additional treatment options being available [31, 32].

Induction and titration vary between different preparations, with some allowing direct start and titration and others requiring establishment on oral buprenorphine before conversion. Therefore, prescribers should take care to familiarize themselves with the products available in their country and their licensed usage.

14.3 Maintenance Therapy

Maintenance therapy with OST is suitable for people who want to stop using illicit opioids but are unable to achieve abstinence from all opioids; it is more appropriate for adults with a significant history of dependence than those moving to rapid detoxification. It forms a key component of MAT. Maintenance therapy supports, rather than prevents patients from returning to a stable lifestyle and improves their physical health, mental health, and social functioning. It should be seen very much as a step within an individual's recovery journey. There should be no time limit placed on how long an individual stays within treatment [9], and each patient should have their own individualized care plan to support them in moving forward in the recovery.

Care plans require regular revisions and routine consideration of the following areas:

- Drug and alcohol misuse
- Physical and mental health
- Participation in rehabilitation, counseling, relapse, and other psychosocial support programs
- Progress with family relationships, training, and employment
- Housing
- Offending and criminal justice involvement

As patients move from the initiation to the maintenance phase of treatment, supervised self-administration of medication should be provided for a length of time appropriate to the patient's individual needs and risks. Relaxation of supervised self-administration can act as an incentive if progress is being made but should only be permitted where:

- A stable dose has been reached.
- Illicit drug and alcohol use have ceased.
- The patient's mental health is stable, and there is no risk of self-harm.
- Medication is stored safely at home, particularly where children are present.
- There is no concern of inappropriate use or diversion of medication.

Random drug testing helps monitor ongoing stability and generally should be considered at least twice a year.

As patients stabilize, there is a risk of "overtreatment" where requirements to attend regularly, frequent medication collections, and supervision may have detrimental effects on their ability to return to or sustain a stable lifestyle and higher levels of social functioning [9]. Therefore, these activities should be part of treatment plans that are not arbitrary but tailored to individual circumstances and based on individual needs and risk assessments. In addition, services should aim to ensure flexibility in appointments, especially for those who may be homeless or have other comorbidities or social issues that may affect their ability to organize their time [8].

14.3.1 Optimal Dosing

A key goal is to provide a dose that leads to complete cessation of opioid use, which may well be higher than the dose at which the patient feels "stable." It may take several weeks to reach the desired optimal dose. To achieve optimization after induction, doses can continue to be increased gradually. A total target dose of between 60 and 120 mg daily of methadone, and occasionally more, may be required. Doses of over 8 mg daily of buprenorphine are associated with better retention in treatment [33] and protection against opioid overdose [29]. Generally, doses of buprenorphine 12–16 mg daily are appropriate; sometimes, up to 32 mg daily may be required.

Caution needs to be exercised, balancing any assessed risk of increasing dose with optimizing treatment where the patient continues to use illicit opioids. For example, a patient may believe that continuing intermittent lapses are due to a lack of willpower when too low a dose is the determining factor. This may need explaining to patients unwilling to increase the dose beyond that which makes them feel comfortable. Additionally, patients' perception of their dose is also significant, with patients perceiving themselves to be on adequate doses of methadone more likely to be retained in treatment than those who see their dose as inadequate [34, 35].

Some patients may be unwilling to increase their dose because they intend to continue to use heroin. This should be addressed, increasing input if needed, but should not stop positive feedback where other significant improvements are achieved. On the other hand, a small minority of patients may persistently seek higher doses during maintenance to seek a psychoactive drug effect. These patients need to be identified and managed without further dose increases.

14.3.2 Less than Daily Dosing with Buprenorphine

Some patients can be maintained successfully taking oral buprenorphine less than daily [36, 37]. This regimen has advantages, especially in allowing ongoing supervised self-consumption for all doses, without the need for daily attendance.

Buprenorphine may also allow for attendance three times weekly, with a 3-day dose to cover the weekend. In some countries, this dosing strategy may fall outside of licensed use, specifying daily dosing. Quantities can be given to cover either a 2- or 3-day period as below. However, the limited amount able to be given as a single dose means the strategy may not be tolerable for those requiring higher daily doses.

Equivalent less than daily doses:

Two-day buprenorphine dose → 2× usually daily dose, maximum 32 mg.
Three-day buprenorphine dose → 3× usual daily dose, maximum 32 mg.

14.3.3 Missed Doses

For patients on oral medication, missing doses for 3 or more consecutive days risks reduction in opioid tolerance, placing patients at an increased risk of overdose when recommencing medication. In these cases, the dose of medication should be withheld and advice sought from the prescriber.

Due to its higher overdose potential, methadone requires particular care. However, the following schedule, in general, can be presumed to be safe:

No. of days missed
1–2: *No change in dose.*

A regular dose may be taken if no evidence of intoxication.

3–4: *Advice must be obtained from the prescriber.*

It is usually safe to continue from half the current dose, and re-titration should be undertaken as required.

5 or more: *Regard as new induction.*

It is not safe to assume any degree of opioid tolerance. Therefore, the existing prescription should be stopped, and a further full assessment and re-titration should be undertaken.

The same principles apply when patients have missed buprenorphine doses. However, the risk of overdose is relatively lower, and experienced prescribers may consider continuing the full dose with 3 days missed or a lesser dose reduction. However, there is the risk of precipitated withdrawal if the patient has been using other opioids, increasing as the time from the last dose of buprenorphine lengthens. Therefore, other opioid use should be clarified with the patient as part of the decision on how to continue treatment.

14.3.4 When Patients Are Not Benefiting from Treatment

When there is a relapse to or continued heroin or other drug use, patients are not fully benefiting from treatment and should be reviewed further. Services should aim to ensure that adequate psychosocial supports are in place and the intensity of these increased, if needed. It should be ensured that OST is at an optimal dose, with dose increases offered if required. For patients already on high doses, alternative substitute medication should be considered.

If a relapse has occurred, the key worker should try to discover what has triggered it. They should aim to go on to support the development of techniques to avoid a breakdown in progress, encouraging participation in relapse prevention, counseling, training, and employment. If supervised self-consumption is not already in place for all doses, reinstating this should be considered.

14.3.5 Excluding Patients from Treatment

The requirement to abstain from all illicit substances to access or continue with MAT should not be in place. In addition, dose reductions or other punitive actions (such as "disciplinary discharge") due to ongoing substance use actively damage engagement and retention in treatment [9].

When patients' behavior toward staff or other interactions with services are hostile, threatening, or otherwise grossly inappropriate, the decision to discharge them and exclude them from MAT should still not be taken lightly. Removing patients from treatment is likely to increase their risk of overdose, offending, and contracting a blood-borne virus. It may also increase the risks to children or other vulnerable adults present in the home environment. Services should explore all possible options, including, where available, treatment within another service or setting and other avenues of conflict resolution, before deciding to discharge.

14.3.6 Supervised Self-Administration

Supervised self-administration by an appropriate professional provides the best guarantee that the patient takes medicine as directed. It is recommended initially and should continue for a time suitable to the patient's individual needs and risks. This can be variable as some countries may have statutory restrictions and requirements before supervised self-administration can be relaxed [17].

Supervised self-administration enhances compliance, reduces the potential for sharing or selling medication, increases prescriber confidence in prescribing higher doses, ensures regular contact with healthcare professionals, and introduces routine due to daily attendance. However, there are also disadvantages such as inconvenience, difficulties in attending for people in employment, complications for patients with child care issues, stigmatizing patients receiving OST, and reduced personal responsibility. It is important that once patients have stabilized, they are trusted to accept some responsibility for their treatment by the introduction of "take-away" doses. Therefore, assessment of suitability for take-away doses should include consideration of the following criteria.

14.3.6.1 Indicators for Supervised Self-Administration

- Recurrent failure to attend appointments and frequently missed doses.
- Continued or return to an irregular pattern of illicit opioid, alcohol, benzodiazepine, or other drug misuse.
- Relevant child protection concerns.
- The patient has not reached a stable dose.
- The patient has significant, unstable psychiatric or physical morbidity or is threatening self-harm.
- Continued or returned concern that prescribed medication is being, or may be, diverted or misused.
- Homelessness or significant social instability.
- Patient recommencing methadone or buprenorphine prescription or a significant increase in the daily dose.

14.3.6.2 Indicators for Take-Away Doses

- Regular attendance at appointments and for medication dispensing.
- Negative drugs tests for illicit drugs (2–3 negative tests usually would be sufficient).
- No child protection concerns (including safe storage of medication being available).
- The patient is prescribed an adequate daily dose.
- There is no significant or destabilizing psychiatric or physical morbidity.
- Positive drug tests are for prescribed medication only.
- Evidence of stable home and social environment.

Figure **14.3** shows the suggested stages of reduction in dispensing frequency and supervision. Supervision less than three times weekly is unlikely to be clinically useful and especially for ***methadone may increase risk, if a patient has not been taking their medication as prescribed and is then supervised a high dose after a gap of several days with loss of tolerance.***

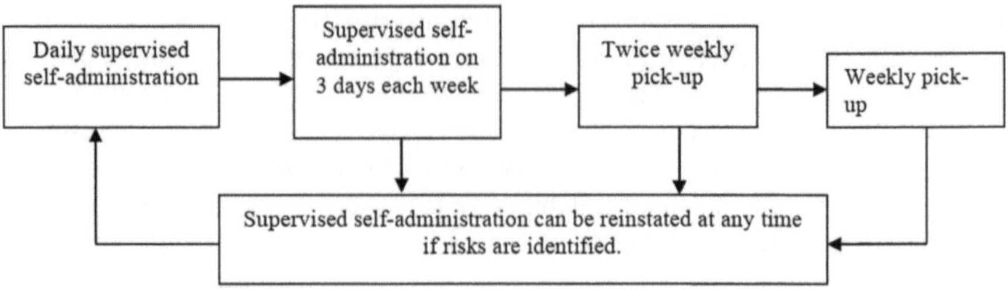

Fig. 14.3 Suggested pathway from supervised self-administration to weekly pick-up

14.3.7 Drug Testing

Drug testing is commonly performed in clinical settings using either urine or oral fluid. Point-of-care testing is now widely available in most countries but may be more limited in the range of substances screened for and relies on immunoassay testing. This limitation can provide less reliable and specific results than gas chromatography-mass spectrometry, which may be available for samples sent for laboratory testing. Nevertheless, access to point-of-care testing is essential if a low threshold approach allowing for same-day assessment and prescribing is adopted. However, access to "instant" results is often not required within the maintenance phase of treatment, and more comprehensive laboratory analyses are preferable.

Urine testing generally has a longer window of detection for most substances (several days), but unless production of the specimen is witnessed, it is more susceptible to adulteration or substitution. Oral fluid is easier to collect, but substances are generally present in lower concentrations, and only more recent use (typically 24–48 hours) is captured. As it is easy to witness samples collection, adulteration or substitution is generally very difficult.

Drug testing should be used for:

- Supporting the initial assessment and diagnosis; confirming drug use
- Ensuring compliance with OST
- Monitoring illicit drug use and returning evidence of progress (or lack of) during treatment to the patient
- Meeting legal requirements

A patient's self-report of illicit substance use should be considered equal to a drug test result and recorded as such. It is not always necessary to confirm self-reported use with testing though this should be done intermittently and more frequently if concerns exist of non-disclosed substance use. Even when stable, patients in maintenance treatment should have at least two random drug tests per year, without the patient knowing in advance. If patients refuse to provide samples for testing, clinicians should interpret this within the overall clinical picture, but this would usually suggest illicit use.

The range of substances tested for will often vary between different areas. Some laboratories may adjust the range of substances tested to reflect patterns of use in their local communities. At a minimum, testing should confirm the presence of any prescribed OST and illicit opioid use. Clinicians should be conscious of the range of substances tested for and those in use in their local community that cannot be detected. Drug tests should always be interpreted as part of the larger clinical picture and not in isolation.

14.4 Physical Monitoring for Those on OST

14.4.1 Cardiovascular Disease and ECG Monitoring for Patients on Methadone

While the evidence for QTc screening strategies in preventing cardiac mortality for those receiving methadone remains limited [38], it remains strongly recommended to assess patients receiving methadone for the risk of QTc prolongation and offer ECG monitoring to all considered at potential risk. Risk factors include:

- Prescribed ≥100 mg methadone
- Prescribed <100 mg methadone daily and taking any other medication that prolongs the QTc interval (e.g., some antidepressants, some antipsychotics)
- Prescribed <100 mg methadone and taking any other medication which inhibits metabolism by the cytochrome P450 pathway
- Family history of cardiac conditions or sudden death
- Patient history of syncope, palpitations, shortness of breath, seizures, or cardiac conditions
- Symptomatic presentation at an appointment – pallor, sweatiness, cyanosis
- Patient using stimulants
- Hypothyroidism
- Liver disease
- Malnourishment
- HIV infection
- Anorexia nervosa
- Alcohol dependence

If an ECG is reported as showing QTc abnormalities, the following actions should be taken:

- Offer to repeat ECG and U&Es to females with QTc > 469 and males with QTc > 439. QTc risk factors should be reviewed and modified where possible.
- Refer patients with QTc > 499 for cardiology opinion. Consider reducing the methadone dose or switching to buprenorphine. Continue with ECG monitoring until QTc normalizes.
- Urgently refer patients with QTc > 550 to cardiology, reduce methadone dose immediately, and give urgent consideration to switching to buprenorphine in an inpatient environment.

14.4.2 Buprenorphine and Liver Function

There have been previous concerns over the potential effect of buprenorphine to elevate liver function tests or exacerbate hepatic pathology. As a result, some guidelines previously recommended checking patients' liver function tests and reviewing blood-borne virus status (to exclude viral hepatitis) before treatment commencing. This approach is not consistent with the low-threshold assessment and initiation model, as it essentially excludes buprenorphine from same-day prescribing. Evidence suggests that buprenorphine does not selectively cause abnormal liver function compared with methadone and that patients known to be hepatitis C positive tolerate buprenorphine treatment [39, 40]. Therefore, buprenorphine can be used in same-day prescribing, though all patients should be offered tests for blood-borne viruses and those found positive offered liver function tests. Prescribers must, however, always also ensure that they are following local guidelines and medication licensing requirements for the country in which they are practicing, which may vary. Should baseline liver function tests be required, services should aim to ensure these are arranged as rapidly as possible to avoid delays in accessing treatment.

It would also be recommended that liver function be monitored for patients receiving buprenorphine:

- With known pre-existing liver enzyme abnormalities
- Who are already known positive for viral hepatitis
- Who use other potentially hepatotoxic medicines/substances such as alcohol

14.5 Management of Opioid Users Admitted to Hospital Settings

Patients requiring admission to acute hospital or psychiatric settings should not be detoxed and maintained on their existing OST. While not life-threatening, opioid withdrawal is highly unpleasant and carries risks of patients seeking to self-discharge and not receiving appropriate medical care if it is not correctly managed. Figure 14.4 gives a suggested approach to the management of opioid-dependent patients following admission, but successful care will require close working with the local addictions team. As described in Chap. 15, symptomatic relief can be used if there is initial uncertainty around assessment and while seeking information, but this is not a substitute for OST.

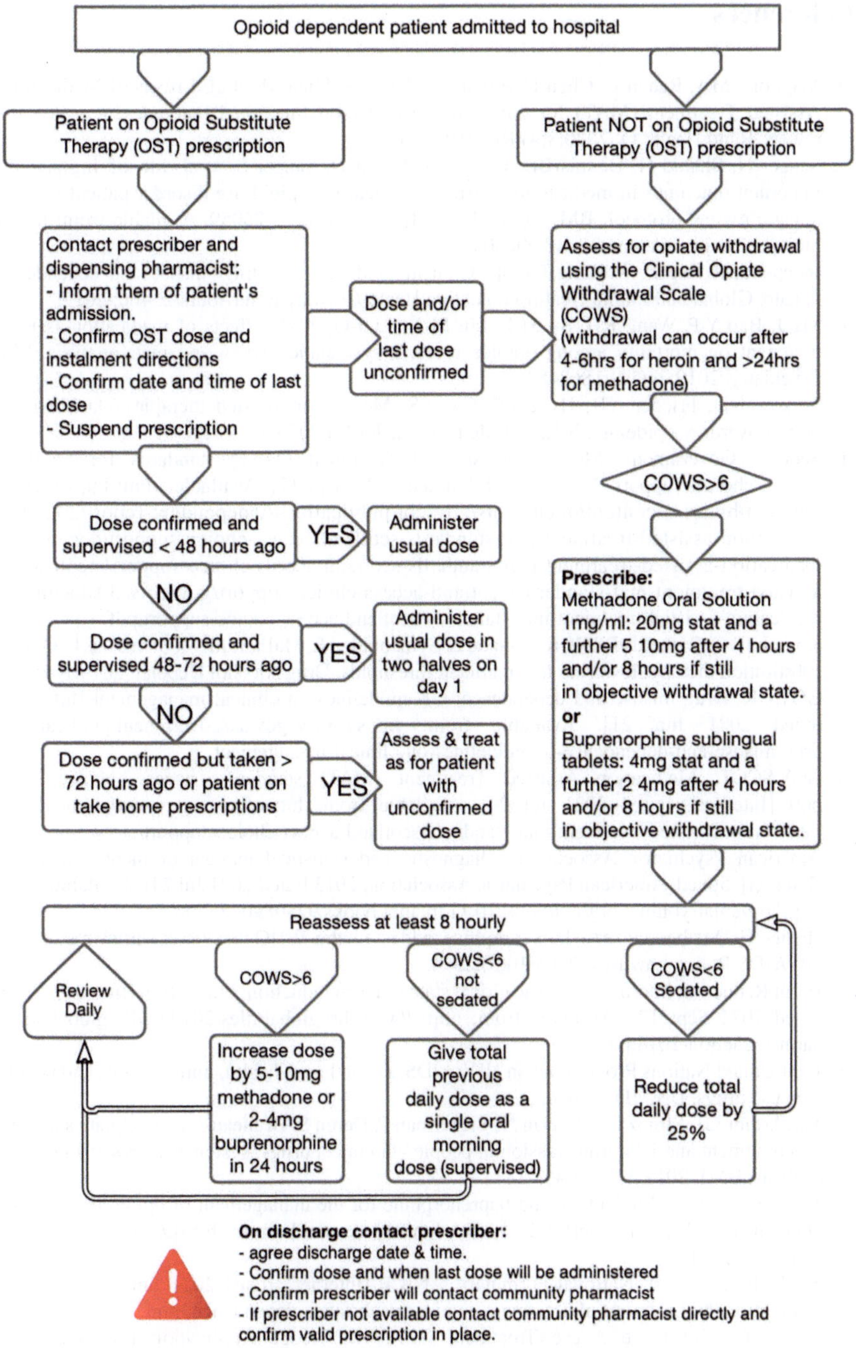

Fig. 14.4 Example pathway for management of an opioid-dependent patient admitted to hospital

References

1. Maglione MA, Raaen L, Chen C, Azhar GS, Nima S, Shen M, et al. Effects of Medication-Assisted Treatment (MAT) for opioid use disorder on functional outcomes: a systematic review. Rand Health Q. 2020;8(4):RR-2108-OSD.
2. Sanger N, Shahid H, Dennis BB, Hudson J, Marsh D, Sanger S, et al. Identifying patient-important outcomes in medication-assisted treatment for opioid use disorder patients: a systematic review protocol. BMJ Open [Internet]. 2018;8(12):e025059. Available from: http://bmjopen.bmj.com/content/8/12/e025059.
3. Independent Expert Working Group. Clinical guidelines on drug misuse and dependence update. Global and Public Health/Population Health/Healthy Behaviours/25460; 2017.
4. Ma J, Bao Y-P, Wang R-J, Su M-F, Liu M-X, Li J-Q, et al. Effects of medication-assisted treatment on mortality among opioids users: a systematic review and meta-analysis. Mol Psychiatry. 2019;24(12):1868–83.
5. Volkow ND, Frieden TR, Hyde PS, Cha SS. Medication-assisted therapies—tackling the opioid-overdose epidemic. N Engl J Med. 2014;370(22):2063–6.
6. Scottish Government. Medication Assisted Treatment (MAT) standards for Scotland access, choice, support [Internet]. 2021 [cited 2021 Sept 12]. Available from: https://www.gov.scot/binaries/content/documents/govscot/publications/independent-report/2021/05/medication-assisted-treatment-mat-standards-scotland-access-choice-support/documents/medication-assisted-treatment-mat-standards-scotland-access-choice-support/medication-assisted-treatment-mat-standards-scotland-access-choice-support/govscot%3Adocument/medication-assisted-treatment-mat-standards-scotland-access-choice-support.pdf.
7. Kourounis G, Richards BDW, Kyprianou E, Symeonidou E, Malliori M-M, Samartzis L. Opioid substitution therapy: lowering the treatment thresholds. Drug Alcohol Depend. 2016;161:1–8.
8. GOV.UK. Drug misuse and dependence: UK guidelines on clinical management [Internet]. [cited 2021 Jun 21]. Available from: https://www.gov.uk/government/publications/drug-misuse-and-dependence-uk-guidelines-on-clinical-management.
9. GOV.SCOT. Medication Assisted Treatment (MAT) standards: access, choice, support [Internet]. [cited 2021 Jul 24]. Available from: https://www.gov.scot/publications/medication-assisted-treatment-mat-standards-scotland-access-choice-support/.
10. American Psychiatric Association. Diagnostic and statistical manual of mental disorders [Internet]. 5th ed. American Psychiatric Association; 2013 [cited 2021 Jul 24]. Available from: http://psychiatryonline.org/doi/book/10.1176/appi.books.9780890425596.
11. Herget G. Methadone and buprenorphine added to the WHO list of essential medicines. HIVAIDS Policy Law Rev. 2005;10(3):23–4.
12. Harm Reduction International. The global state of harm reduction: 6th edition [Internet]. 2018 [cited 2021 Sept 12]. Available from: https://www.hri.global/files/2019/02/05/global-state-harm-reduction-2018.pdf.
13. Joint United Nations Programme on HIV/AIDS. Do no harm: health, human rights and people who use drugs. UNAIDS; 2016.
14. MacArthur GJ, Minozzi S, Martin N, Vickerman P, Deren S, Bruneau J, et al. Opiate substitution treatment and HIV transmission in people who inject drugs: systematic review and meta-analysis. BMJ. 2012;345:e5945.
15. NICE. Overview | Methadone and buprenorphine for the management of opioid dependence | Guidance | NICE [Internet]. [cited 2021 Jun 21]. Available from: https://www.nice.org.uk/guidance/ta114.
16. NICE. BNF: British National Formulary – NICE [Internet]. [cited 2021 Jun 21]. Available from: https://bnf.nice.org.uk/treatment-summary/substance-dependence.html.
17. Center for Substance Abuse Treatment (US); Knowledge Application Program (US). Medications for opioid use disorder: for healthcare and addiction professionals, policymakers, patients, and families (SAMHSA/CSAT treatment improvement protocols) [Internet].

Rockville: Substance Abuse and Mental Health Services Administration (US); 2018 [cited 2021 Jul 26]. Available from: http://www.ncbi.nlm.nih.gov/books/NBK535268/.
18. Kristensen K, Christensen CB, Christrup LL. The mu1, mu2, delta, kappa opioid receptor binding profiles of methadone stereoisomers and morphine. Life Sci. 1995;56(2):PL45–50.
19. Joseph H, Stancliff S, Langrod J. Methadone maintenance treatment (MMT): a review of historical and clinical issues. Mt Sinai J Med. 2000;67(5–6):347–64.
20. Eap CB, Buclin T, Baumann P. Interindividual variability of the clinical pharmacokinetics of methadone: implications for the treatment of opioid dependence. Clin Pharmacokinet. 2002;41(14):1153–93.
21. Wolff K, Rostami-Hodjegan A, Hay AW, Raistrick D, Tucker G. Population-based pharmacokinetic approach for methadone monitoring of opiate addicts: potential clinical utility. Addiction. 2000;95(12):1771–83.
22. Chugh SS, Socoteanu C, Reinier K, Waltz J, Jui J, Gunson K. A community-based evaluation of sudden death associated with therapeutic levels of methadone. Am J Med. 2008;121(1):66–71.
23. Lutfy K, Cowan A. Buprenorphine: a unique drug with complex pharmacology. Curr Neuropharmacol. 2004;2(4):395–402.
24. Walsh SL, Eissenberg T. The clinical pharmacology of buprenorphine: extrapolating from the laboratory to the clinic. Drug Alcohol Depend. 2003;70(2, Supplement):S13–27.
25. Schottenfeld RS, Pakes J, O'Connor P, Chawarski M, Oliveto A, Kosten TR. Thrice-weekly versus daily buprenorphine maintenance. Biol Psychiatry. 2000;47(12):1072–9.
26. Paone D, Tuazon E, Stajic M, Sampson B, Allen B, Mantha S, et al. Buprenorphine infrequently found in fatal overdose in New York City. Drug Alcohol Depend. 2015;155:298–301.
27. Wesson DR, Ling W. The Clinical Opiate Withdrawal Scale (COWS). J Psychoactive Drugs. 2003;35(2):253–9.
28. NHS Fife – Methadone tolerance testing procedure (2011) [Internet]. [cited 2021 Aug 31]. Available from: https://www.fifeadtc.scot.nhs.uk/media/7032/a1-procedure-for-tolerance-testing-final-april-2011.pdf.
29. Buprenorphine | Prescribing information | Opioid dependence | CKS | NICE [Internet]. [cited 2021 Jun 27]. Available from: https://cks.nice.org.uk/topics/opioid-dependence/prescribing-information/buprenorphine/.
30. Tay Wee Teck J, Baldacchino A, Gibson L, Lafferty C. Using microdosing to induct patients into a long-acting injectable buprenorphine depot medication in low threshold community settings: a case study. Front Pharmacol. 2021;12:294.
31. Chappuy M, Trojak B, Nubukpo P, Bachellier J, Bendimerad P, Brousse G, et al. Prolonged-release buprenorphine formulations: perspectives for clinical practice. Therapie. 2020;75(5):397–406.
32. Parsons G, Ragbir C, D'Agnone O, Gibbs A, Littlewood R, Hard B. Patient-reported outcomes, experiences and satisfaction with weekly and monthly injectable prolonged-release buprenorphine. Subst Abus Rehabil. 2020;11:41–7.
33. Mattick RP, Breen C, Kimber J, Davoli M. Buprenorphine maintenance versus placebo or methadone maintenance for opioid dependence. Cochrane Database Syst Rev [Internet]. 2014;(2):CD002207. Available from: https://www.cochranelibrary.com/cdsr/doi/10.1002/14651858.CD002207.pub4/full.
34. Roux P, Lions C, Michel L, Cohen J, Mora M, Marcellin F, et al. Predictors of non-adherence to methadone maintenance treatment in opioiddependent individuals: implications for clinicians. Curr Pharm Des. 2014;20(25):4097–105.
35. Bao Y-P, Liu Z-M, Epstein DH, Du C, Shi J, Lu L. A meta-analysis of retention in methadone maintenance by dose and dosing strategy. Am J Drug Alcohol Abuse. 2009;35(1):28–33.
36. Amass L, Bickel WK, Crean JP, Blake J, Higgins ST. Alternate-day buprenorphine dosing is preferred to daily dosing by opioid-dependent humans. Psychopharmacology. 1998;136(3):217–25.

37. Johnson RE, Eissenberg T, Stitzer ML, Strain EC, Liebson IA, Bigelow GE. Buprenorphine treatment of opioid dependence: clinical trial of daily versus alternate-day dosing. Drug Alcohol Depend. 1995;40(1):27–35.
38. Pani PP, Trogu E, Maremmani I, Pacini M. QTc interval screening for cardiac risk in methadone treatment of opioid dependence. Cochrane Database Syst Rev. 2013;(6):CD008939.
39. Lofwall MR, Stitzer ML, Bigelow GE, Strain EC. Comparative safety and side effect profiles of buprenorphine and methadone in the outpatient treatment of opioid dependence. Addict Disord Their Treat. 2005;4(2):49–64.
40. Bruce RD, Altice FL. Case series on the safe use of buprenorphine/naloxone in individuals with acute hepatitis C infection and abnormal hepatic liver transaminases. Am J Drug Alcohol Abuse. 2007;33(6):869–74.
41. Ostgathe C, Voltz R, Van Aaaken A, Klein C, Sabatowski R, Nauck F, et al. Practicability, safety, and efficacy of a "German model" for opioid conversion to oral levo-methadone. Support Care Cancer. 2012;20(9):2105–10.
42. Carrieri MP, Amass L, Lucas GM, Vlahov D, Wodak A, Woody GE. Buprenorphine use: the international experience. Clin Infect Dis. 2006;43(Supplement_4):S197–215.
43. Strang J, Reed K, Bogdanowicz K, Bell J, van der Waal R, Keen J, et al. Randomised comparison of a novel buprenorphine oral lyophilisate versus existing buprenorphine sublingual tablets in opioid-dependent patients: a first-in-patient phase II randomised open label safety study. Eur Addict Res. 2017;23(2):61–70.
44. Orman JS, Keating GM. Buprenorphine/naloxone. Drugs. 2009;69(5):577–607.
45. Lofwall MR, Walsh SL, Nunes EV, Bailey GL, Sigmon SC, Kampman KM, et al. Weekly and monthly subcutaneous buprenorphine depot formulations vs daily sublingual buprenorphine with naloxone for treatment of opioid use disorder: a randomized clinical trial. JAMA Intern Med. 2018;178(6):764–73.
46. Minozzi S, Amato L, Vecchi S, Davoli M, Kirchmayer U, Verster A. Oral naltrexone maintenance treatment for opioid dependence. Cochrane Database Syst Rev. 2011;2011(4):CD001333.
47. Krupitsky E, Zvartau E, Blokhina E, Verbitskaya E, Wahlgren V, Tsoy-Podosenin M, et al. Randomized trial of long-acting sustained-release naltrexone implant vs oral naltrexone or placebo for preventing relapse to opioid dependence. Arch Gen Psychiatry. 2012;69(9):973–81.
48. Ferri M, Minozzi S, Bo A, Amato L. Slow-release oral morphine as maintenance therapy for opioid dependence. Cochrane Database Syst Rev. 2013;(6):CD009879.
49. Haasen C, Verthein U, Degkwitz P, Berger J, Krausz M, Naber D. Heroin-assisted treatment for opioid dependence: randomised controlled trial. Br J Psychiatry. 2007;191(1):55–62.
50. Oviedo-Joekes E, Brissette S, MacDonald S, Guh D, Marchand K, Jutha S, et al. Safety profile of injectable hydromorphone and diacetylmorphine for long-term severe opioid use disorder. Drug Alcohol Depend. 2017;176:55–62.
51. Wedam EF, Bigelow GE, Johnson RE, Nuzzo PA, Haigney MCP. QT-interval effects of methadone, levomethadyl, and buprenorphine in a randomized trial. Arch Intern Med. 2007;167(22):2469–75.

Chapter 15
Medication-Assisted Treatment for Opioid Use Disorders 2: Detoxification

James Tidder, Alexander M. Baldacchino, and Joseph Tay Wee Teck

Abstract This chapter is the second of two parts describing the advantages, standards, and goals of medication-assisted treatment (MAT) for opioid use disorders. MAT is a complex biopsychosocial intervention with the provision of opioid substitution treatment (OST) at its core. These chapters aim to provide the clinician with charts, tables, and clinical guidance around the prescribing of OST within MAT. This second part (this chapter) covers the process of detoxification from OST as well as aftercare including the use of naltrexone. This chapter also looks at the care of special patient groups requiring MAT, such as those who are pregnant or who have multiple substance dependencies and gives brief guidance around the management of comorbid mental health problems. These chapters focus on the prescribing issues within MAT and are not able to cover the full scope of the social and psychological aspects of this complex intervention; clinicians are recommended to read broadly to ensure they understand the full scope of MAT, which is far more than just the prescribing of OST.

Data presented here is based on clinical guidelines and the experience in treating persons with OUDs in Scotland, United Kingdom (UK), and reflects evolution in practice in response to the high levels of drug deaths and Scottish national MAT standards published in May 2021. Within the UK, the National Health Service (NHS) exists as a single-payer healthcare system, providing universal care to all residents. This of course is not the case worldwide, and there is substantial variation in availability of services between countries. Clinicians should always ensure that they are familiar with what services are available locally as well as local licensing

J. Tidder (✉)
NHS Borders Addiction Service, Galashiels, Scotland, UK
e-mail: james.tidder@borders.scot.nhs.uk

A. M. Baldacchino
University of St Andrews, St Andrews, Scotland, UK

NHS Fife Addiction Services, Leven, Scotland, UK

J. Tay
Forward Leeds, Leeds, UK

S. L. Cruz (ed.), *Opioids*, https://doi.org/10.1007/978-3-031-09936-6_15

and legal requirements around prescribing OST, as substitute medications are usually controlled drugs.

Keywords MAT · OST · OAT · OUD · Methadone · Buprenorphine · Lofexidine · Naltrexone · Detoxification

15.1 Introduction

MAT is a complex intervention designed to consider the full range of biological, psychological, and social interventions needed for high standard care of those with opioid use disorders (OUDs). The effective delivery of opioid substitution therapy (OST) sits at the heart of MAT and is the focus of Chap. 14 and this chapter. Chapter 14 covers the assessment of patients with opioid use disorders (OUDs) and introduces the low threshold model. It also describes the initiation, titration, and maintenance of OST, as well as physical monitoring and the treatment of those with OUDs requiring admission to hospital. This chapter covers detoxification and aftercare for those ceasing OST, as well as the care for special patient groups requiring MAT, such as those who are pregnant, have mental health problems, or have multiple dependencies. While the psychosocial aspects of MAT are briefly touched on, it has not been possible to detail all of the aspects of MAT within these two chapters. Clinicians are recommended to explore the full range of therapeutic options available and to always consider that while effective OST provision is the essential starting point, MAT is more than the provision of OST.

15.1.1 Before Detoxification

Services should not impose any time limit on how long a patient can continue to be prescribed OST, and enforced detoxification is likely to increase the risk of relapse. Equally, services should ensure that patients can detox from opioids in a controlled and supported way when this is right for them within their own recovery journey.

Opioid detoxification should be part of a package including preparation and post-detoxification support to prevent relapse. Where possible, patients should be encouraged not to attempt detoxification until they have been fully stable on their prescribed medication for a period adequate to allow them to address other psychological and social factors that may have driven their substance use. The decision to detoxify must then be taken jointly with the patient to allow informed consent.

Before detoxification, patients should receive information on:

- The physical and psychological aspects of detoxification, the duration and intensity of symptoms, and how these may be managed
- The use of non-pharmacological approaches to manage and cope with withdrawal symptoms

- The importance of continued support during detoxification, to maintain abstinence and reduce the risk of adverse outcomes

Psychosocial interventions and key working should continue to be delivered alongside pharmacological interventions throughout detoxification. If detoxification is unsuccessful, patients should have rapid access back into maintenance treatment with OST and other interventions. In addition, the patient must understand that, as the dose of opioid is reduced, tolerance to previous doses is lost, and any relapse into drug-taking will carry a high risk of overdose. It should be ensured that overdose awareness training has been delivered and naloxone offered.

15.1.2 Preparation Process for Detoxification

During the preparation process for detoxification, the following areas should be carefully considered together with the patient:

- Lessons learned from previous treatments, detoxifications, and rehab programs
- Expectations and acknowledgment of positive outcomes
- Motivation and readiness for a detoxification program
- Methods of detoxification, including choice of medication
- Coping skills to deal with detoxification programs and strategies to maintain abstinence
- Support network during and after detoxification program
- Creation of a care plan aimed at relapse prevention
- The setting or location of detoxification (in-patient or community detox)

Community-based detoxification is suitable for most patients, but exceptions may include:

- Those who have not benefited from previous care-planned community detoxifications
- Those who need medical and nursing care due to significant mental or physical health problems
- Those who require complex polydrug detoxification
- Those who have significant social problems, such as homelessness, that may limit the success of community-based detoxification

15.1.3 Supervised Self-Administration During Detoxification

A return to supervised self-administration may have an advantage in managing dose reduction, particularly if patients are putting extra pressure on themselves by reducing too quickly or having difficulty coping with reductions and tempted to take doses too early. However, the decision must also take into account the patients' social circumstances. For instance, a requirement to once again attend daily may impact work or childcare responsibilities. Any decision to reinstate supervised self-administration must be discussed with the patient, giving clearly the reasons why this has been taken and highlighting that this is to ensure safety and not as a punitive measure.

15.2 Detoxification from OST

Clinicians should ensure that they are familiar with the approaches to detoxification with both methadone and buprenorphine and should aim to come to a shared decision with the patient clearly setting out their individualized plan for detoxification. The sections below offer guidance around common approaches.

15.2.1 Detoxification with Methadone

- Negotiate a structured rate of reduction with the patient and set an end date.
- Aim to reduce the dose initially by about 5 mg every 2 weeks.
- Patients are likely to tolerate a reasonably rapid dose reduction at the beginning until reaching 30 mg of methadone.
- Reduction to 0 is likely to be more successful if slowed to 1–2 mg fortnightly over the last few weeks.
- Prolonged, slow reductions should not be endorsed – longer detoxes are associated with higher risks of relapse.
- When the patient has reached a dose of 30 mg methadone, they may opt for transfer to buprenorphine to complete the detox, which may be better tolerated.

15.2.2 Detoxification with Oral Buprenorphine

- Negotiate a structured rate of reduction with the patient and set an end date.
- A common regime is reducing by 2–4 mg every 2 weeks.
- When the dose is reduced below 2 mg, it may be necessary to change to a smaller tablet size to continue the reduction (Table 15.1 gives some examples on how to reduce buprenorphine doses).

Buprenorphine detoxification often causes less significant or prolonged withdrawal features than full agonists such as methadone; patients struggling to detoxify from methadone can be offered to transfer to buprenorphine and to use this for detoxification. However, patients should be counseled around the risk of precipitated withdrawal emerging during the transfer.

Table 15.1 Reduction rates of buprenorphine doses

Daily buprenorphine dose (mg)	Reduction rate
Above 16	4 mg every 1–2 weeks
8–16	2–4 mg every 1–2 weeks
2–8	2 mg every 12 weeks
Below 2	0.4–0.8 mg every 1–2 weeks

Table 15.2 Dose transfer example

Last methadone dose (mg)	Buprenorphine day 1 (mg)	Buprenorphine day 2 (mg)
20–30	4	6–8
10–20	4	4–6
<10	2	2–6

15.2.2.1 Transfer from 30 mg or Less of Methadone to Buprenorphine

Due to the risk of precipitated withdrawal, the first dose of buprenorphine should be administered at least 24–36 hours after the last use of methadone and preferably with the onset of mild to moderate withdrawal symptoms. Increasing the time interval between the last dose of methadone and the first dose of buprenorphine reduces the incidence and severity of precipitated withdrawal (Table 15.2) gives example dosing.

15.2.2.2 Transfer from Higher Doses of Methadone to Buprenorphine

It is possible to transfer from a higher dose of methadone to buprenorphine, but the risk of experiencing precipitated withdrawal symptoms is significantly higher. Therefore, the first dose of buprenorphine must be delayed until there are clear signs of withdrawal. Consideration can be given to a starting dose of 2 mg with a second dose an hour later if there is no evidence of precipitated withdrawal – though even then, there is a risk of precipitated withdrawal emerging. Withdrawal symptoms may be treated with lofexidine or managed with other agents for symptomatic relief.

15.2.3 Detoxification with Lofexidine

Lofexidine is a non-opioid, alpha-adrenergic agonist drug that may be used to relieve symptoms of withdrawal in patients undergoing opioid detoxification and is widely licensed across Europe and in the United States for this purpose. It is a structural analog of clonidine, another alpha-adrenergic receptor agonist, which has also been used similarly in some areas. While clonidine has been extensively used in a similar way, it may have more adverse effects, and a worse risk/benefit profile, compared to lofexidine [1–3], is not licensed for use in managing opioid withdrawal symptoms, and its use is not recommended by major clinical guidelines [4]. Lofexidine may effectively relieve symptoms such as chills, sweating, stomach cramps, muscle pain, and rhinorrhea but is less effective at suppressing symptoms of subjective discomfort and will not stop cravings to use opioids.

Due to the hypotensive effect of lofexidine, it is necessary to closely monitor blood pressure (BP) and pulse. Therefore, baseline BP reading should be obtained before treatment and BP monitored at least once daily for the first 3 days or until reaching the peak dosing phase.

Lofexidine is a treatment option for patients who:

- Are using smaller amounts of opioids
- Have shorter drug and treatment histories
- Are at the last stages of methadone or buprenorphine detoxification
- Do not want or have been previously unsuccessful with methadone or buprenorphine
- Are not at risk of harming themselves or others as a result of detoxification
- Are strongly motivated to stop using opioids

Lofexidine should be used with caution in patients with severe coronary insufficiency, recent myocardial infarction, cerebrovascular disease or chronic renal failure, bradycardia, hypotension, or risk of QT prolongation.

Clinicians should be aware that lofexidine may enhance the effects of:

- Alcohol, barbiturates, and other sedatives
- Anti-hypertensive drugs
- Drugs known to prolong the QT interval, such as erythromycin/clarithromycin, some antidepressants, some antipsychotics

Lofexidine is initiated and the dose built within an induction phase, aiming to reach a peak dosing phase and maintained at that dose during the period where opioid withdrawals are expected to be at their peak before being reduced and withdrawn. The amount of lofexidine should be titrated according to the patient's response, and it will not always be necessary to reach the maximum daily dose of 2.4 mg.

Some patients will feel comfortable with the symptoms of withdrawal being controlled by a lower daily dose. At the same time, those undergoing acute detoxification will usually require the highest recommended dose and dosage increments to provide optimum relief at the time of expected peak withdrawal symptoms. The length of treatment will vary depending on when the patient becomes opioid-free and their individual response to withdrawal.

The initial dosage of lofexidine should be 0.8 mg/day, in divided doses (usually four times daily). The dosage may have increments of 0.4–0.8 mg/day to a maximum of 2.4 mg daily. However, the maximum single dose taken at one time should not exceed 0.8 mg. Table 15.3 gives suggested dosing schedules, given as total daily doses.

Table 15.3 Suggested dosing schedules for lofexidine depending on previous opioid treatment

Days since last opioid use	Short $t_{1/2}$ opioid[a]	Modified release opioid[b]	Methadone <30 mg	Methadone 30–80 mg	Methadone >80 mg	Bupre-norphine <8 mg	Bupre-norphine >8 mg
	0.8 mg						
1	1.6 mg	0.8 mg	0.8 mg	0.8 mg	0.8 mg	0.8 mg	0.8 mg
2	2.4 mg	1.6 mg	1.6 mg	1.2 mg	0.8 mg	1.2 mg	1.2 mg
3	2.4 mg	2.4 mg	2.4 mg	1.6 mg	1.2 mg	1.6 mg	1.6 mg
4	2.4 mg	2.4 mg	2.4 mg	2.4 mg	1.6 mg	1.6 mg	1.6 mg
5	1.6 mg	2.4 mg	2.4 mg	2.4 mg	2.4 mg	1.6 mg	1.6 mg
6	1.2 mg	2.4 mg	2.4 mg	2.4 mg	2.4 mg	1.6 mg	1.6 mg
7	0.8 mg	1.6 mg	2.4 mg	2.4 mg	2.4 mg	1.6 mg	1.6 mg
8	0.4 mg	1.2 mg	1.6 mg	2.4 mg	2.4 mg	1.2 mg	1.6 mg
9	0.2 mg	0.8 mg	1.2 mg	2.4 mg	2.4 mg	0.8 mg	1.2 mg
10		0.4 mg	0.8 mg	2.4 mg	2.4 mg	0.4 mg	0.8 mg
11		0.2 mg	0.4 mg	1.6 mg	2.4 mg	0.2 mg	0.4 mg
12			0.2 mg	1.2 mg	2.4 mg		0.2 mg
13				0.8 mg	2.4 mg		
14				0.4 mg	1.6 mg		
15				0.2 mg	1.2 mg		
16					0.8 mg		
17					0.4 mg		
18					0.2 mg		

Induction Phase	Peak dosing phase: this should coincide with the anticipated onset & duration of peak withdrawal symptoms	Reduction Phase

[a]**Heroin, codeine, dihydrocodeine, morphine, tramadol, pethidine, oxycodone** – *start lofexidine 6–12 hours after last dose*

[b]Modified release preparations of morphine, dihydrocodeine, tramadol, or oxycodone – *start lofexidine about 12 hours after last dose, when withdrawal symptoms develop*

15.3 Symptomatic Relief in Detoxification

It is possible to attempt detoxification using only symptomatic relief, though many patients will find this difficult to tolerate and it is generally not recommended. In the last phase of detoxification with methadone and buprenorphine, patients may experience withdrawal symptoms, and symptomatic relief can also be helpful in this setting.

The following medications may be helpful:

Diarrhea: Loperamide 4 mg stat, then 2 mg after each loose stool (max. 16 mg/day)
Stomach cramps: Hyoscine butylbromide* 10–20 mg four times daily when required or mebeverine 135 mg three times daily, 20 minutes before meals
Nausea and vomiting: Metoclopramide 10 mg three times daily when required. Prochlorperazine 5–10 mg two or three times daily when required
Agitation and Anxiety: Propranolol 40 mg once daily, increasing to three times daily if required. Diazepam* 2–10 mg up to three times daily when required
Muscular pain/headaches: Paracetamol 1 g four times daily. Ibuprofen 400 mg three times daily
Insomnia: Zopiclone* 7.5 mg at night. Trazodone 50 mg at night

*These drugs have the potential for abuse or dependence – prescribe for no more than 14 days.

15.4 Aftercare

Newly detoxified patients remain at increased risk of relapse. For this reason, they should retain their treatment place (i.e., they are not discharged) for at least 4 weeks, while they are in the after-care phase of their opioid replacement therapy. Relapse to opioid use during this post medication phase will usually mean automatic re-induction into OST for another maintenance period before another attempt at reduction. Keyworkers should continue to provide weekly support to the patient for 4 weeks during this phase, focusing on relapse prevention. Emphasis should also be on engagement in work-related activity and meaningful occupations or activities to replace drug-using lifestyles and increase the likelihood of staying drug-free.

Patients should be encouraged to access aftercare services such as rehabilitation, group services such as SMART Recovery, Narcotics Anonymous, and counseling services, depending on what is available locally. In addition, consideration should be given to information sharing with child and family services to ensure that parents and children receive additional support if needed.

Patients remain at high risk of relapse (often for years) after they have detoxified, so the after-care phase of their treatment is an essential part of their Recovery Care Plan. Some patients may feel more encouraged to attempt reduction if they know that their treatment "slot" will still be there for them should they relapse.

15.4.1 Naltrexone for Relapse Prevention

Naltrexone is a long-acting opiate antagonist. If taken by an individual continuing to take opioids, it will precipitate opiate withdrawal symptoms. However, taken regularly after detoxification, it can assist in relapse prevention by blockade of opioid receptors and is licensed for this indication in multiple jurisdictions. Supervision by either a family member, pharmacy, or another dispensing setting can aid success. Naltrexone should be used only as an adjunct to other forms of support and treatment for patients who have recently come off opiates.

15.4.1.1 Transfer to Naltrexone

- Wait for at least 72 hours after the last dose of oral buprenorphine administration or at least 7 days after the last dose of methadone to initiate naltrexone treatment.
- An instant negative urine test to confirm the patient's opioid-free status should be obtained within 12 hours of initiation of naltrexone if there is any doubt.
- The initial dose of naltrexone is 25 mg, followed by 50 mg daily. A three-times-a-week dosing schedule may be considered if it is likely to result in better compliance (e.g., 100 mg on Monday and Wednesday and 150 mg on Friday).
- Continue with naltrexone for at least 6–12 months.

15.4.1.2 Cautions and Monitoring with Naltrexone

- Liver function tests should be carried out before starting, 1-month post-transfer, and then 6-monthly. Naltrexone should be discontinued if there is evidence of progressive hepatic impairment.
- Naltrexone does not prevent the use of other classes of drugs, but there is evidence for reduced alcohol consumption for problematic drinkers.
- The absence of documented evidence means that naltrexone should only be given to pregnant or breastfeeding women when the potential benefits outweigh the possible risks.

15.4.1.3 Long-Acting Naltrexone

In some countries, such as the United States, naltrexone is available as a monthly intramuscular injection. This may allow for the benefit of naltrexone treatment with the advantage of patients not needing to take oral medication and potentially greater compliance. Prescribers should familiarize themselves with local licensing and prescribing guidelines.

15.5 Psychosocial Interventions Within MAT

When delivered correctly, MAT is a complex psychosocial intervention and far more than just prescribing OST. Treatment for drug misuse should always involve a psychosocial component to help support an individual's recovery. Changing

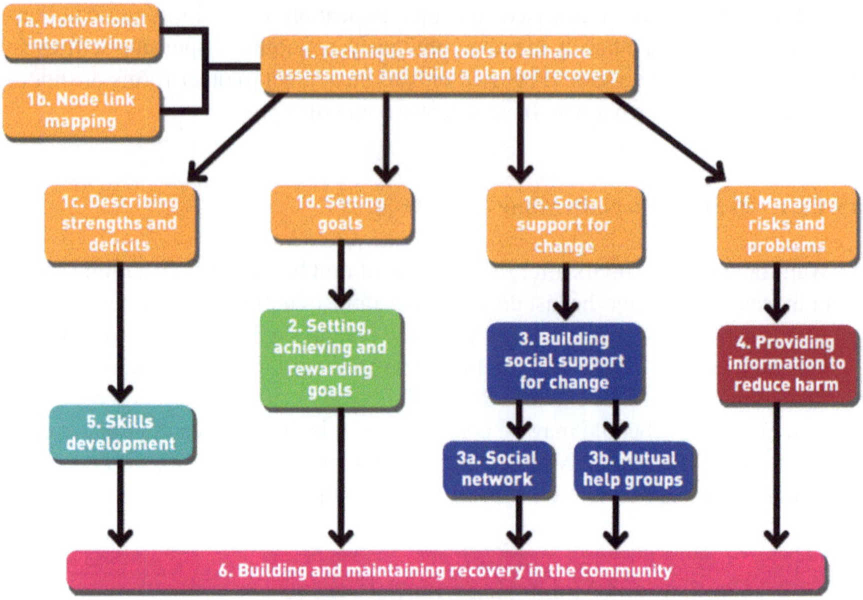

Fig. 15.1 Node link map illustrating the psychosocial components of treatment to support an individual's recovery. (Modified from Day [47]; [5])

entrenched patterns of drug-using behavior is complex, and concerted efforts with psychological and social interventions are crucial to this process.

Psychosocial interventions should be integrated into an individual's recovery care plan and delivered closely with medical and pharmacological interventions. Access to a range of interventions at different stages of the recovery journey should be offered and supported, and these should reflect local availability of services and the cultural context. Stabilizing OST provides powerful support for behavior changes and allows meaningful engagement with psychosocial therapy; however, an individual does not need, and is unlikely to be, completely drug- or alcohol-free before commencing treatment or engaging in broader interventions.

The node-link mapping approach and associated manual [5] developed by Public Health England gives one example of a framework that clinicians can use with patients to develop a broad recovery care plan; this approach is illustrated in Fig. 15.1.

15.6 Use of Take-Home Naloxone

Naloxone is an opioid/opiate antagonist and may be used for complete or partial reversal of central nervous system depression and especially respiratory depression caused by natural or synthetic opioids and treatment of suspected acute opioid overdose or intoxication. Provision of naloxone to patients can be effective in reversing opioid overdoses, and there is also evidence for the effectiveness of training family members or peers in how to administer the drug [6–9]. Naloxone has no dependence forming potential or intoxicating effects, and side effects are rarely reported [10]. Increasing the availability of naloxone to those who may be likely to witness an opioid overdose is recognized as a key measure by the World Health Organization (WHO) [11].

The legal status and prescription requirements around the distribution of naloxone vary between countries and require different approaches to naloxone provision. However, all providers should aim to have takeaway naloxone provision and appropriate training in its use alongside overdose awareness training available as soon as possible within the assessment process. This should remain available through all stages of treatment, and keyworkers should be regularly confirming that patients remain aware and familiar with how to use naloxone and offer further supply if the original is used, lost, or expires.

Availability of naloxone and associated training should, wherever possible, not be limited to drug treatment settings but also be available in broader healthcare and community settings. These include hospital emergency departments, primary care, prisons, community pharmacies, and non-healthcare drug support and treatment services. There is also increasing evidence of the success of making naloxone available for administration by non-medical first responders such as police officers and firefighters in reducing opioid overdose deaths [12–14].

15.7 Special Patient Groups

15.7.1 MAT, Pregnancy, and Breastfeeding

Pregnant women dependent on opioids are at high risk of experiencing complications generally due to inadequate antenatal care and lifestyle factors, including smoking, poor nutrition, high levels of stress, and deprivation. In addition, repeated cycles of intoxication and withdrawal can harm the fetus or precipitate premature labor or miscarriage.

Opiate-dependent women entering treatment due to pregnancy should be offered methadone as preparation of choice unless special circumstances apply (e.g., recurrent failure on methadone). Methadone carries the potential risk of respiratory depression in neonates and neonatal withdrawal syndrome. However:

- Respiratory depression is not a significant problem in babies born to mothers on methadone maintenance treatment.
- Babies may experience neonatal withdrawal syndrome. Occurrence is unpredictable, with no relationship between the maternal methadone dose and the severity of the neonatal withdrawal syndrome.
- The benefits of methadone maintenance treatment for both the mother and baby outweigh any risks from neonatal withdrawal syndrome.

Those who are already stable on buprenorphine should be maintained on it. Buprenorphine provokes a similar incidence, compared to methadone, of neonatal abstinence syndrome, but this tends to be less severe, needing less and shorter treatment.

15.7.1.1 Management in Pregnancy

Attending regular antenatal care is of high priority, and liaison between OST prescriber and maternity service is essential. Pregnant women should be maintained on adequate doses of methadone to achieve stability and prevent relapse or continued illicit opioid drug use. Women already in methadone treatment who become pregnant can generally be safely maintained on their current dose. Pregnant women should be considered as priority cases and titrated onto methadone as soon as possible. It may be necessary to divide the daily dose or possibly to increase the dose in the third trimester of pregnancy to avoid withdrawal symptoms due to the reduced bioavailability of methadone in the later stages of pregnancy.

15.7.1.2 Dose Reductions or Detoxification During Pregnancy

Opioid withdrawal in the first trimester of pregnancy is associated with an increased risk of miscarriage, while in the third trimester, it may be related to fetal distress and death. Therefore, pregnant women must not be exposed to withdrawal during these two trimesters. Dose reductions should only occur in the second trimester if the pregnancy is stable and should be flexible, with withdrawal symptoms avoided as much as possible as they cause considerable distress to the fetus. If the decision is made to reduce the dose, careful monitoring of the fetus should be undertaken. In most instances, dose reductions of 2–3 mg of methadone every 3–5 days (or less frequently) are considered safe.

15.7.1.3 Breastfeeding

Breast milk contains only small amounts of methadone, and mothers can be encouraged to breastfeed regardless of methadone dose provided they are not using other drugs. Breastfeeding may reduce the severity of neonatal withdrawal syndrome. Women on high doses of methadone should be advised to wean their babies slowly to avoid withdrawal in the infant.

15.7.2 Patients with Comorbid Mental Disorders

Co-occurring mental health problems are common among people who use drugs. For some patients, dual-focused treatment will be appropriate. An approach such as cognitive-behavioral therapy or motivational interviewing may be adapted to address both the mental health issue and the drug dependency. For other patients, where the mental health disorder is the primary diagnosis and clinical priority, the patient should receive treatment with a specialist from the appropriate adult mental health service. Table 15.4 gives details around approaches in common mental health comorbidities.

15.7.3 Treatment of Pain in Those on OST

It is important to recognize that opioid users experience the same sources of pain as others and will have similar needs for pharmacological and other interventions to address pain. Opioid users may have previously self-medicated to relieve pain and psychological distress and may have a poor acceptance of non-pharmacological

Table 15.4 Additional treatment considerations in people with comorbid health conditions

Comorbidity	Additional considerations
Depression	Patients may present with a depressed mood following recent substance use, intoxication, or withdrawal. However, primary, mild, or moderate depressive disorder is also pervasive. For less severe and less complex problems (often described as "mild" or "moderate"), online therapy, therapeutic group work, and 1–2–1 guided self-help can be beneficial
Bipolar affective disorder	Some evidence suggests that dual-focused psychological interventions based on cognitive-behavioral principles have a better impact on reducing substance use than individual treatments [45, 46]
Post-traumatic stress disorder (PTSD)	High rates of PTSD have been reported in those attending substance misuse treatment services. Stabilization of drug misuse should allow initial psychological interventions focused on reducing risky behaviors and emotional regulation, moving toward specific interventions for trauma. Services should aim to ensure that staff has been trained in low-level interventions to support this and that the service delivery model is "trauma-informed"
Anxiety spectrum disorders	Diagnosis of a comorbid anxiety disorder is common in those with substance misuse problems. Sometimes, confirmation of diagnosis and treatment planning needs to await stabilization of substance use, although advice on anxiety management can often be given at assessment.
Borderline personality disorder	Patients with borderline personality disorder (BPD) and drug/alcohol dependence should be assessed on a case-by-case basis to determine how best to meet their needs. Patients may remain under the care of mental health services – for treatment of BPD and with addiction services for assessment and management of their drug dependence. However, some aspects of their psychological care might be integrated depending on the setup of individual services. For example, many patients with BPD will often have underlying trauma, and initial low-level interventions around this may be possible in an addictions setting
Psychotic illnesses	At present, there is not sufficient evidence to recommend dual-focused treatment for the management of psychosis and substance use disorder. Patients should therefore be offered specific interventions for each disorder

interventions for pain control. Detailed assessment of the pain, the dependence, and any comorbid mental health problems are essential, particularly for chronic pain.

Patients dependent on opioids require empathic communication and reassurance that their pain will be taken seriously and managed. Patients abstinent in recovery risk relapse from re-exposure to opioids or undertreatment of pain. It is important to discuss treatment options with the patient and respect the patient's decisions, being transparent and open with the patient as to the rationale if any treatment options are not considered appropriate. Effective non-opioid acute pain regimens should be used where possible in preference to additional opioids for those with substance misuse problems. Figure 15.2 shows an approach to pain management in patients on OST.

15.7.4 Patients with Respiratory Disease

The prevalence and mechanisms involved in respiratory disease in the opioid-dependent population are not well understood. While there is evidence of an increased burden of respiratory diseases in people who use illicit opioids, it is difficult to gain accurate estimates of the prevalence of respiratory disease in this population due to the heterogeneity of study design and samples in publications [15]. However, in contrast with findings from studies of marijuana or crack cocaine smokers, studies of heroin smokers recruited from drug treatment services have shown a higher prevalence of COPD than tobacco smokers of comparable age, with a high mortality rate from COPD in young heroin smokers [16]. It is virtually impossible to disentangle the different inhaled drugs that someone who uses heroin smokes (e.g., one estimate suggests 90% of heroin smokers also smoke tobacco), and it is difficult to be sure of the real impact of inhaled heroin [16]; however, there is evidence of a high burden of respiratory disease in this patient group [17].

During the ongoing assessment, patients should be asked about:

- Current and previous levels of smoking and current quit status for all substances
- Current or recent history of cough, shortness of breath, wheeze, or other signs of respiratory disease and any consequent impairment in activity such as walking
- Previous respiratory diagnoses and any treatment for existing lung disease
- The desire, now or in the future, to quit tobacco smoking and experience of previous quit attempts

The patient should be referred and supported to attend for investigation and treatment, where symptoms suggest potential respiratory disease. In addition, the patient should be supported to continue treatment for respiratory problems by respiratory services. If diagnosed with respiratory disease (or other qualifying conditions), the patient should be encouraged to receive annual vaccinations for influenza and pneumococcus if suggested by local policies.

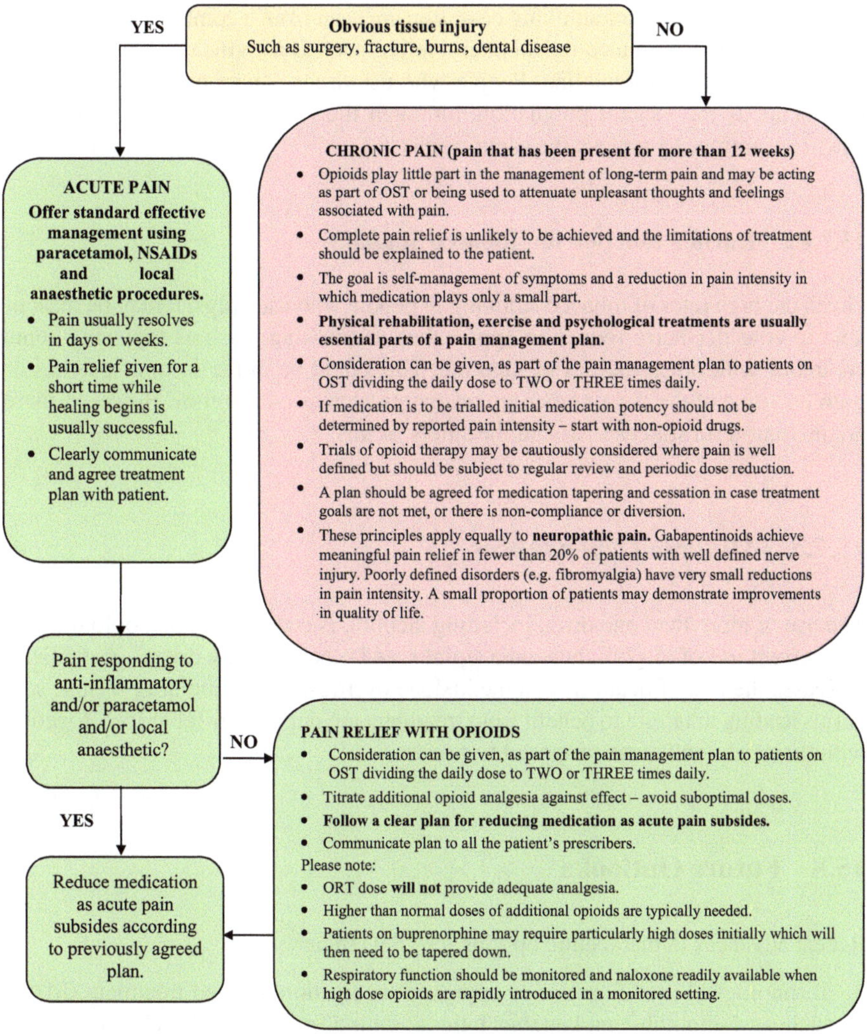

Fig. 15.2 Approach to pain management in patients on OST

15.7.4.1 Smoking Cessation

While smoking cessation might not seem a priority, engagement with smoking cessation support has been associated with improved drug treatment outcomes for patients in treatment. There is no reason to delay a discussion around smoking cessation as evidence suggests most patients express the desire to quit. Given that different patients may wish to help with smoking at varying stages of their treatment journey, repeated brief advice for smoking cessation should be offered as treatment progresses.

The best outcomes for smoking cessation are seen from a combination of behavioral support and pharmacological interventions such as nicotine replacement therapies, bupropion, or varenicline. People who use drugs can respond to these same treatments as the general population, although they may need more intensive or extensive options to achieve the same results.

15.7.4.2 Harm Reduction for Tobacco Smoking

Given the high rates of tobacco smoking in people who use drugs, it may be reasonable to consider harm reduction approaches to smoking, such as replacing some cigarettes with other sources of nicotine. This could be in the form of patches or gum for some of the day or other replacements such as e-cigarettes. However, there are no long-term studies to assure the safety of this.

15.7.5 Multiple Dependences

The use of more than one drug, including alcohol, is common among opioid users. Concurrent use of alcohol, benzodiazepines, and other sedating drugs substantially increases the risk of death from a methadone overdose. Common drug misuse scenarios leading to failure to benefit from treatment are outlined in Table 15.5, together with their risks and some proposed responses.

15.8 Future Outlooks

Recent advances in therapeutic options for OUD include:

1. Technological modifications to approved medication-assisted treatment (MAT) such as 6-monthly implantable buprenorphine, monthly injectable buprenorphine, and extended-release injectable naltrexone [18–21]
2. Heroin vaccines [22–24]
3. Gene-targeted therapy in (a) pharmacogenomics and (b) gene splicing to optimize and personalize OUD [25–29]
4. Epigenetic informed opioid treatment [30–35]
5. Medical cannabinoids [36, 37]
6. Neuroscience informed psychoeducation and metacognitive training [38–40]
7. Biased agonism at the G protein-coupled receptor (GPCR) function (see Chap. 9) [41, 42]
8. Technology-based neuromodulation interventions [43, 44]

A few have undergone the necessary clinical trial stages required to be considered as an evidence-based intervention (e.g., 1). However, most are either in early clinical trial stages (e.g., 5) or have shown promising results when using animal models (e.g., 2, 3).

Table 15.5 Responses to drug and alcohol misuse on top of an opioid prescription

Scenario	Risks	Possible response
Alcohol or benzodiazepine misuse on top of an opioid prescription	Overdose or "near misses" Drug interactions Alteration of methadone metabolism Deterioration of hepatic functioning in those with hepatitis C Street drinking Intoxicated presentations	Review evidence of alcohol/benzodiazepine dependence and the need for alcohol-focused key working support or assisted withdrawal Increase frequency of key working and psychosocial interventions and medical review Reintroduce daily supervised consumption, carefully titrating up the proportion supervised as appropriate, and agree on the progress needed before relaxing the arrangements Do not reduce opioid dose simply because of alcohol/benzodiazepine use but for review opioid tolerance and any evidence of opioid intoxication Consider whether breathalyzer testing can be helpful in monitoring progress (e.g., to confirm no evidence of recent alcohol use)
Opioid misuse on top of an opioid prescription	Overdose Blood-borne viruses and other infections if injecting Continued offending and involvement in drug misusing lifestyle Impaired engagement	Increase dose, if inadequate Divide dose, in addition, if fast metabolizer Offer to change OST medication If a patient is an on reducing regimen, re-stabilize the patient on a higher dose, and review support and patient goals Reintroduce daily supervised consumption, carefully titrating up the proportion supervised if appropriate, and monitor successful progress before relaxing this arrangement Consider increasing other psychosocial interventions (e.g., increase the frequency of key working and motivational support or medical review or provide more formal contingency management) Ensure access to safer injecting advice and supplies Reinforce advice and support for overdose prevention Confirm suitability of medication collection and review arrangements
Crack cocaine and cocaine misuse on top of an opioid prescription	Blood-borne viruses and other infections if injecting More chaotic drug misuse Increased crime Psychological problems Overdose	Confirm adequate stability on the current dose of OST Increase frequency of key working or other psychosocial interventions Ensure access to safer injecting advice and supplies Review understanding of overdose risk and reinforce advice on reducing risk Review for any comorbid mental health problems Review level of instability and the possible need for daily supervised consumption of OST

15.9 Conclusion

Medication-assisted treatment (MAT) when delivered well allows the intervention plans of individuals with substance misuse disorders alongside diverse comorbid issues to be tailored according to their specific biopsychosocial needs. Chapter 14 and this chapter have focused primarily on the effective prescribing of OST, the medication component of MAT. Rapid and low barrier provision of OST is essential and allows patients to be engaged in treatment and to start to stabilize. However, given the additional health issues prominent in this population including chronic conditions such as chronic obstructive airways disease (COAD), cardiovascular disease, alcohol liver diseases, hepatitis C, and HIV, often alongside trauma and other significant mental health difficulties, OST alone is not an adequate intervention, and these chapters cannot capture all the details of MAT as a complex intervention. MAT involves far more than OST and it is clear that the starting point of a productive and supportive relationship between the patient, the prescriber, and the rest of the clinical team is key in building trust and engagement, allowing broader interventions to then be delivered. Services will look different across the world as locally they seek to reflect the needs and culture of the population they are serving. But all services should be striving to better understand the diversity and complexity of this patient population.

References

1. Kuszmaul AK, Palmer EC, Frederick EK. Lofexidine versus clonidine for mitigation of opioid withdrawal symptoms: a systematic review. J Am Pharm Assoc. 2020;60(1):145–52.
2. Crumb MW, Atkinson TJ, et al. Comparing lofexidine and clonidine for opioid withdrawal management [Internet]. Pract Pain Manag. 20(4). [cited 2021 Nov 30]. Available from: https://www.practicalpainmanagement.com/treatments/addiction-medicine/comparison-alpha-2-adrenergic-receptor-agonists-managing-opioid-withdr.
3. Carnwath T, Hardman J. Randomised double-blind comparison of lofexidine and clonidine in the out-patient treatment of opiate withdrawal. Drug Alcohol Depend. 1998;50(3):251–4.
4. NICE. Overview | Drug misuse in over 16s: opioid detoxification | Guidance | NICE [Internet]. [cited 2021 Nov 30]. Available from: https://www.nice.org.uk/guidance/cg52.
5. GOV.UK. Routes to recovery from substance addiction [Internet]. [cited 2021 Sept 14]. Available from: https://www.gov.uk/government/publications/routes-to-recovery-from-substance-addiction.
6. McDonald R, Strang J. Are take-home naloxone programmes effective? Systematic review utilizing application of the Bradford Hill criteria. Addiction. 2016;111(7):1177–87.
7. Irvine MA, Buxton JA, Otterstatter M, Balshaw R, Gustafson R, Tyndall M, et al. Distribution of take-home opioid antagonist kits during a synthetic opioid epidemic in British Columbia, Canada: a modelling study. Lancet Public Health. 2018;3(5):e218–25.
8. Walley AY, Xuan Z, Hackman HH, Quinn E, Doe-Simkins M, Sorensen-Alawad A, et al. Opioid overdose rates and implementation of overdose education and nasal naloxone distribution in Massachusetts: interrupted time series analysis. BMJ. 2013;346:f174.
9. Tzemis D, Al-Qutub D, Amlani A, Kesselring S, Buxton JA. A quantitative and qualitative evaluation of the British Columbia Take Home Naloxone program. CMAJ Open. 2014;2(3):E153–61.

10. GOV.UK. Naloxone: a review [Internet] [cited 2021 Jun 27]. Available from: https://www.gov.uk/government/publications/naloxone-a-review.
11. World Health Organization. Community management of opioid overdose [Internet]. Geneva: World Health Organization; 2014 [cited 2021 Jun 27]. Available from: https://apps.who.int/iris/handle/10665/137462.
12. Rando J, Broering D, Olson JE, Marco C, Evans SB. Intranasal naloxone administration by police first responders is associated with decreased opioid overdose deaths. Am J Emerg Med. 2015;33(9):1201–4.
13. Fisher R, O'Donnell D, Ray B, Rusyniak D. Police officers can safely and effectively administer intranasal naloxone. Prehosp Emerg Care. 2016;20(6):675–80.
14. Kitch BB, Portela RC. Effective use of naloxone by law enforcement in response to multiple opioid overdoses. Prehosp Emerg Care. 2016;20(2):226–9.
15. Hulin J, Brodie A, Stevens J, Mitchell C. Prevalence of respiratory conditions among people who use illicit opioids: a systematic review. Addiction (Abingdon, England). 2020;115(5):832–49.
16. Tashkin DP. Heroin smoking and COPD: a case for targeted screening spirometry. Chest. 2019;155(2):247–8.
17. Nightingale R, Mortimer K, Giorgi E, Walker PP, Stolbrink M, Byrne T, et al. Screening heroin smokers attending community drug clinics for change in lung function. Chest. 2020;157(3):558–65.
18. Rosenthal RN, Lofwall MR, Kim S, Chen M, Beebe KL, Vocci FJ, et al. Effect of buprenorphine implants on illicit opioid use among abstinent adults with opioid dependence treated with sublingual buprenorphine: a randomized clinical trial. JAMA. 2016;316(3):282–90.
19. Tanum L, Solli KK, Latif Z-E-H, Benth JŠ, Opheim A, Sharma-Haase K, et al. Effectiveness of injectable extended-release naltrexone vs daily buprenorphine-naloxone for opioid dependence: a randomized clinical noninferiority trial. JAMA Psychiat. 2017;74(12):1197–205.
20. Haight BR, Learned SM, Laffont CM, Fudala PJ, Zhao Y, Garofalo AS, et al. Efficacy and safety of a monthly buprenorphine depot injection for opioid use disorder: a multicentre, randomised, double-blind, placebo-controlled, phase 3 trial. Lancet Lond Engl. 2019;393(10173):778–90.
21. Soyka M. Recent advances in the treatment of opioid use disorders. Tech Neurosurg Neurol. 35(5):1–5. https://doi.org/10.31031/TNN.2020.03.000572.
22. Hwang CS, Bremer PT, Wenthur CJ, Ho SO, Chiang S, Ellis B, et al. Enhancing efficacy and stability of an antiheroin vaccine: examination of antinociception, opioid binding profile, and lethality. Mol Pharm. 2018;15(3):1062–72.
23. Bremer PT, Janda KD. Conjugate vaccine immunotherapy for substance use disorder. Pharmacol Rev. 2017;69(3):298–315.
24. Bremer PT, Schlosburg JE, Banks ML, Steele FF, Zhou B, Poklis JL, et al. Development of a clinically viable heroin vaccine. J Am Chem Soc. 2017;139(25):8601–11.
25. Nielsen DA, Ji F, Yuferov V, Ho A, Chen A, Levran O, et al. Genotype patterns that contribute to increased risk for or protection from developing heroin addiction. Mol Psychiatry. 2008;13(4):417–28.
26. Bart G, Heilig M, LaForge KS, Pollak L, Leal SM, Ott J, et al. Substantial attributable risk related to a functional mu-opioid receptor gene polymorphism in association with heroin addiction in central Sweden. Mol Psychiatry. 2004;9(6):547–9.
27. Mouly S, Bloch V, Peoc'h K, Houze P, Labat L, Ksouda K, et al. Methadone dose in heroin-dependent patients: role of clinical factors, comedications, genetic polymorphisms and enzyme activity. Br J Clin Pharmacol. 2015;79(6):967–77.
28. Ahmad T, Valentovic MA, Rankin GO. Effects of cytochrome P450 single nucleotide polymorphisms on methadone metabolism and pharmacodynamics. Biochem Pharmacol. 2018;153:196–204.
29. Pan Y-X, Xu J, Xu M, Rossi GC, Matulonis JE, Pasternak GW. Involvement of exon 11-associated variants of the mu opioid receptor MOR-1 in heroin, but not morphine, actions. Proc Natl Acad Sci U S A. 2009;106(12):4917–22.

30. Doehring A, Oertel BG, Sittl R, Lötsch J. Chronic opioid use is associated with increased DNA methylation correlating with increased clinical pain. Pain. 2013;154(1):15–23.
31. Chorbov VM, Todorov AA, Lynskey MT, Cicero TJ. Elevated levels of DNA methylation at the OPRM1 promoter in blood and sperm from male opioid addicts. J Opioid Manag. 2011;7(4):258–64.
32. Nielsen DA, Yuferov V, Hamon S, Jackson C, Ho A, Ott J, et al. Increased OPRM1 DNA methylation in lymphocytes of methadone-maintained former heroin addicts. Neuropsychopharmacology. 2009;34(4):867–73.
33. Hwang CK, Kim CS, Kim DK, Law P-Y, Wei L-N, Loh HH. Up-regulation of the mu-opioid receptor gene is mediated through chromatin remodeling and transcriptional factors in differentiated neuronal cells. Mol Pharmacol. 2010;78(1):58–68.
34. Oertel BG, Kettner M, Scholich K, Renné C, Roskam B, Geisslinger G, et al. A common human micro-opioid receptor genetic variant diminishes the receptor signaling efficacy in brain regions processing the sensory information of pain. J Biol Chem. 2009;284(10):6530–5.
35. Ferrer-Alcón M, La Harpe R, García-Sevilla JA. Decreased immunodensities of micro-opioid receptors, receptor kinases GRK 2/6 and beta-arrestin-2 in postmortem brains of opiate addicts. Brain Res Mol Brain Res. 2004;121(1–2):114–22.
36. Ren Y, Whittard J, Higuera-Matas A, Morris CV, Hurd YL. Cannabidiol, a nonpsychotropic component of cannabis, inhibits cue-induced heroin seeking and normalizes discrete mesolimbic neuronal disturbances. J Neurosci. 2009;29(47):14764–9.
37. Hurd YL, Yoon M, Manini AF, Hernandez S, Olmedo R, Ostman M, et al. Early phase in the development of cannabidiol as a treatment for addiction: opioid relapse takes initial center stage. Neurotherapeutics. 2015;12(4):807–15.
38. Verdejo-Garcia A, Lorenzetti V, Manning V, Piercy H, Bruno R, Hester R, et al. A roadmap for integrating neuroscience into addiction treatment: a consensus of the neuroscience interest group of the International Society of Addiction Medicine. Front Psych. 2019;10:877.
39. Verdejo-García A, Pérez-García M. Profile of executive deficits in cocaine and heroin polysubstance users: common and differential effects on separate executive components. Psychopharmacology. 2007;190(4):517–30.
40. Rezapour T, DeVito EE, Sofuoglu M, Ekhtiari H. Perspectives on neurocognitive rehabilitation as an adjunct treatment for addictive disorders: from cognitive improvement to relapse prevention. Prog Brain Res. 2016;224:345–69.
41. Soergel DG, Subach RA, Burnham N, Lark MW, James IE, Sadler BM, et al. Biased agonism of the μ-opioid receptor by TRV130 increases analgesia and reduces on-target adverse effects versus morphine: a randomized, double-blind, placebo-controlled, crossover study in healthy volunteers. Pain. 2014;155(9):1829–35.
42. Schmid CL, Kennedy NM, Ross NC, Lovell KM, Yue Z, Morgenweck J, et al. Bias factor and therapeutic window correlate to predict safer opioid analgesics. Cell. 2017;171(5):1165–1175.e13.
43. Luigjes J, Segrave R, de Joode N, Figee M, Denys D. Efficacy of invasive and non-invasive brain modulation interventions for addiction. Neuropsychol Rev. 2019;29(1):116–38.
44. Ekhtiari H, Tavakoli H, Addolorato G, Baeken C, Bonci A, Campanella S, et al. Transcranial electrical and magnetic stimulation (tES and TMS) for addiction medicine: a consensus paper on the present state of the science and the road ahead. Neurosci Biobehav Rev. 2019;104:118–40.
45. Weiss RD, Griffin ML, Kolodziej ME, Greenfield SF, Najavits LM, Daley DC, et al. A randomized trial of integrated group therapy versus group drug counseling for patients with bipolar disorder and substance dependence. Am J Psychiatry. 2007;164(1):100–7.
46. Weiss RD, Griffin ML, Jaffee WB, Bender RE, Graff FS, Gallop RJ, et al. A "community-friendly" version of integrated group therapy for patients with bipolar disorder and substance dependence: a randomized controlled trial. Drug Alcohol Depend. 2009;104(3):212–9.
47. Day E. Routes to recovery via the community. London: Public Health England; 2013.

Chapter 16
Synthetic Opioids as New Psychoactive Substances (NPS)

Silvia L. Cruz and Raúl Martín-del-Campo

Abstract The last decade has seen a significant increase in synthetic opioids among the new psychoactive substances (NSP). The first two sections of this chapter review the different opioids pertaining to this group and the main effects of the opioid drugs recently incorporated into the NPS or currently in use. The third section examines the precursor and pre-precursors needed to synthesize opioids and how they are regulated. The fourth section analyzes the role of the public Internet, the darknet, cryptocurrencies, and postal services in NPS trading. Finally, the last section presents some challenges these substances pose to prevention and regulation policies and the strategies proposed to face them.

Keywords New psychoactive substances (NPS) · Darknet · Cryptocurrencies · Regulation · Prevention

16.1 The Rapidly Evolving Field of NPS

According to United Nations Office on Drugs and Crime (UNODC), new psychoactive substances (NPS) are "substances of abuse, either in a pure form or a preparation, that are not controlled by the 1961 Single Convention on Narcotic Drugs or the 1971 Convention on Psychotropic Substances, but which may pose a public health

S. L. Cruz (✉)
Department of Pharmacobiology, Center for Research and Advanced Studies (Cinvestav), Calzada de los Tenorios, Mexico City, Mexico

Opioids Working Group, Global Studies Seminar, Faculty of Medicine, National Autonomous University of Mexico, Mexico City, Mexico
e-mail: slcruzm@cinvestav.mx

R. Martín-del-Campo
Opioids Working Group, Global Studies Seminar, Faculty of Medicine, National Autonomous University of Mexico, Mexico City, Mexico

National Institute of Psychiatry Ramón de la Fuente Muñiz (INPRFM), Mexico City, Mexico

International Narcotics Board Control, United Nations, Vienna, Austria

S. L. Cruz (ed.), *Opioids*, https://doi.org/10.1007/978-3-031-09936-6_16

threat" [1]. Within this group, "new" refers to the drug's presence in the market and not to the date of synthesis.

Although the use of uncontrolled synthetic drugs is not a new phenomenon, NPS became a matter of global concern at the beginning of this century. The UNODC presented the first report on "The challenge of new psychoactive substances" in 2013, summarizing data from 80 countries and territories collected through the Global *S*ynthetics *M*onitoring: *A*nalyses, *R*eporting, and *T*rends (SMART) program. This program identified different groups of NPS, with the synthetic cannabinoids synthetic psychostimulant comprising approximately two-thirds of all the substances then available. In the same year, the annual World Drug Report highlighted an increase in NPS from 166 in 2009 to 251 in 2012. At that point, NPS already surpassed the 234 drugs under international control [1].

The NPS market is highly dynamic, with substances constantly emerging and disappearing. For example, the UNODC identified 541 NPS based on data gathered from 95 countries and territories in 2015 [2], but 2 years later, there were "only" 492. This occurred because 78 NPS had newly emerged, but several dozen identified in previous years were no longer available.

By December 2019, the UNODC reported more than 950 NPS from 120 places from all over the world [3]. Synthetic stimulants and cannabinoids accounted for 50% of the total in 2020. However, the number of synthetic opioids changed from 1 in 2009 to 56 in 2019, and 87 in 2020 [2, 4]. Opioids now account for approximately 8% of the total NPS, and their presence in the market contributes to the increasing opioid-related fatal overdose cases (see Chap. 5).

Synthetic opioids can be marketed as "legal highs" and "research chemicals" or labeled as prescription drugs, like OxyContin®, and sold, like the rest of NPS, mainly via the Internet.

16.2 Main Opioid NSP Classes and Effects

Synthetic new psychoactive opioids are analogs of existing opioids or "failed drugs" developed as potential medications to treat pain, coughing, or diarrhea but were discarded due to serious adverse effects, high dependence liability, or lack of advantages over already marketed drugs. These opioids are often synthesized in concealed laboratories using patents or information available in the scientific literature. From the chemical point of view, opioid NPS include 4,5-epoxymorphinans (morphine-like drugs), diphenyl-heptyl-amines (methadone-like drugs), and benzomorphans, but the majority are phenylpiperidines or fentanyl-like opioids (see Chap. 8 for opioid chemical classification). Recently, a new family of synthetic opioids, the benzimidazoles, has become available and is gaining momentum [5].

16.2.1 Fentanyl and Its Analogs

Fentanyl analogs are highly selective μ-opioid receptor agonists of the phenyl-piperidine group of opioids (Chap. 8). As such, they produce morphine-like effects but are much more potent. A single administration of fentanyl causes euphoria, analgesia, miosis, and constipation. At high doses, fentanyl produces chest rigidity ("wooden chest"), respiratory depression, and death. Repeated administration causes dependence so that withdrawal occurs upon drug discontinuation. Fentanyl and its analogs are common heroin adulterants, and users can use them inadvertently, which significantly increases the risk of fatal overdoses [6, 7].

Fentanyl and some derivatives have been controlled via class-wide scheduling rather than individual substances as an emergency measure [8]. However, they were the first opioids to enter the NPS market, the emergency measure to control their marketing is temporary, and new closely related alternatives are constantly emerging [9]. Moreover, according to a recent review, most NPS opioids remain in the market from 6 months to 1 year on average. As a result, only a few compounds stay in circulation long enough to be detected and controlled [10, 11].

Fentanyl was synthesized by Paul Janssen in 1959 and is a clinically effective analgesic and anesthetic adjuvant. It is 50–100 times more potent than morphine and has a shorter duration of action and higher lipophilicity. Potency refers to the amount of drug needed to produce a response. For example, the effective dose of morphine to produce maximal analgesia (antinociception) in rodents is 10 mg/kg, but the fentanyl dose to achieve the same effect is only 0.1 mg/kg. In this example, morphine and fentanyl can cause the same actions because they have similar efficacy, but this is not always the case. Some drugs have the same potency; for example, they can produce their maximal effects with 10 mg/kg, but one would completely abolish pain, and the other would only diminish it because they have different efficacy. Fentanyl and its analogs are highly potent and effective opioids [12].

Fentanyl is available in solutions, tablets, spray lollipops, spray, film, and transdermal patches for clinical purposes. Because it has legitimate uses and high dependence liability, fentanyl is a Drug Enforcement Administration (DEA) controlled (schedule II) substance [9]. From the chemical point of view, fentanyl is a member of the phenyl-piperidine group and the prototype of several drugs that share its core structure (Fig. 16.1) and pharmacological effects. The main risk of this opioid group is its high potency, which has caused numerous fatalities around the globe. Therefore, any person using fentanyl or its analogs is at high risk, especially if the presence of these drugs is unknown to the user. Table 16.1 summarizes basic information about some representative fentanyl-like synthetic opioids.

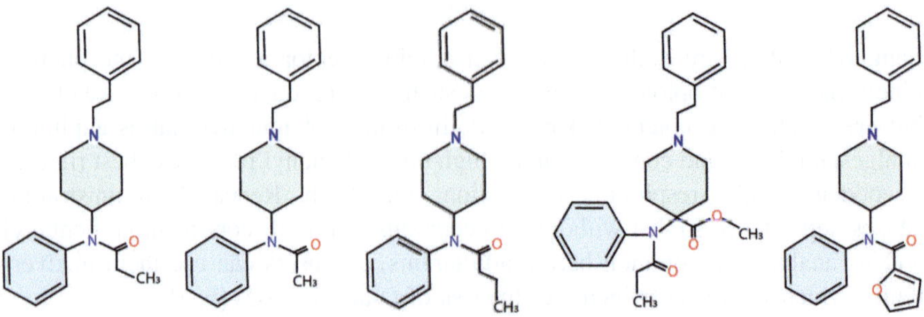

Fig. 16.1 Chemical structure of fentanyl and representative analogs. All these compounds are potent phenyl-piperidine opioids (see Chap. 8)

Table 16.1 Selected fentanyl analogs sold as NPS

Compound	Potency relative to morphine (Mor)	Routes of administration and presentations	Others
Acetylfentanyl	15× > Mor	i.v., p.o., rectal, snorted, vaped. Sold as tablets, in powder form, in herbal products, in blotter papers, nasal spray, and e-liquids	(Desmethylfentanyl). Synthetized in 1968. Involved in fatal ODs. DEA Schedule I controlled substance
Butyrfentanyl	7× > Mor	Rectal, nasal, i.v., transdermal, and sublingual. Sold as white/yellow powder, nasal spray and blotter papers	Synthetized in 1961. Metabolized by CYP3A4 and CYP2D6. Involved in fatal ODs. DEA Schedule I controlled substance
Furanylfentanyl	50–100× > Mor	p.o., insufflation, nasal sprays, i.v. Sold as tablets, in powder form, or injectable solution	Patented in 1986. Involved in fatal ODs. DEA Schedule I controlled substance
Ocfentanyl	100–200× > Mor	i.v., p.o., snorting and smoked. Sold as white or brown powder	Patented in 1984. Involved in fatal ODs. DEA Schedule I controlled substance
Carfentanyl	10,000× > Mor	p.o., insufflation. Sold as powder, tablets, blotter papers, patches, and aerosol	Synthetized in 1979. It is the most potent fentanyl analog. DEA Schedule II controlled substance

Modified from [46], with information from [47] and [6]

16.2.2 *Non-fentanyl Opioids*

16.2.2.1 Isotonitazine and Analogs (Benzimidazoles)

Isotonitazine, etonitazene, metonitazene, and protonitazene are part of the benzimidazole group of synthetic opioids unrelated to fentanyl. These compounds, collectively called nitazenes, were synthesized in the 1950s by the pharmaceutical company CIBA. Despite their somewhat unusual structure (Fig. 16.2),

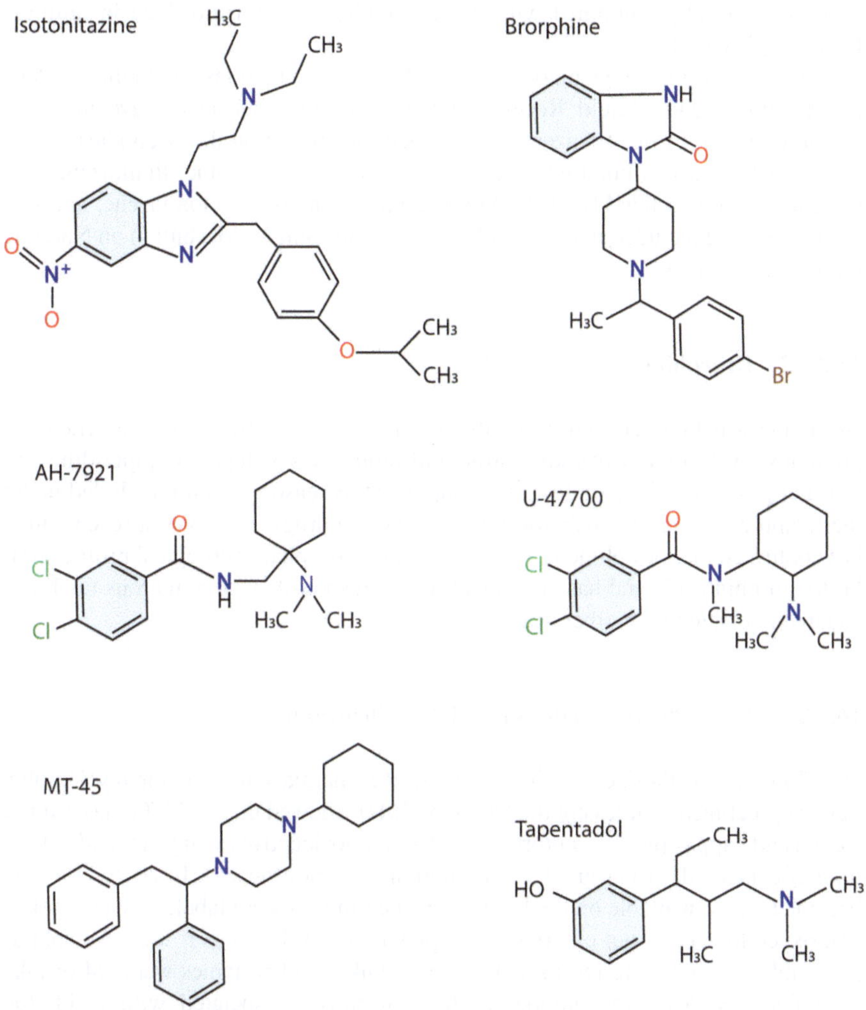

Fig. 16.2 Chemical structures of non-fentanyl opioids sold as NPS. These drugs include members of all the opioid chemical groups. For example, isotonitazine is a benzimidazole, brorphine is a phenyl piperamide, and MT-45 is a methadone-like drug (see Chap. 8)

benzimidazoles are potent analgesics with mu-opioid receptor affinity. However, they were never developed into medicines because they lack advantages over morphine as painkillers, have similar adverse effects, and are very potent, contributing to unintentional overdoses. Seventy years later, benzimidazole opioids re-emerged as NPS after fentanyl analogs scheduling.

Isotonitazine, a compound 50–100 times more potent than morphine, was identified in the black market for the first time in Europe in March 2019. It was initially sold as "etonitazene," which is, in fact, an analog. In 2020, the analysis of some samples revealed that isotonitazine was undiluted [5]. Given its high potency and that of its active metabolite (N-desethylisotonitazene), it is not surprising that isotonitazene, usually combined with other opioids, was soon involved in multiple fatalities [4, 10, 11].

Isotonitazine has been marketed in different presentations, including tablets, powder, liquid, and e-liquid. Recent efforts to control this substance have prompted the introduction of metonitazene, a compound identified on darkweb sites by the end of 2018 that is a hundred times more potent than morphine, flunitazene, and etazene, to name a few [10, 11]. Only recently, etonitazene, clonitazene, and isotonitazine were included in the 1961 United Nations Single Convention on Narcotic Drugs (see Chap. 4).

16.2.2.2 Brorphine

Brorphine was first detected in the illicit drug market in 2019. From the chemical point of view, brorphine has some structural similarities with phenyl-piperidine opioids but possesses a benzimidazole group. For this reason, it is not included under the temporary generic legislation for fentanyl-like drugs. However, a recent study has confirmed that brorphine is a potent opioid with a pharmacological profile similar to morphine [13] and has been identified as responsible for numerous fatalities, usually associated with other drugs [14].

16.2.2.3 U-47700 and Analogs (Cyclohexylbenzamides)

U-47700 was synthesized in the 1970s by the pharmaceutical company Upjohn. Several preclinical studies conducted in the 1980s showed that U-47700 has affinity for mu and kappa opioid receptors, as well as antinociceptive (analgesic) and behavioral effects similar to or greater than morphine (7.5× efficacy). In 2016, U-47700 became readily available on the Internet. Some samples were labeled as and looked like oxycodone pills, but others were in powder form. Because it was also sold as pink tablets, people referred to U-47700 as "pinky." Other names were U4 or fake morphine. By 2016, the number of fatal overdoses associated with U-447700 increased in several countries, and this substance was brought under international control in 2017 [15].

U-49900 is a drug of the same group, also developed by Upjohn, of which there is minimal information. It was advertised on the Internet as a "research chemical" with higher potency than U-47700. U-49900 does not have good reviews in drug forums because of its foul odor and little efficacy [16]. A few reports include U-48800 as a drug seized recently responsible for several fatalities [17]. It is expected that other members of this group will soon be incorporated into the growing number of NPS. Table 16.2 shows other non-fentanyl opioids sold as NPS.

16.2.3 Plant-Derived Compounds with Opioid Activity

16.2.3.1 Mitragynine (Kratom)

Mitragynine is the most abundant of the kratom's active compounds. Kratom, also known as biak or ketum, is a native tropical tree found in Southeast Asia and Africa. People have chewed kratom leaves to alleviate pain and reduce fatigue for centuries. Reports of the medical use of kratom in Thailand and Malaysia include pain, diarrhea, and coughing treatment. The coincidence with the traditional medicinal use of opium is because mitragynine and 7-hydroxymitragynine, also present in kratom leaves, have affinity for μ- and κ-opioid receptors despite lacking structural similarity to opioids. In addition, mitragynine acts on α_2 adrenergic, D_2 dopamine, and serotonin receptors, which could explain its use to increase sexual desire and as a natural alternative to treat opioid withdrawal [18].

Adverse effects associated with high doses of kratom are similar to those produced by opioids (itching, constipation, respiratory depression) and others related to other receptors' activation, including dry mouth, seizures, and hallucinations [19].

Table 16.2 Selected non-fentanyl synthetic opioids sold as NPS

Compound	Potency relative to morphine (Mor)	Route of administration	Others
AH-7921	1× Mor	i.v., p.o., nasal, rectal, sublingual. Powder presentation	*Cyclohexylbenzamide* Synthetized in the early 1970s. Known as *doxylam*. Involved in fatal overdoses. DEA Schedule I controlled substance
MT-45	1× Mor	p.o., nasal, rectal. Sold as a white powder	*Diphenylpiperazine* Synthetized in the 1970s. It has opioid and non-opioid effects (dissociative-like). Can produce hearing loss and cataracts. DEA Schedule I controlled substance
Tapentadol	0.5× Mor	p.o. (normal and extended-release presentations); i.v.	Approved for clinical use in 2008. Weak opioid analgesic structurally related to tramadol. Non-medical use has been associated with several fatalities

DEA Drug Enforcement Administration. For additional information, see Refs. [7, 47, 48]

Preclinical and clinical studies have shown that kratom, mitragynine, and 7-hydroxymitrogynine have addiction potential, and their repeated use produces tolerance and withdrawal after cessation [20, 21]. In addition, fatal overdose cases have been reported usually among people who combined kratom with other drugs. Although several countries have issued restrictive measures to kratom products (leaves, tablets, extracts, powder, or capsules), they are still marketed as NPS due to their psychoactive effects and popularity as opioid alternatives [20, 22].

16.2.3.2 Salvinorin A (*Salvia divinorum*)

Salvia divinorum also called the "diviner's sage" or "magic mint" is a traditional plant used in ancient cultures from Oaxaca, Mexico. Salvinorin A is the active compound responsible for *Salvia*'s effects. Despite its chemical structure unrelated to opioids, salvinorin A is a selective and potent agonist of k-opioid receptors. Activation of this opioid receptor subtype produces psychotomimetic (psychosis-like) effects and analgesia, but not respiratory depression [22]. *Salvia divinorum* fresh leaves can be chewed or used to prepare a drink. As a psychoactive NPS, it is usually smoked to experience short but intense hallucinations. There are several reports of acute psychosis after *Salvia* exposure [19, 23].

16.3 Synthetic Opioid Precursors

The continued proliferation of precursors, pre-precursors, and designer precursors of all illegal drug classes has a global impact and is a critical challenge for the international drug control system. However, the context surrounding this issue has changed dramatically over the past three decades. The pharmaceutical industry has gone global, there have been significant advances in chemistry, the Internet has eased access to substances, and the world drug trade has quadrupled. Moreover, drug conventions did not anticipate the emergence of designer precursors developed solely for illicit drug-related activities and are not yet adequately addressed in international and local legislation and policies [24–27].

The traditional approach to drug precursors assumes that substances have a legitimate use. Because many design precursors do not have a lawful use, they cannot be monitored in the same way. Furthermore, design precursors must gradually be subjected to international control [24, 26, 27]. Manufacturers of design precursors will often claim that they were unaware that their chemicals were used for illegal purposes. Since the substance is not regulated, prosecutors will find it challenging to conduct an expert investigation to prove this.

16.3.1 The International Precursor Control System

According to article 12 of the United Nations Convention against Illicit Traffic in Narcotic Drugs and Psychotropic Substances of 1988, the international precursor control system (which is part of INCB) is responsible for monitoring and promoting government measures to prevent the diversion of substances frequently used in illicit drug manufacture. In addition, INCB also evaluates precursors to determine whether it is necessary to add new compounds to the tables included as an Annex in the 1988 United Nations Convention against Illicit Traffic in Narcotic Drugs and Psychotropic Substances (see Chap. 4). INCB supports the efforts of governments around the world in the areas of monitoring legal international trade, precursor-related investigations, and real-time intelligence sharing on precursor incidents [25, 28].

16.3.2 Synthetic Opioid Precursor for Fentanyl and Fentanyl Analogs

NPP (N-phenethyl-4-piperidone) and *ANPP* (4-anilino-N-phenethylpiperidine) are the main precursors for fentanyl and fentanyl-like drugs. ANPP is also a minor and pharmacologically inactive metabolite of fentanyl. Both NPP and ANPP belong to the phenyl-piperidine class of opioids (Chap. 8) and have limited use in research and analytical laboratories [24, 26, 27, 29].

NPP and ANPP are precursors to several fentanyl analogs, including acetyl-fentanyl (under international control in Schedules I and IV of the 1961 Convention), acryl-fentanyl, butyryl-fentanyl, furanyl-fentanyl, and valery fentanyl. In addition, NPP and ANPP are also used to synthesize 4-fluoro-fentanyl, 4-fluoro-butyr-fentanyl, 4-methoxy-fentanyl, ocfentanil, and carfentanil [29].

Various chemical pathways exist to synthesize NPP and ANPP. For example, some pharmaceutical companies use benzyl cyanide to synthesize phenethylamine, while others start directly from phenethylamine. In the early 2000s, the DEA reported two primary synthetic procedures used in the illicit manufacture of fentanyl: the "Janssen" and "Siegfried" methods. The "Janssen" method, created in the 1960s for the pharmaceutical manufacture of fentanyl, is considered the more difficult and time-consuming of the two, as specialized knowledge of chemistry is required. The much simpler "Siegfried" method was first published on the Internet in the 1990s under a pseudonym and made it possible to refine alternative synthesis procedures that were previously published (in the 1980s). Based on the impurities found in seized drugs, the "Siegfried" method (see below) for NPP and ANPP appears to be the route most used by drug cartels [24, 29]. Although it is beyond the scope of this chapter to provide details on synthesis methods, listing the compounds associated with the chemical synthesis of NPP and ANPP evidences the complexity of regulating them all (Table 16.3).

Table 16.3 Reagents used in the synthesis of the precursors of fentanyl-like opioids NPP and ANPP

Siegfried method	Janssen method
4-Piperidone1 (usually in the form of 4-piperidone hydrochloride monohydrate) Phenethyl bromide (typically) or phenethyl chloride Phenethylamine or benzyl cyanide Methyl-acrylate or ethyl-acrylate Phenethyl halides Phenethylamine/(m) ethacrylate	4-Piperidone N-Benzyl-4-piperidone Norfentanyl Aniline Benzyl halides Phenethyl halides Propionyl chloride Sodium borohydride

The only industrial use of NPP and ANPP is the manufacture of pharmaceutical fentanyl as an analgesic and anesthetic adjuvant. The theoretical yield is 1 kg of NPP and 1.2 kg of ANPP per 1.6 kg of fentanyl base. INCB is also aware of at least one unrelated Active Pharmaceutical Ingredient (API) not under international control. Most companies offering NPP and ANPP for illegal fentanyl synthesis appear to be suppliers or resellers of pharmaceutical chemicals. Therefore, NPP and ANPP are often sold (in gram quantities) for "use as analytical standards" and "for research."

16.3.3 Use of Pre-precursors and Non-scheduled Designer Precursors

As the INCB itself has reported, after the approval of the United Nations Convention against Illicit Traffic in Narcotic Drugs and Psychotropic Substances of 1988, three overlapping shifts in the market occurred:

- The replacement of scheduled precursors by non-scheduled precursors
- The generation of new designer precursors
- The growing versatility of illicit manufacturers to innovate in terms of precursors and synthesis procedures

As a result, criminal groups alternate between using scheduled precursors, non-controlled precursors, and designer precursors to synthesize drugs [24, 26, 27, 30].

The emergence of an increasing number of designer precursors in recent years is of great concern to the international community. Generally speaking, designer precursors are chemicals expressly manufactured to facilitate the production or recovery of scheduled precursors or drugs. Design precursors have been classified as either "masked" precursors or intermediates. Masked precursors are chemicals specifically designed to cover up scheduled precursors later recovered with simple chemical changes. Chemical intermediates are compounds produced during the manufacture of drugs from precursors that do not have a use by themselves.

Chemicals in both categories lack licit uses; they were designed to circumvent current legislation or avoid detection and identification. Design precursors, especially disguised precursors, pose a significant challenge for scheduling measures because there is an almost infinite number of ways to mask or hide the known precursors already included in international lists [24, 26, 29, 30].

In 2017 and 2018, the Commission on Narcotic Drugs decided to control NPP and ANPP [31], forcing drug cartels to develop novel methods to synthesize fentanyl and its analogs. Since then, the most commonly seized precursor has been 4-anilinopiperidine (4-AP), a product used in the pharmaceutical industry for the synthesis of medical fentanyl [29]. Besides having a legitimate use, the synthesis of fentanyl from 4-AP is more straightforward than from NPP. Benzylfentanyl (N-1-benzyl-4-piperidyl) is another precursor increasingly detected through seizures and drug profiling. In addition, in 2020, the INCB was informed of a shipment of propionyl chloride used to manufacture fentanyl illicitly. Propionyl chloride is not the subject of international control, but it is included in the Limited Special International Surveillance List [28, 29].

The other development of international relevance on the regulation of fentanyl precursors and their analogs is the adoption in China of local control measures applicable to many fentanyl precursors. This restriction, which began in May 2019, increased the number of attempts to smuggle precursors out of China, particularly uncontrolled alternative precursors, and manufacture fentanyl in the destination countries. However, there are still gaps in the available intelligence on various aspects of this new situation, such as trafficking routes, modus operandi, the nature of alternative chemicals, and the location of fentanyl synthesis laboratories [26–28, 30, 31].

Although China made all forms of fentanyl illegal, fentanyl's precursors remained uncontrolled. In May 2019, the marketing of at least four non-legislatively controlled substitutes of fentanyl precursors appeared on the public Internet. These substitutes included "covert" precursors designed to disguise their relationship with fentanyl. Then in September 2019, the Alibaba e-commerce page started selling "99918-43-1," which is the CAS (Chemical Abstracts Service) Registry Number of 4-AP. This product appeared on more than 100 ads linked to 29 different companies [29, 32].

Many illicit synthetic opioid manufacturers offer a diverse range of chemicals, including legal products for the pharmaceutical industry. This variety not only allows manufacturers to move quickly from illegal to legal chemical production and circumvent controls, but it also offers them opportunities to conceal illegal chemical production.

On the other hand, the results of the forensic profiling of impurities in fentanyl samples seized in the United States have shown that the illicit opioid market has adapted to the tightening of the control measures of NPP and ANPP in several countries. The results of the profiling analyses carried out in 2019 indicate that the use of the Siegfried method for the illicit manufacture of fentanyl, using NPP or ANPP, had decreased again. In contrast, samples of fentanyl manufactured with the Janssen method increased. Faced with this new situation, the United States placed 4-AP and two chemical substances linked to the Janssen method under control [26, 28, 30, 31, 33].

It is necessary to achieve international consensus and collaboration agreements. Action is imperative. There seems to be a race in which governments and international agencies compete to control and legislate on precursors versus organized crime organizations that devise new drugs and new methods to synthesize illicit drugs as soon as a substance is classified as illegal. If the regulatory mechanisms of chemical precursors continue with regulatory schemes that go from substance to substance, drug cartels will win this race. On the other hand, chemical products with legal and justified uses must be available to the proper extent without diverting them to illegal drug manufacturing.

16.4 Synthetic Opioid Purchase via the Web

Life as we know it cannot be conceived without the Internet; however, new communication technologies make it easier to buy and sell psychoactive substances online. The Internet promotes business on a global scale, with access to potential clients anywhere in the world. Internet transactions can provide anonymity and security for sellers and buyers, complicating the government authorities' investigations in the fight against drug trafficking. In addition, virtual platforms act as discussion forums on the characteristics of the substances and serve to redirect potential clients to pages on the darknet to place orders and make purchases.

The darknet is the part of the Internet that standard browsers cannot access. As a result, illegal drug transactions through the darknet have multiplied since 2015. English is the main language used between sellers and buyers, making it challenging to identify the country where transactions occur. Home deliveries and using cryptocurrencies as a form of payment appear to be trends in individual transactions [34–36]. These behavioral changes, once established, may persist in the long term. Drug traffickers and their organizations remain resilient to the current COVID-19 pandemic and progressive regulations in various countries and adapt their modus operandi to the current situation. Between 2019 and 2020, more than 100 online commerce and social media platforms were detected on the public Internet dedicated to selling illegal drugs and chemical precursors with no legitimate uses in the industry [28, 30, 31, 35, 36].

The sale of synthetic opioids is, after cannabis and stimulant drugs, the one that occurs most frequently in transactions carried out by digital platforms in the world. These cyber drug trafficking networks respond quickly to the pressure of new laws by adjusting advertising techniques or changing chemical formulas to develop substances that mimic the desired effect and fall outside the existing drug controls [32, 33, 36]. Furthermore, because synthetic opioids do not rely on poppy plants' availability, drug traffickers can buy precursors on the Internet and synthesize what they need as long as they have the right laboratory equipment [24, 31].

16.4.1 Drug Trading on the Public Internet

The US Drug Enforcement Agency (DEA) has tracked electronic sales of fentanyl, its analogs, and other synthetic opioids since 2010 [35, 36]. Although darknet use has increased in recent years, most of the transactions carried out over the Internet still occur on public Internet platforms. For example, more than 100 websites for online sales have been detected in social networks such as Facebook, Twitter, and Reddit and e-commerce platforms including eBay, Amazon, Alibaba, and Mercado Libre. In addition, chemicals used to make synthetic drugs, including precursors, pre-precursors, adulterants, NPS, and other controlled substances, are currently sold on independent or well-known websites [30, 35, 37].

E-commerce pages and social media applications create forums of tens to hundreds of users who share information and refer to encrypted websites or the darknet. These private chats may have names that clearly describe their purpose, such as "Research Group on Chemical Drugs" [30, 32].

Chemical companies associated with these social networks have been identified as Chinese "shell companies," suggesting that several suppliers of synthetic drugs connect with customers through Facebook [30].

Images of illicit drugs and precursors are available through simple image search engines [32]. An analysis of more than a thousand websites with drug images from January 2019 to March 2021 found some e-commerce platforms specialized in selling chemical products online. More than 50 companies within the "bio-technology" branch in China are presumed to trade synthetic opioids or their precursors [30, 33, 36].

Another marketing strategy of online vendors is to use drugs' chemical names and identifiers (CAS numbers) instead of their commonly known names. As previously mentioned, a CAS number is a unique numerical code that refers to a specific chemical. For example, fentanyl hydrochloride has the CAS number 437-38-7. The International Union of Pure and Applied Chemistry (IUPAC) also provides standard rules for identifying chemicals by their nomenclature. In this example, fentanyl hydrochloride has the IUPAC name "N-phenyl-N- [1- (2-phenylethyl) piperidin-4-yl] propanamide; hydrochloride." Online sellers use these codes and chemical names that are little known to the general public to offer both drugs and precursors in a veiled way and thus avoid prosecution from the authorities [29, 32].

Fentanyl and other synthetic opioids are advertised as "research chemicals," a term easily recognized by prospective clients. Similarly, some chemical companies offer "custom synthesis" of chemicals, whereby customers can order substances not included in a list of products available on their website, including illegal substances [31, 32].

16.4.2 New Synthetic Opioids and Fentanyl Precursors on the Internet

Internet marketers adapt rapidly to legislative changes that bring different fentanyl analogs and their precursors under control. For example, offers on the public Internet for selling fentanyl analogs such as carfentanil or sufentanil have decreased since they were controlled, but new drugs have replaced them. In particular, the INCB alerted governments on the availability of new synthetic opioids: first came isotonitazine, whose potency is similar to or greater than fentanyl. When isotonitazine became controlled in the United States and the European Union, brorphine and metonitazene appeared [28, 30, 31, 33, 36].

16.4.3 The Use of Darknet for Drugs' Trading

The darknet offers a more discreet option for people involved in illegal drug deals because it can only be accessed through an anonymous browser, the most common of which is Tor. Between 74% and 90% of commerce transactions on the darknet are related to the sale of drugs and their precursors. Although some darknet drug sales are distributor-to-distributor, the majority are distributor-to-user. Like the public Internet, marketplaces and forums are the leading platforms for advertising illicit products on the darknet. There, sellers post advertisements, often without disguising the unlawful nature of the products for sale [30, 35, 36].

Darknet platforms are highly volatile. For example, an analysis of pages between 2010 and 2017 showed that platforms were active on average for just over 8 months. Moreover, of more than 110 pages selling drugs on the darknet, only ten remained fully operational in 2019. During the 2017–2020 period, drug darknet sales amounted to 315 million dollars per year, representing only 0.2% of the illicit drug sales estimated by traditional media. Nevertheless, overall sales of illegal drugs using this system quadrupled between 2011 and 2020 [30].

In an analysis of active darknet platforms in 2014 and 2015, 93 countries worldwide were identified as source countries and 164 as destination countries for transactions [30, 37].

Each year, the Global Drug Survey is conducted on the Internet in a non-representative convenience sample with 100,000–500,000 respondents in more than 50 countries. According to the 2020 survey, "although half of the users who buy drugs on the darknet continue to use the same type of drugs they used before obtaining them through the darknet, some have changed their drug use habits. More than a quarter of those who started using drugs before buying them on the darknet report using a more comprehensive range of drugs." "Furthermore, the darknet may increasingly become an alternative to other traditional sources of drug supply, such as friends, acquaintances, or drug dealers. It is striking that the percentage of people whose first drug purchase was via the darknet doubled, going from 4.5% in 2015 to 9.3% in 2020" [38].

In the specific case of heroin users, buying and selling through the darknet is not attractive since most heroin users need a daily ration and often lack the financial means to purchase large quantities of the drug. However, qualitative research with heroin users in Austria found that they buy heroin via the darknet for special occasions and parties to ensure the availability of high-quality drugs. At the same time, daily purchases of smaller quantities still take place on the street [30].

Fentanyl posts on darknet pages increase constantly but are rarely detailed enough to identify the specific people involved in potential transactions. Instead, these posts usually include some contact information, such as a username, a confidentiality ID, an email address, or a WhatsApp number, which customers can use to communicate directly with sellers and conclude sales transactions [32].

Usernames, usually pseudonyms that sometimes also appear on the public Internet, can be used on darknet platforms, suggesting that some fentanyl providers maintain a presence on the public Internet to communicate with the broader community of synthetic drug users.

Despite the general unpredictability of the darknet and its penchant for fraudulent activity, its encryption, anonymity, and higher barriers to entry continue to offer advantages for those who wish to participate in a comparatively unregulated platform [32, 35–37].

Darknet sales parallel new synthetic opioids' emergence on the open Internet, offering the same NPS, such as isotonitazine (since 2020), or products that are hard to find anywhere else [28].

16.4.4 Purchase of Drugs on the Internet with Cryptocurrency

The use of cryptocurrencies to purchase synthetic drugs over the web reduces the risk of detection by buyers and sellers. It presents a particular challenge to authorities seeking to dismantle these networks. According to different investigations, around two-thirds of all transactions involving the acquisition of drugs through the darknet use cryptocurrencies [35, 36]. The best-known example is Bitcoin, introduced in 2008 by an anonymous individual using the pseudonym Satoshi Nakamoto. Bitcoins are self-managed by an online community and, although not initially conceived for illegal activities, are the currency of choice for most transactions involving illegal drugs. While Bitcoin is still the preferred cryptocurrency, there are other alternatives, such as Litecoin, Dogecoin, Zcash, Ethereum, Darkcoin, and Monero [31, 34–36].

16.5 Drug Distribution by Mail and Express Messaging Services

Various international (Interpol, UNODC, INCB), regional (Europol and CICAD), and national drug control agencies have identified a growing local and cross-border illegal drug trafficking through public postal services (UPS or FedEx), private couriers, and express mail services. Almost all transnational drug shipments by these services arise from the previously discussed transactions. In addition, because fentanyl, its analogs, and other synthetic opioids are highly potent, they can be trafficked in minimal quantities, including them in what could appear to be personal letters [31, 39, 40].

The two most significant limitations that drug control authorities face regarding the possibility of inspecting packages for the presence of drugs are, firstly, the number of postal shipments that can amount to tens of thousands per day within a particular territory and, secondly, the local laws that protect the privacy of the content of letters and packages. Indeed, under most national laws, customs and postal officials are prohibited from opening packages suspected of containing illicit drugs without first obtaining a search warrant [36, 40].

Due to the volume of correspondence, it is necessary to determine which parcels may or may not be suspicious. Therefore, international and national agencies are working on profiling criteria for screening suspicious packages [28, 39, 40].

Like any postal service user who makes a shipment, drug traffickers can send their packages by leaving them in a mailbox or presenting them at a post office or through a sender authorized by third parties. Postal services transport packages from any of these origins to sorting facilities, where the processing equipment scans them and routes them to the delivery unit that will take the package to its final destination [36, 39].

Once a postal item arrives at the destination delivery unit, it is delivered by a mail carrier to a business, residence, PO box, private mailbox, or any location with a personal address. People who engage in illegal activities often rent private mailboxes to disguise their identity, location, or business name. One of the greatest ironies of this phenomenon is that by sending drugs through the postal services of each country, drug traffickers are using government resources to perpetrate a crime [40].

Governments have generally made efforts to combat the use of their postal services to facilitate the distribution of illicit drugs by creating law enforcement agencies within their postal services. In addition to the specific case of fentanyl trafficking and its analogs, these agencies have dedicated part of their efforts to training and equipping their employees with the security equipment to prevent the dangers of handling packages containing these substances, with a high potency that can compromise the health and life of those in contact with them [31, 36]. The US Postal Service (USPS), for example, has reported that drug seizures have increased since 2014, with more than 40,000 pounds seized in 2017 [39, 40].

In a recent (2020) investigation, the DEA detected that of 125 Internet sites illegally selling drugs; 32% of public Internet sites and 92% of darknet sites used the

US Postal Service to make shipments [31, 36]. Most traffickers who have provided information report that only 1 out of every 100 shipments was usually intercepted [31, 34, 40].

Several web pages offer instructions on how to package illicit substances to avoid detection. These how-to websites often cited the limited ability of postal services to detect and intercept packages due to the high volume of mail they handle [31, 39].

Specifically, regarding the trafficking of synthetic opioids via mail, the INCB reports that of the 3298 seizures of these substances reported worldwide in the first 5 months of 2020, 86% were related to postal services, compared to 46% in all of 2019. In its 2020 Report, the INCB also reports that several countries expressed concern regarding the increase in the seizure of synthetic opioids and their precursors through postal services and the use of cryptocurrencies to pay for them. For example, China reported almost 7000 seizures in drug-related mailings in 2019. However, trafficking of synthetic opioids by postal services also occurs in India, Mexico, and the Golden Triangle region of Southeast Asia [28].

16.6 Regulation and Attention Challenges

The regulation of NPS faces several challenges, beginning with the definition of NPS itself. As previously mentioned, many of these substances are not new, but are marketed as such. Also, not everybody agrees on including plant-derived drugs in the NPS category. Another problem is the legal status of these drugs. NPS are not scheduled under the international drug conventions of 1961 and 1971, but some countries have local legislation prohibiting or regulating them, while others do not [41]. On the other hand, the NPS market does not affect all countries simultaneously or in the same way. Reaching international consensus under these circumstances is not an easy task.

The dynamism of the NPS market requires rapid responses. From the mid-2000s to 2019, more than 950 drugs emerged; many of them rapidly disappeared and were replaced with chemical analogs with minor modifications. The current approach of controlling individual substances does not match the market volatility and cannot provide timely interventions. Additional difficulties arise from Internet transactions and postal services for drug delivery because of the massive mail volume and the difficulty of searching for suspicious packages without violating people's privacy [28].

Strategies to face the challenges posed by NPS include:

- Strengthening drug monitoring organizations that gather and exchange information on NPS, such as the European Monitoring Centre for Drugs and Drug Addiction [42], the UNODC Early Warning Advisory, and the Global SMART (Synthetics Monitoring: Analyses, Reporting, and Trends) program [27]
- Sharing information on:

- Drugs seized in different countries
- Identification and detection of NPS in biological samples (usually with unique spectra profiles)
- Signs of drugs intoxication gathered from patients arriving at emergency services
- Preclinical and clinical data of closely-related drugs (e.g., pharmacological effects and possible antagonists)

- Developing clinical guides to treat NPS intoxication such as the one prepared by NEPTUNE, the Novel Psychoactive Treatment United Kingdom Network [43, 44]
- Adopting flexible policies to control classes of substances (e.g., fentanyl analogs) instead of, or in addition to, controlling individual drugs

The dissemination of information to clinicians, policymakers, and people involved in drug prevention in the constantly evolving field of NPS is the first step to provide timely responses to the many challenges posed by these drugs [45].

Governments concerned about the potential trafficking of illicit drugs via web and postal services must enhance cooperation with their national e-commerce industry to prevent and respond to such trafficking. Authorities may also adopt new technologies to strengthen their capacity to identify and monitor trafficking attempts on e-commerce platforms and the darknet via private and public postal services. Unfortunately, many governments have failed to monitor malicious vendors and suspicious platforms offering dangerous substances with no legitimate use, including non-medical synthetic opioids, fentanyl, new psychoactive substances, and precursors.

References

1. United Nations Office on Drugs and Crime (UNODC). The challenge of new psychoactive substances: Global SMART Programme [Internet]. Vienna: UNODC; 2013. Available from: https://www.unodc.org/documents/scientific/NPS_Report.pdf.
2. United Nations Office on Drugs and Crime. World Drug Report 2022 [Internet]. (United Nations publication. Sales No. E.20.XI.6). Vienna: UNODC; 2020 Jun [cited 2021 Aug 15]. Available from: https://www.unodc.org/res/wdr2022/MS/WDR22_Booklet_4.pdf
3. United Nations Office on Drugs and Crime (UNODC). New psychoactive substances (NPS) [Internet]. 2020 [cited Aug 29]. Available from: https://www.unodc.org/documents/scientific/NPS-Leaflet_WEB_2020.pdf.
4. Ujváry I, Christie R, Evans-Brown M, Gallegos A, Jorge R, De Morais J, et al. DARK classics in chemical neuroscience: etonitazene and related benzimidazoles. ACS Chem Neurosci. 2021;12(7):1072–92.
5. Blanckaert P, Balcaen M, Vanhee C, Risseeuw M, Canfyn M, Desmedt B, et al. Analytical characterization of "etonitazepyne," a new pyrrolidinyl-containing 2-benzylbenzimidazole opioid sold online. Drug Test Anal. 2021;13(9):1627–34.
6. Frisoni P, Bacchio E, Bilel S, Talarico A, Gaudio RM, Barbieri M, et al. Novel synthetic opioids: the pathologist's point of view. Brain Sci. 2018;8(9):1–17.

7. Ventura L, Carvalho F, Dinis-Oliveira RJ. Opioids in the frame of new psychoactive substances network: a complex pharmacological and toxicological issue. Curr Mol Pharmacol. 2018;11(2):97–108.
8. Lovrecic B, Lovrecic M, Gabrovec B, Carli M, Pacini M, Maremmani AGI, et al. Non-medical use of novel synthetic opioids: a new challenge to public health. Int J Environ Res Public Health. 2019;16(2):1–21.
9. Weedn VW, Elizabeth Zaney M, McCord B, Lurie I, Baker A. Fentanyl-related substance scheduling as an effective drug control strategy. J Forensic Sci. 2021;66(4):1186–200.
10. Vandeputte MM, Krotulski AJ, Papsun DM, Logan BK, Stove CP. The rise and fall of isotonitazene and brorphine: two recent stars in the synthetic opioid firmament. J Anal Toxicol [Internet]. 2021. Available from: https://academic.oup.com/jat/advance-article/doi/10.1093/jat/bkab082/6317457.
11. Vandeputte MM, Van Uytfanghe K, Layle NK, St. Germaine DM, Iula DM, Stove CP. Synthesis, chemical characterization, and μ-opioid receptor activity assessment of the emerging group of "nitazene" 2-benzylbenzimidazole synthetic opioids. ACS Chem Neurosci. 2021;12(7):1241–51.
12. Karila L, Marillier M, Chaumette B, Nicolas F, Amine B. New synthetic opioids: part of a new addiction landscape. Neurosci Biobehav Rev. 2019;106:133–40.
13. Verougstraete N, Vandeputte MM, Lyphout C, Cannaert A, Hulpia F, Van Calenbergh S, et al. First report on brorphine: the next opioid on the deadly new psychoactive substance horizon? J Anal Toxicol [Internet]. 2021;44(9):937–46. Available from: https://academic.oup.com/jat/article/44/9/937/5879253.
14. Krotulski AJ, Papsun DM, Noble C, Kacinko SL, Logan BK. Brorphine—Investigation and quantitation of a new potent synthetic opioid in forensic toxicology casework using liquid chromatography-mass spectrometry. J Forensic Sci. 2021;66(2):664–76.
15. Beardsley PM, Zhang Y. Synthetic opioids. In: Handbook of experimental pharmacology [Internet]. Cham: Springer; 2018. p. 353–81. Available from: https://link.springer.com/chapter/10.1007/164_2018_149.
16. Alzghari SK, Amin ZM, Chau S, Fleming SW, Cho K, Fung V. On the horizon: the synthetic opioid U-49900. Cureus [Internet]. 2017;9(9):1–3. Available from: https://pubmed.ncbi.nlm.nih.gov/29152436/.
17. Fogarty MF, Mohr ALA, Papsun DM, Logan BK. Analysis of the illicit opioid U-48800 and related compounds by LC-MS-MS and case series of fatalities involving U-48800. J Anal Toxicol [Internet]. 2020. Available from: https://academic.oup.com/jat/advance-article/doi/10.1093/jat/bkaa180/6006236.
18. Boyer EW, Babu KM, Adkins JE, McCurdy CR, Halpern JH. Self-treatment of opioid withdrawal using kratom (Mitragynia speciosa korth). Addiction. 2008;103(6):1048–50.
19. Rech MA, Donahey E, Cappiello Dziedzic JM, Oh L, Greenhalgh E. New drugs of abuse. Pharmacotherapy. 2015;35(2):189–97.
20. Meireles V, Rosado T, Barroso M, Soares S, Gonçalves J, Luís Â, et al. Mitragyna speciosa: clinical, toxicological aspects and analysis in biological and non-biological samples. Medicines (Basel). 2019;6(1):35.
21. Stanciu CN, Gnanasegaram SA, Ahmed S, Penders T. Kratom withdrawal: a systematic review with case series. J Psychoactive Drugs [Internet]. 2019;51(1):12–8. Available from: https://doi.org/10.1080/02791072.2018.1562133.
22. Feng LY, Battulga A, Han E, Chung H, Li JH. New psychoactive substances of natural origin: a brief review. J Food Drug Anal [Internet]. 2017;25(3):461–71. Available from: https://doi.org/10.1016/j.jfda.2017.04.001.
23. European Monitoring Centre for Drugs and Drug Addiction (EMCDDA). Salvia divinorum drug profile [Internet]. Available from: https://www.emcdda.europa.eu/publications/drug-profiles/salvia_en.
24. International Narcotics Control Board (INCB). Precursors and chemicals frequently used in the illicit manufacture of narcotic drugs and psychotropic substances: report of the International

Narcotics Control Board for 2018 on the Implementation of Article 12 of the United Nations Convention against Illicit Traffic in Narcotic Drugs and Psychotropic Substances of 1988. Vienna: United Nations; 2019.
25. International Narcotics Control Board (INCB). Report of the International Narcotics Control Board for 2020 [Internet]. Vienna: United Nations; 2021. 142 p. (Report of the International Narcotics Control Board). Available from: https://www.un-ilibrary.org/content/books/9789210056380.
26. United Nations Office on Drugs and Crime (UNODC). Global Smart Update: fentanyl and its analogues-50 years on [Internet], vol. 17. Vienna: UNODC; 2017. Available from: https://www.unodc.org/unodc/en/scientists/global-smart-update-2017-vol-17.html.
27. United Nations Office on Drugs and Crime (UNODC). The growing complexity of the opioid crisis. Global Smart Update, vol. 24. Vienna: UNODC; 2020.
28. International Narcotics Control Board (INCB). Precursors and chemicals frequently used in the illicit manufacture of Narcotic Drugs and Psychotropic Substances: report of the International Narcotics Control Board for 2020 on the Implementation of Article 12 of the United Nations Convention against Illicit Traffic in Narcotic Drugs and Psychotropic Substances of 1988 [Internet]. Vienna: United Nations; 2021. Available from: https://www.incb.org/incb/en/precursors/technical_reports/2020.html.
29. International Narcotics Control Board (INCB). Precursor chemical monographs 2020. Vienna: United Nations Office on Drugs and Crime (UNODC); 2020.
30. United Nations Office on Drugs and Crime (UNODC). World Drug Report 2021 [Internet]. 2021. Available from: https://www.unodc.org/unodc/en/data-and-analysis/wdr2021.html.
31. Drug Enforcement Administration (DEA). 2020 National drug threat assessment, vol. 3. Johnstown: National Drug Intelligence Center; 2021.
32. Lohmuller M, Cook N, Pauley L. Lethal exchange: synthetic drug networks in the digital exchange era [Internet]. Washington, DC: Center for Advanced Defense Studies; 2020. Available from: https://www.c4reports.org/lethal-exchange.
33. Prekupec MP, Mansky PA, Baumann MH. Misuse of novel synthetic opioids: a deadly new trend. J Addict Med. 2017;11(4):256–65.
34. European Monitoring Centre for Drugs and Drug Addiction (EMCDDA) and Europol. Drugs and the darknet. Perspectives for enforcement, research and policy [Internet]. Luxembourg: Europol Joint Publications, Publications Office of the European Union; 2017. Available from: https://www.emcdda.europa.eu/publications/joint-publications/drugs-and-the-darknet_en.
35. Horton-Eddison M, Di Cristofaro M. Hard interventions and innovation in crypto-drug markets: the escrow example [Internet]. Swansea: Global Drug Policy Observatory; 2017. Available from: http://www.swansea.ac.uk/media/Escrow_PB11_GDPO_AUGUST2017.pdf.
36. United States Departament of Justice. Fentanyl and related threats. US Atty Bull [Internet]. 2018;66(4):111. Available from: https://www.justice.gov/usao/page/file/1083791/download.
37. Kruithof K, Aldridge J, Décary-Hétu D, Sim M, Dujso E, Hoorens S. Internet-facilitated drugs trade: an analysis of the size, scope and the role of the Netherlands [Internet]. RAND Corporation Europe; 2016. Available from: https://www.rand.org/pubs/research_reports/RR1607.html.
38. Global Drug Survey. GDS 2020 [Internet]. 2020. Available from: https://www.globaldrugsurvey.com/gds-2020/.
39. United States Postal Service. Use of postal service networks to facilitate illicit drug distribution. Report Number SAT-AR-18-002 [Internet]. 2018. Available from: https://www.uspsoig.gov/sites/default/files/document-library-files/2018/SAT-AR-18-002.pdf.
40. World Customs Organization (WCO). Illicit trade report 2019 [Internet]. 2019. Available from: https://c4ads.org/wco-2019.
41. Peacock A, Bruno R, Gisev N, Degenhardt L, Hall W, Sedefov R, et al. New psychoactive substances: challenges for drug surveillance, control, and public health responses. Lancet. 2019;394(10209):1668–84.

42. United Nations Office on Drugs and Crime (UNODC). The Global Smart Programme [Internet]. Available from: https://www.unodc.org/unodc/en/scientists/smart-new.html.
43. Abdulrahim D, Bowden-Jones O. Guidance on the clinical management of acute and chronic harms of club drugs and novel psychoactive substances [Internet]. London: Novel Psychoactive Treatment UK Network (NEPTUNE); 2015. Available from: http://neptune-clinical-guidance.co.uk/wp-content/uploads/2015/03/NEPTUNE-Guidance-March-2015.pdf.
44. Abdulrahim D, Bowden-Jones O. The misuse of synthetic opioids: harms and clinical management of fentanyl, fentanyl analogues and other novel synthetic opioids. Information for clinicians. London: Novel Psychoactive Treatment UK Network (NEPTUNE); 2018.
45. Vicknasingam B, Narayanan S, Singh D, Corazza O. Global strategy for new psychoactive substances: an update. Curr Opin Psychiatry. 2020;33(4):295–300.
46. Cruz SL, Granados-Soto V. Opioids and opiates: pharmacology, abuse, and addiction. In: Pfaff DW, Volkow ND, editors. Neurosci 21st century. New York: Springer; 2016. p. 3625–57. https://doi.org/10.1007/978-1-4939-3474-4_156.
47. Tabarra I, Soares S, Rosado T, Gonçalves J, Luís Â, Malaca S, et al. Novel synthetic opioids–toxicological aspects and analysis. Forensic Sci Res. 2019;4(2):111–40.
48. Peacock A, Gisev N, Memedovic S, Larance B, Brown J, Cairns R, et al. Opioid use and harms associated with a sustained-release tapentadol formulation: a post-marketing surveillance study. Drug Alcohol Depend [Internet]. 2020;206:107697. Available from: https://doi.org/10.1016/j.drugalcdep.2019.107697.

Index

S. L. Cruz (ed.), *Opioids*, https://doi.org/10.1007/978-3-031-09936-6

MIX
Papier aus verantwortungsvollen Quellen
Paper from responsible sources
FSC® C105338

FSC
www.fsc.org

If you have any concerns about our products,
you can contact us on
ProductSafety@springernature.com

In case Publisher is established outside the EU,
the EU authorized representative is:
Springer Nature Customer Service Center GmbH
Europaplatz 3, 69115 Heidelberg, Germany

Printed by Libri Plureos GmbH
in Hamburg, Germany